International Space Law in the New Space Era

International Space Law in the New Space Era

Principles and Challenges

Edited by

SANDEEPA BHAT B.

Professor of Law and Director, Centre for Aviation and Space Laws
The WB National University of Juridical Sciences, Kolkata, India

DILIP UKEY

Vice-Chancellor and Professor of Law
Maharashtra National Law University, Mumbai, India

ADITHYA VARIATH

Assistant Professor of Law and Coordinator of the Centre for Research in Air and Space Law
Maharashtra National Law University, Mumbai, India

OXFORD
UNIVERSITY PRESS

OXFORD
UNIVERSITY PRESS

Great Clarendon Street, Oxford, OX2 6DP,
United Kingdom

Oxford University Press is a department of the University of Oxford.
It furthers the University's objective of excellence in research, scholarship,
and education by publishing worldwide. Oxford is a registered trade mark of
Oxford University Press in the UK and in certain other countries

Published in the United States of America by Oxford University Press
198 Madison Avenue, New York, NY 10016, United States of America

British Library Cataloguing in Publication Data

Data available

Library of Congress Control Number: 2024931147

ISBN 9780198909385

DOI: 10.1093/9780198909415.001.0001

Printed and bound by
CPI Group (UK) Ltd, Croydon, CR0 4YY

Contents

List of Abbreviations

ARNS	aeronautical radio navigation services
ARRA	agreement on the Rescue of Astronauts, the Return of Astronauts and the Return of Objects Launched into Outer Space
ASAT	anti-satellite
ATM	air traffic management
BITs	bilateral investment treaties
CCDev	commercial crew development
CD	conference on disarmament
CEO	chief executive officer
CHM	common heritage of mankind
CIA	central intelligence agency
CNSA	Chinese space administration
COMSAT	Communications Satellite Corporation
COPUOS	United Nations Committee of the Peaceful Uses of Outer Space
COSPAR	Committee on Space Research
COTS	Commercial Orbital Transportation Service
CRS	commercial resupply services
ECHO	European Concept for Higher Airspace Operations
EMP	electromagnetic pulse
ENDC	Eighteen Nation Committee on Disarmament
ESA	European Space Agency
EU	European Union
EVP	Executive Vice President
FAA	Federal Aviation Administration
FAI	Fédération aéronautique internationale
FCC	Federal Communications Commission
FIR	Flight Information Regions
FL	flight level
FOBS	Fractional Orbital Bombardment System
FSIA	Foreign Sovereign Immunities Act (US)
GEO	geosynchronous earth orbit
GGE	Group of Governmental Experts
GHz	gigahertz
GLONASS	Russia's global navigation satellite system
GNSS	Global Navigation Satellite System
GPS	Global Positioning System
GSO	Geostationary Satellite Orbit
HAPS	High-Altitude Platform Stations
HF	high frequency

HGV	hypersonic glide vehicle
Hz	hertz
IADC	Inter-Agency Space Debris Coordination Committee
IAEA	International Atomic Energy Agency
ICAO	International Civil Aviation Organization
ICBM	intercontinental ballistic missile
ICC	International Chamber of Commerce
ICJ	International Court of Justice
ICSID	International Centre for Settlement of Investment Disputes
ICSU	Scientific Committee of International Council for Science
IGA	Intergovernmental Agreement
IGO	intergovernmental organization
IGY	International Geophysical Year
IHL	International humanitarian law
ILA	International Law Association
ILRS	International Lunar Research Station
IMO	International Maritime Organization
INTELSAT	International Telecommunication Satellite Organization
IRNSS	Indian Regional Navigation Satellite System
ISC	International Science Council
ISS	International Space Station
ITU	International Telecommunications Union
ITU-D	Telecommunication Development Sector
ITU-R	Radiocommunication Sector
ITU-T	Telecommunication Standardization Sector
LEMNOS	Laser-Enhanced Mission Communications Navigation and Operational Services pipeline project
LEO	Low Earth Orbit
LSC	Legal Subcommittee
LTE	long-term evolution
MFIR	Master Frequency International Register
MSL	mean sea level
NAFTA	North American Free Trade Agreement
NAM	Non-Aligned Movement
NASA	National Aeronautics and Space Administration
NATS	National Air Traffic Services
NAVSTAR GPS	United States Global Positioning System
NIEO	New International Economic Order
NOS	Nuclear Power Sources
NSA	US National Security Administration
NTIA	National Telecommunications and Information Administration
O2O	which is developing the Orion Artemis II Optical Communications System
OEWG	Open-Ended Working Group
OOS	On Orbit Servicing
PAROS	Prevention of an Arms Race in Outer Space

PCA	Permanent Court of Arbitration
PNT	position, navigation and timing
PPWT	Treaty on Prevention of the Placement of Weapons in Outer Space, the Threat or Use of Force against Outer Space Objects
PTBT	1963 Treaty Banning Nuclear Weapon Tests in the Atmosphere, in Outer Space and Under Water
RF	Radio frequency
RF	Radio wave frequencies
RNSS	Radio Navigation Satellite Systems
ROSCOSMOS	Russian Space Agency
RR	Radio Regulations
SAA	Space Arbitration Association
SALT II	Strategic Arms Limitation Talks II
SARPs	Standards and Recommended Practices
SASTIND	State Administration of Science, Technology and Industry for National Defense (China)
satnav	space satellite navigation
SDI	strategic defence initiative
START I	Strategic Arms Reduction Treaty I
STS	UNCOPUOS Science and Technology Subcommittee
STSC	Scientific and Technical Subcommittee
TCBMs	Transparency and confidence-building measures
TT&C	Tracking, Telemetry, and Control
TWAIL	Third World Approaches to International Law
UAE	United Arab Emirates
UK	United Kingdom
UN	United Nations
UNCITRAL	United Nations Commission on International Trade Law
UNCOPUOS	United Nations Committee on the Peaceful Uses of Outer Space
UNESCO	United Nations Education, Scientific and Cultural Organization
UNGA	United Nations General Assembly
UNIDIR	United Nations Institute for Disarmament Research
UNIDROIT	International Institute for the Unification of Private Law
UNISPACE	United Nations Conference on the Exploration and Peaceful Uses of Outer Space
UNOOSA	United Nations Office for Outer Space Affairs
UNSG	UN Secretary-General
US	United States
USAF	United States Air Force
USMCA	US-Mexico-Canada Agreement
USSR	Union of Soviet Socialist Republics
UTC	Coordinated Universal Time
VCLT	Vienna Convention on the Law of Treaties
VLCs	Very large constellations
VSS	Virgin Space Ship
WIPO	World Intellectual Property Organization

WMD	weapons of mass destruction
WMO	World Meteorological Organization
WRC	World Radiocommunication Conferences
WTO	World Trade Organization

List of Contributors

Setsuko Aoki, Professor of Law, Keio University Law School, Keio University, Tokyo, Japan.

Sandeepa Bhat B., Professor of Law and Director, Centre for Aviation and Space Laws, The WB National University of Juridical Sciences, Kolkata, India.

Jennifer A Brobst, Assistant Professor, Cecil C. Humphreys School of Law, University of Memphis, Memphis, TN, United States.

Melissa de Zwart, Professor, Space Law and Governance, Andy Thomas Centre for Space Resources, University of Adelaide, Adelaide, Australia. https://orcid.org/0000-0002-9372-1530

Steven Freeland, Emeritus Professor, Western Sydney University & Bond University, Sydney, Australia. https://orcid.org/0000-0002-0287-8509

Hamza Hameed, Senior Policy Manager, Access Partnership, Singapore.

Michelle LD Hanlon, Executive Director, Center for Air and Space Law at the University of Mississippi, Oxford, MS, United States. https://orcid.org/0009-0005-2118-7307

Mahulena Hofmann, SES Chair in Space, SatCom and Media Law, University of Luxembourg, Luxembourg.

Joel Lisk, Research Fellow (Space and Regulation), Jeff Bleich Centre for Democracy and Disruptive Technologies, Flinders University, Adelaide, Australia. https://orcid.org/0000-0002-6180-2352

Anne-Sophie Martin, Visiting Researcher in International Law and Space Law, Institute for International Legal Studies of the National Research Council (ISGI-CNR), Rome, Italy.

Christopher J Newman, Professor of Space Law and Policy, Northumbria Law School, Northumbria University, Newcastle upon Tyne, United Kingdom. https://orcid.org/0000-0001-5131-4665

Mark J Sundahl, Director of the Global Space Law Center, Cleveland State University College of Law, Cleveland, United States. https://orcid.org/0000-0002-0678-776X

Fabio Tronchetti, Associate Professor of Law, School of Law, Northumbria University, Newcastle, United Kingdom.

Dilip Ukey, Vice Chancellor and Professor of Law, Maharashtra National Law University Mumbai, India.

Adithya Variath, Assistant Professor of Law and Coordinator, Centre for Research in Air and Space Law, Maharashtra National Law University Mumbai, India. https://orcid.org/0000-0001-7693-1010

Kiran Mohan Vazhapully, Senior Legal Officer, Asian-African Legal Consultative Organization (AALCO), New Delhi, India.

Yun Zhao, Henry Cheng Professor in International Law and Associate Dean (Mainland Affairs), Faculty of Law, The University of Hong Kong, Hong Kong.

1
AN INTRODUCTION TO INTERNATIONAL SPACE LAW

Sandeepa Bhat B.

1 Introduction

Since soon after the entry of Sputnik into outer space, one of the most significant debates in the domain of outer space has been on the issue of applicable law for outer space activities. The absence of clarity on the legal regulation of space activities carried the risk of converting the scientific achievement into a domain of chaos. Fortunately, the United Nations (UN) stepped onto the scene to prevent a possible disaster. The development of international space law started with the UN General Assembly passing its first resolution on space in 1958.[1] This was followed by multiple other UN General Assembly resolutions to address the foundational aspects of space activities. The most significant among them has been the UN General Assembly Resolution No 1962 of 1963,[2] which formed the basis for the development of the Outer Space Treaty[3] by outlining the fundamental principles of space law.

Since the UN General Assembly resolutions are not binding, the first phase of the development of space law is characterised as the soft law stage.[4] With the increased activities in space, it was realised that binding norms are required in regulating space activities.[5] This resulted in the second phase of the development of space law in the form of binding treaties, starting with the Outer Space Treaty 1967. The subsequent years witnessed the development of four more international treaties relating to activities in outer space and celestial bodies. All these treaties were negotiated at the United Nations Committee on the Peaceful Uses of Outer Space

[1] UNGA Res 1348 (XII) (13 December 1958).

[2] UNGA Res 1962 (XVIII) (13 December 1963).

[3] Treaty on Principles Governing the Activities of States in the Exploration and Use of Outer Space, Including the Moon and Other Celestial Bodies (adopted 27 January 1967, entered into force 10 October 1967) 610 UNTS 205 (Outer Space Treaty).

[4] See Bin Cheng, 'United Nations Resolutions on Outer Space: "Instant" International Customary Law' (1965) 5(1) IJIL 23.

[5] Paul G Dembling and Daniel M Arons, 'The Evolution of the Outer Space Treaty' (1967) 33 JAirL&Com 419, 427.

Sandeepa Bhat B., *An Introduction to International Space Law* In: *International Space Law in the New Space Era.*
Edited by: Sandeepa Bhat B., Dilip Ukey, and Adithya Variath, Oxford University Press. © Sandeepa Bhat B. 2024.
DOI: 10.1093/9780198909415.003.0001

(UNCOPUOS), which acts on the basis of consensus.[6] Unfortunately, the treaty-making phase ended in 1979 with a lack of consensus among the States.[7] Thus, several proposals at UNCOPUOS for the further development of international space law did not conceptualise and, consequently, concerns started to mount, especially in light of big power politics.

With the gates of treaty-making being closed, resorting back to the UN General Assembly remained the only option, since a two-thirds majority is sufficient to pass any resolution at the UN General Assembly. Accordingly, the agenda items were moved from the UNCOPUOS to the UN General Assembly, and the 1980s and 1990s witnessed the third stage of development of space law, with three set of principles and a declaration being part of that. The set of principles on the subject of: (i) Direct Television Broadcasting,[8] (ii) Remote Sensing of Earth from Outer Space,[9] and (iii) Use of Nuclear Power Sources in Outer Space,[10] are the three different sets of principles adopted at this stage. The twentieth-century efforts at the international level ended with the Space Benefits Declaration of 1996,[11] which is also considered to be the last major endeavour in the line of international space law developments under the aegis of the United Nations.

Even though the UN General Assembly continues to pass resolutions relating to various aspects of space activities in the twenty-first century,[12] their effectiveness is seriously in question. On the fiftieth anniversary of the Outer Space Treaty in 2017, the UN General Assembly also adopted a Declaration[13] to showcase the continuing relevance of the principles laid down under the Outer Space Treaty, as well as the need for the 'widest adherence to the Treaty'.[14] Unfortunately, the current scenario

[6] Steven Freeland and Anja NakaradaPecujilic, 'Do You Like Your Regulation—Hard or Soft: The Antarctic Treaty and the Outer Space Treaty Compared' (2018) 30(1) NatlLSchIndRev 11, 21–22.

[7] Nandasiri Jasentuliyana, 'Treaty Law and Outer Space: Can the United Nations Play an Effective Role?' (1986) XI AnnAir&SpaceL 219, 224–25.

[8] Principles Governing the Use by States of Artificial Earth Satellites for International Direct Television Broadcasting, UNGA Res 37/92 (10 December 1982).

[9] Principles Relating to Remote Sensing of the Earth from Outer Space, UNGA Res 41/65 (3 December 1986).

[10] Principles Relevant to the Use of Nuclear Power Sources in Outer Space, UNGA Res 47/68 (14 December 1992).

[11] Declaration on International Cooperation in the Exploration and Use of Outer Space for the Benefit and in the Interest of All States, Taking into Particular Account the Needs of Developing Countries, UNGA Res 51/122 (13 December 1996).

[12] Between 2000 and 2022, the UN General Assembly has passed fifty resolutions on aspects such as international cooperation, peaceful uses, prevention of arms race, application of the concept of launching State, transparency and confidence building, disaster management and emergency response, registration of space objects, national space legislation, the UN programme on space applications, declaration on the fiftieth anniversary of the Outer Space Treaty, sustainable development and global health. See UNOOSA, 'Documents and Resolutions Database' <www.unoosa.org/oosa/documents-and-resolutions/search.jspx?&view=resolutions> accessed 20 July 2023.

[13] Declaration on the Fiftieth Anniversary of the Treaty on Principles Governing the Activities of States in the Exploration and Use of Outer Space, including the Moon and Other Celestial Bodies, UNGA Res 72/78 (7 December 2017).

[14] ibid para 17.

does not reflect in practice adherence of the States to the true spirit of the principles of the space treaties.

2 The Five Treaties

Before delving into the areas of operation of international space law, one needs to understand the object and purposes of the five UN space treaties, as all binding commitments of the international community emanate from them. The Outer Space Treaty provided the foundation for a legal regime by incorporating the fundamental principles of space law outlined in UN General Assembly (UNGA) Resolution 1962 of 1963 into the realm of hard law. With some of its provisions and principles attaining the status of customary international law,[15] the Outer Space Treaty is being called the Magna Carta[16] of international space law. While the Outer Space Treaty laid down the framework, it was found essential to supplement it with other treaties to make it practically significant. This resulted in the drafting of four more treaties.

The Rescue Agreement 1968[17] was the second international agreement in the field of outer space. This was enacted to supplement Article V of the Outer Space Treaty, which declares astronauts to be envoys of mankind in outer space. Being the representatives of mankind, they are entitled to receive emergency assistance from all States. The Rescue Agreement carries forward this notion of emergency assistance being available to space personnel, along with a detailed procedure for rescue and return. The procedure for recovery and return of space objects and component parts thereof is also laid down under the Rescue Agreement.

The Liability Convention 1972[18] was adopted as the next step to supplement Article VII of the Outer Space Treaty. The Outer Space Treaty distinguishes State responsibility from international liability by laying down two separate provisions. While Article VI stipulates the former, Article VII deals with the latter. Despite the coverage of international liability for damage caused by space activities under Article VII of the Outer Space Treaty, failure to mention the basis of liability[19] as well as to deal with the situation of joint operations necessitated a supplementary treaty to clarify the position. The Liability Convention not only addresses the

[15] See Valentina Vecchio 'Customary International Law in the Outer Space Treaty: Space Law as Laboratory for the Evolution of Public International Law' (2017) 66(3) GerJAir&SpaceL 491.

[16] Detlev Wolter, *Common Security in Outer Space and International Law* (UN Institute for Disarmament Research 2006) 19.

[17] Agreement on the Rescue of Astronauts, the Return of Astronauts and the Return of Objects Launched into Outer Space (adopted 22 April 1968, entered into force 3 December 1968) 672 UNTS 119 (Rescue Agreement).

[18] Convention on International Liability for Damage Caused by Space Objects (adopted 29 March 1972, entered into force 1 September 1972) 961 UNTS 187 (Liability Convention).

[19] Absolute liability or fault-based liability.

substantive gaps of the Outer Space Treaty but also provides for a dispute settlement mechanism.

As space activities started to increase, the need for space situational awareness and space traffic management were found to be absolutely essential. Even to identify the State that is liable to pay compensation for damage caused by space objects, there was a requirement that a system be brought in force. In order to cater to these needs, the Registration Convention 1975[20] came into existence.[21] The Registration Convention seeks to create transparency in carrying on space activities by mandating the launching State to undertake the process of double registration of space objects, one at the national level and the other at the international level. It also goes on to ask those States with space monitoring and tracking facilities to assist other States in identifying the space object that has caused damage to any person or property.[22]

Finally, the Moon Agreement 1979[23] came into existence as the last international treaty in the domain of outer space. It attempts to establish a stronger regime to govern the activities on the celestial bodies as well as in their orbits and other trajectories.[24] Most of the provisions of the Outer Space Treaty are incorporated in the Moon Agreement, with necessary improvisation. Since the celestial bodies contain tangible resources, the Moon Agreement incorporates separate provisions to govern activities such as resource mining. Unfortunately, the stronger provisions of the Moon Agreement have little practical significance as only a few States have ratified it.[25] In addition, Saudi Arabia notified its decision to withdraw from the Moon Agreement on 5 January 2024.[26] This withdrawal has taken effect from 5 January 2024, as per Article 20 of the Moon Agreement.

Even though the Moon Agreement suffers from limited acceptance, the other four treaties have a substantial number of ratifications.[27] The Outer Space Treaty is the most popular among them, with almost every space-faring nation having signed up to it. While acceptance of the Rescue Agreement and the Liability Convention

[20] Convention on Registration of Objects Launched into Outer Space (adopted 6 June 1976, entered into force 15 September 1976) 1023 UNTS 15 (Registration Convention).

[21] See Stefan A Kaiser, 'Legal and Policy Aspects of Space Situational Awareness' (2015) 31 SpacePoly 5, 11.

[22] Registration Convention, art VI.

[23] Agreement governing the Activities of States on the Moon and Other Celestial Bodies (adopted 5 December 1979, entered into force 11 July 1984) 1363 UNTS 3 (Moon Agreement).

[24] Moon Agreement, art 1.

[25] As it stands in 2023, eighteen States have become parties to the Moon Agreement. See United Nations Treaty Collections, 'Chapter XXIV, Outer Space: Agreement governing the Activities of States on the Moon and Other Celestial Bodies' <https://treaties.un.org/Pages/ViewDetails.aspx?src=IND&mtdsg_no=XXIV-2&chapter=24&clang=_en>accessed 21 July 2023.

[26] Depositary Notification C.N.4.2023.TREATIES-XXIV.2 of 5 January 2023 (Withdrawal: Saudi Arabia) <https://treaties.un.org/doc/Publication/CN/2023/CN.4.2023-Eng.pdf> accessed 21 July 2023.

[27] See United Nations Office for Outer Space Affairs, 'Status of International Agreements relating to Activities in Outer Space' <www.unoosa.org/oosa/en/ourwork/spacelaw/treaties/status/index.html> accessed 21 July 2023.

is fast catching up with the Outer Space Treaty, the Registration Convention is also not far too behind. Hence, the key aspects dealt with under these treaties form the foundation of international space law, which are still significant in the modern era of space commercialisation. Accordingly, the subsequent sections of this chapter outline those key aspects that have been dealt with under international space law over the years.

3 Demilitarisation and Peaceful Uses

The first aspect to be considered in international space law has been the issue of space militarisation. It is significant to note here that the entry into outer space has been used to gain military advantage.[28] Neither the erstwhile Soviet Union nor the United States had any better reason to invest in the costly affair of space entry during the Cold War period, which was a widely known fact. Accordingly, the need to prevent the evil effects of space militarisation was felt immediately after the beginning of the space race in 1950s.

Responding to this need, the UN General Assembly devoted its first resolution on outer space to the domain of peaceful uses.[29] It recognised the common interest of mankind in outer space and stressed that the use of outer space should be for 'peaceful purposes only'. While reiterating the sacrosanct obligations of States under the Charter of the United Nations, the UN General Assembly also appealed to the member States 'to avoid the extension of present national rivalries' into the domain of outer space. The UN General Assembly went ahead with the establishment of an adhoc Committee on Peaceful Uses of Outer Space (AdHoc COPUOS) under this first resolution to further the cause of peaceful uses. Although the AdHoc COPUOS hit a roadblock due to political differences,[30] the UN General Assembly responded to the crisis by establishing UNCOPUOS as a permanent body.[31]

Since it started operating in 1961, the UNCOPUOS has established two subcommittees: (i) the Scientific and Technical Subcommittee; and (ii) the Legal Subcommittee.[32] All five treaties governing space activities are developed by the UNCOPUOS on the basis of its mandate of upholding peaceful uses. This requirement of peaceful uses of outer space was found to be so significant that even after

[28] Mircea Mateesco-Matte, 'Space Militarisation and Space Law at a Time of Non-Peaceful Coexistence' (1984) IX AnnAir&SpaceL 355, 364.

[29] UNGA Res 1348 (XII) (13 December 1958).

[30] See Philip C Jessup and Howard J Taubenfeld, 'The United Nations Ad Hoc Committee on the Peaceful Uses of Outer Space' (1959) 53(4) AJIL 87.

[31] UNGA Res 1472 (XIV) (12 December 1959).

[32] See C Wilfred Jenks, *Space Law* (Stevens & Sons 1965) 56.

the establishment of the UNCOPUOS, the UN General Assembly has passed a series of resolutions on the requirement of peaceful uses of outer space.[33]

In terms of treaty norms, the Outer Space Treaty and the Moon Agreement incorporate provisions on peaceful uses and the prohibition on certain kinds of weapons. The need for the maintenance of international peace and security is highlighted in both agreements by mandating adherence to international law, including the UN Charter, while carrying on activities in outer space or on celestial bodies.[34] Article IV of the Outer Space Treaty prohibits the placement of nuclear weapons and any other kinds of weapons of mass destruction in outer space or on celestial bodies. It mandates the States to use the Moon and other celestial bodies exclusively for peaceful purposes and forbids the establishment of military bases, installations and fortifications; the testing of weapons; and conduct of military manoeuvres on celestial bodies. Article 3 of the Moon Agreement, while incorporating all elements of Article IV of the Outer Space Treaty, goes a step further to prohibit the threat or use of force or any other hostile act or threat of such act in connection with any celestial activity.[35]

Despite addressing the concern of military space activities extensively under international space law, the process of space militarisation has never been halted.[36] In fact, it has increased over time with the establishment of space forces in different States.[37] The provisions of the space treaties are not adhered to in the right spirit and States have made it evident that they continue to use space for military advantage. The most crucial aspect that has led to this defiance from States is the loose wording of the Outer Space Treaty. The prohibition of weapons, for example, is limited to nuclear weapons and weapons of mass destruction only. Conventional weapons or tools of modern warfare, such as laser weapons or cyberattack, are not explicitly mentioned under Article IV's prohibition. In the wake of the ongoing Russia–Ukraine War, the extent of damage that can be inflicted by using these weapons in outer space has become apparent.[38] Hence, the blanket prohibition on any kind of weapon in space has become the need of the hour.

[33] More than 90 per cent of the UN General Assembly resolutions mention 'peaceful uses' or 'prevention of arms race' in the title itself. See UNOOSA, 'Documents and Resolutions Database' (n 12).

[34] Outer Space Treaty, art III and Moon Agreement, art 2.

[35] Moon Agreement, art 3(2).

[36] See Ryan M Esparza, 'Event Horizon: Examining Military and Weaponization Issues in Space by Utilizing the Outer Space Treaty and the Law of Armed Conflict' (2018) 83(2) JAirL&Com 333.

[37] The United States established the US Space Force as an independent wing of the armed forces in December 2019. See Robert Farley, 'Space Force Ahead of Its Time, or Dreadfully Premature?' (2020) 904 PolyAnal 1. The United Kingdom and Australia have subsequently followed the US path by establishing their military space command. See 'Australia Announces New "Space Command" Defence Agency' *BBC News* (22 March 2022) <www.bbc.com/news/world-australia-60835136> accessed 26 July 2023.

[38] See NC Bipindra, 'STAR WARS! Russia-Ukraine Conflict is World's 1st Commercial Space War as Moscow's EW Tries to Cripple Ukraine' *The Eurasian Times* (23 April 2023) <www.eurasiantimes.com/star-wars-russia-ukraine-conflict-is-worlds-first-commercial/>accessed 26 July 2023; see also 'How the War in Ukraine is Affecting Space Activities: New Challenges and Opportunities' (*OECD Policy Responses on the Impacts of the War in Ukraine*, 15 November 2022) <www.oecd-ilibrary.org/docserver/

In addition, the absence of a definition of 'peaceful uses' has swept away the effectiveness of efforts to prevent the evils of space militarisation. Though the initial understanding of 'peaceful uses' had been non-military uses, over the years States have shifted to interpret it as non-aggressive uses.[39] Consequently, States have used outer space for different military activities under the pretence of defensive uses. The list of such uses of outer space includes not just indirect uses for some military advantage but also direct military uses such as anti-satellite missile testing (ASAT), strategic defence initiative (SDI/Star Wars), Rods from God (Project Thor) etc. Given the dual-use nature of space technology,[40] it is also difficult to contain space militarisation unless there is a collective will on the part of the international community. Hence, in the practical sense, efforts to divert from military uses to scientific or commercial uses of outer space have not been entirely successful.

4 Status of Global Commons

Since the beginning of space exploration, one of the crucial questions has been about its legal status. While the airspace was already subject to the sovereign rights of the States below it,[41] there was no clarity on the application of State sovereignty in outer space. Sensing the possible concerns, the UN General Assembly, in its first resolution on outer space, has recognised the 'common interest of mankind' in outer space.[42] This was the first step towards the application of a common rights regime to outer space, rather than a regime of individual rights, whether in the form of sovereignty or in the form of private rights.

4.1 Sovereignty v Freedom

The Latin maxim *cujus est solum ejus est usque ad coelum et ad inferos*, meaning whoever owns the surface of the land owns the area above it to an unlimited height as well as the area below it to an unlimited depth, characterised the early legal status of the column above the Earth's surface.[43] Though the maxim remained

ab27ba94-en.pdf?expires=1690353668&id=id&accname=guest&checksum=196B11A290058EF58 40D68C0E60FE890>accessed 26 July 2023.

[39] Robert L Bridge, 'International Law and Military Activities in Outer Space' (1980) 13(4) AkronLRev 649, 658.

[40] See Yasuo Otani and others, 'Dual-Use Concept on Civil and Defense Uses of Outer Space'(2012) 10(ists 28) Trans JSASS AerospTechnolJpn Tv_1 - Tv_5.

[41] Convention on International Civil Aviation (adopted 7 December 1944, entered into force 4 April 1947) (1994) 15 UNTS 295 (Chicago Convention) art 1.

[42] UNGA Res 1348 (XII) (13 December 1958).

[43] See Herbert David Klein, 'Cujus Est Solum Ejus Est . . . Quousque Tandem' (1959) 26(3) JAirL&Com 237.

unchallenged for a long period of time, developments in aviation have meant that it is now disregarded. Since the beginning of the twentieth century, United States jurisprudence started to question the validity of this maxim. While the United States judiciary decided against the maxim in several cases, the most notable among them has been the case of *US v Causby*,[44] which was decided by the Supreme Court of the United States.[45]

The logic of *US v Causby* has influenced the legal developments in different corners of the world.[46] Lord McNair also led the charge against the maxim by arguing against the possibility of claiming ownership up to an unlimited height.[47] Disregard of the *usque ad coelum* theory is clearly evident in State practices since the beginning of the space era. While the Soviet Union and the United States did not seek permission from other States during the orbiting of their satellites over several States, neither did the States over which those satellites orbited object to space activities.[48] In addition, neither of the early entrants to outer space claimed sovereignty over the domain that they had entered. These approaches by States have been a clear indicator of the fact that neither the space-faring nations nor others believed in the lateral extension of State sovereignty into outer space.

The development of State practice in this regard got conceptualised under the Outer Space Treaty. Article II of the Outer Space Treaty forbids the 'national appropriation by claim of sovereignty by means of use or occupation or by any other means'. This was necessitated by the fact that the assertion of State sovereignty in outer space would have resulted in unending conflicts. Maurice N. Andem sums up the danger of celestial appropriation as follows:

> The very nature of the moon and other celestial bodies, it should be borne in mind, would place the existence of mankind in constant danger, if the legal principles which were adopted in the appropriation of foreign lands during the seventeenth to nineteenth centuries were to be applied by those states which are now in a position to occupy this new domain of man's activity.[49]

As a corollary to the absence of sovereignty, the principle of freedom has been adopted under Article I of the Outer Space Treaty. This freedom is available only for exploration, use and scientific investigation to be carried on in outer space or celestial bodies. It does not extend to the 'exploitation' of celestial resources, which is

[44] *US v Causby* 328 US 256 (1946).
[45] See Yehuda Abramovich, 'The Maxim "Cujus Est Solum Ejus Usque Ad Coelum" as Applied in Aviation' (1962) 8(4) McGillLJ 247.
[46] See Stuart Randolph Hayes, 'United States v. Causby: An Extension Thereof' (1961) 3(1) William&MaryLRev 36.
[47] Arnold McNair, *The Law of the Air* (2nd edn, Stevens & Sons 1953) 31.
[48] S Bhatt, *Legal Controls of Outer Space—Law, Freedom and Responsibility* (S Chand & Co 1973) 95.
[49] Maurice N Andem, *International Legal Problems in the Peaceful Exploration and Use of Outer Space* (University of Lapland 1992) 95.

dealt with separately under the Moon Agreement.[50] Moreover, the exercise of such freedom is based on the principle of equality and the application of international law, as well as ensuring respect for all other provisions of the UN space treaties. The principle of freedom is further strengthened by expressly guaranteeing free access to all areas of celestial bodies to everyone.[51]

4.2 Res nullius v Res communis

Along with the debate on State sovereignty, the nature of outer space, celestial bodies and their resources from the dimension of property rights has also been in question since the beginning of space activities. The tussle between the notion of individual rights and collective rights was evident in the then prevailing Roman law concepts of *res nullius*[52] and *res communis*.[53] Some of the early writers argued in favour of *res nullius* by equating the celestial bodies to an unoccupied land or newly discovered island on the Earth.[54] However, the concerns of applying the concept of *res nullius* were soon realised by the international community, especially against the backdrop of the evils of colonisation.[55]

The application of *res nullius* to the domain of outer space would have resulted in the two superpowers of the Cold War era claiming everything in outer space and celestial bodies. Hence, *res nullius* was discarded.[56] However, a strict adherence to the concept of *res communis* was also found to be problematic. Since it allowed unrestricted use without mandating benefit sharing, the evils of *res nullius* would have been reflected in *res communis* when applied to outer space. This is because only the Soviet Union and the United States would have used and exploited outer space and celestial bodies as others were still developing their space technology. By the time other States developed their technology, the two space-faring nations could have drained the celestial resources in a race similar to the one that was witnessed in the era of the gold rush.[57]

In addition to the above concerns related to Roman law concepts, space exploration started during the period of decolonisation. Given the economic and political

[50] Moon Agreement, art 11.

[51] See Outer Space Treaty, art I.

[52] Eike-Henner W Kluge, 'Res Nullius, Res Communis and Res Propria: Patenting Genes and Patenting Life-Forms' (2005) 14 AnnRevL&Eth 543, 551.

[53] See Charlotte Ku, 'The Concept of Res Communis in International Law' (1990) 12(4) HistEurIdeas 459.

[54] See Ernst Fasan, 'Dominium Lunae, Proprietas Lunae' (1997) 39 ProceColloqLOuterSpace 1, 1–3.

[55] Harry H Almond, Jr, 'The Legal Status of Property on the Moon and Other Celestial Bodies' (1997) 39 ProceColloqLOuterSpace 20, 27.

[56] Gyula Gal, 'Acquisition of Property in the Legal Regime of Celestial Bodies' (1997) 39 ProceColloqLOuterSpace 45, 46.

[57] See Andrea G McDowell, 'From Commons to Claims: Property Rights in the California Gold Rush' (2002) 14(1) YaleJL&Human 1.

tensions prevailing in the newly formed States, these nations expected space exploration to be in the common interest and for the common benefit.[58] Accordingly, there was a requirement to develop a new concept or principle beyond the regime of *res communis*. This is conceptualised in the form of the principle of the province of all mankind, which is enshrined in Article I of the Outer Space Treaty. Christol observes that the principle of the province of all mankind is a starting point of legal reasoning in space law, which seeks the voluntary sharing of benefits.[59]

4.3 Province of All Mankind v Common Heritage of Mankind

While the Outer Space Treaty was being adopted, the concept of the common heritage of mankind (CHM) was spreading its roots in the law of the sea negotiations.[60] However, the Outer Space Treaty did not absorb the concept of CHM into the domain of space law. It was only the Moon Agreement that incorporated CHM and set out in detail its application to the moon and other celestial bodies under eight subparagraphs of Article 11. Hence, it is pertinent to note here that outer space per se is not declared as part of CHM, rather it is only the moon and other celestial bodies that are sought to be governed by the CHM regime. More significantly, the practical significance of the application of CHM to the moon and other celestial bodies is also debatable, since the Moon Agreement has a limited number of State acceptances.

The principle of the province of all mankind applies to the exploration and use of outer space as well as the celestial bodies since the Outer Space Treaty also covers celestial bodies. Some authors attempt to distinguish the province of all mankind from CHM by stating that the former is only connected with territory and responsibility over territory, while the latter has property connotations.[61] However, accepting this distinction does not seem appropriate in light of the above-discussed background of the emergence of the province of all mankind as a development in preference to the *res communis* regime. Moreover, such an interpretation goes against the text of the Outer Space Treaty, since Article I stipulates 'exploration and use' and not any 'territory' as the province of all mankind.

[58] Carl Q Christol, 'Important Concepts for the International Law of Outer Space' (1998) 40 ProceColloqLOuterSpace 73, 74.

[59] ibid 79.

[60] See UNGA, 'Note verbale dated 17 August 1967 from the Permanent Mission of Malta to the United Nations addressed to the Secretary-General' (17 August 1967) DocA/6695 (1967 Note verbale from the Permanent Mission of Malta).

[61] See JES Fawcett, *Outer Space—New Challenges to Law and Policy* (Clarendon Press 1984) 4; See also B Maiorsky, 'A Few Reflections on the Meaning and the Interpretation of "Province of All Mankind" and "Common Heritage of Mankind" Notions' (1987) 29 ProceColloqLOuterSpace 58, 59.

The CHM carries forward the essence of the province of all mankind and brings in a more concrete regime. As advocated by Arvid Pardo[62] and reflected in the law of the sea negotiations, CHM embodies five novel elements. Prohibition on individual appropriation, establishing an international management system, equitable sharing of benefits, conservation for future generations and peaceful uses of the resources are the crux of CHM.[63] Hence, though both are rooted strongly in the concept of *res communis*, CHM has gone beyond the province of all mankind. Accordingly, using CHM and the province of mankind interchangeably in the legal characterisation of outer space is incorrect.

5 Application of International Law

As already mentioned, the beginning of space activities intensified the question of the applicable law in outer space. With the lateral extension of domestic laws to outer space being ruled out, along with the rejection of State sovereignty in outer space, the risk of a legal vacuum in space loomed. Hence, the UN General Assembly suggested bringing space activities within the domain of international law, including the Charter of the United Nations. This was done initially through asking the States to be 'guided by' international law, including the Charter of the United Nations, in the exploration and use of outer space.[64] However, in the next resolution, the UN General Assembly asserted this in more specific terms by stipulating that 'the activities of States in exploration and use of outer space should be carried out in conformity with international law including the Charter of the United Nations'.[65]

The Outer Space Treaty and the Moon Agreement have incorporated the stricter version of compliance with international law, including the Charter of the United Nations, as is evident in UN General Assembly Resolution 1802.[66] This application of international law has its own significance even after space law has emerged as an independent discipline in the modern period. Wherever the principles or provisions are absent or weak under space law, we can fall back on public international law or the Charter of the United Nations to prevent any misuse. However, it should also be understood that the application of international law should not result in making UN space treaties redundant. In essence, international law should be used only to supplement the space treaties whenever required.

[62] Arvid Pardo, Malta's representative to the United Nations, is considered to be the originator of the CHM principle. See John E Noyes, 'The Common Heritage of Mankind: Past, Present, and Future' (2011) 40(1) DenvJIntL&Pol 447, 456.

[63] See1967 Note verbale from the Permanent Mission of Malta (n 60). See also RP Anand, *Legal Regime of the Sea-Bed and the Developing Countries* (Thomson Press (India) Ltd 1975) 212.

[64] See UNGA Res 1721 (XVI) (20 December 1961).

[65] See UNGA Res 1802 (XVII) (14 December 1962).

[66] Outer Space Treaty, art III and Moon Agreement, art 2.

6 International Cooperation

Another aspect of significance that is dealt with under the first UN General Assembly resolution on space is the requirement of international cooperation. Insistence on international cooperation also stemmed from the success of State cooperation during the International Geophysical Year as well as under the Antarctic Treaty of 1959.[67] Though it seems to be a general aspect, international cooperation is of tremendous significance in the space sector, especially due to the fact that the less developed States can enjoy their freedom in space only with the cooperation of space-faring nations.[68] Cooperation between States is also found to be essential from the perspectives of space traffic management, ensuring peaceful uses, preventing environmental degradation, space research, sharing high cost of space activities, etc.

Given its tremendous importance, the need for cooperation between States is reflected in one form or another under all five UN space treaties. While Articles I, III and IX call for State cooperation in general, Articles V, X, XI and XII seek cooperation on specific aspects under the Outer Space Treaty. The Rescue Agreement is fully based on cooperation between the States since the effective system of rescue and return of personnel or space objects is possible only with cooperation among the States involved. The Liability Convention and the Registration Convention also reflect the requirement of cooperation as their implementation is not subject to a strict mandate of either a supervising body or a strong dispute settlement system. Though the Liability Convention envisages a dispute settlement system, it is not strong enough to provide a binding decision in the absence of the willingness of States.[69] Finally, the Moon Agreement carries forward the spirit of cooperation stipulated under the Outer Space Treaty and contains stronger provisions on cooperation between the States.[70]

Though the extent and specific nuances of cooperation between the States are not mentioned, the spirit of cooperation under the UN space treaties has yielded practical results. Cooperation in terms of allocation of orbits for space activities through the International Telecommunication Union (ITU) is one such example at the global level.[71] While there are several bilateral or plurilateral cooperative space

[67] Wei-chin Lee, 'China and Antarctica' (1990) XXX(6) AsianSurv 576, 576.

[68] Megumu Nakamura, 'Review of Article I of the Outer Space Treaty' (1997) 39 ProceColloqLOuterSpace 132, 133.

[69] See Liability Convention, art XIX(2).

[70] The requirement of cooperation under arts 5, 11 and 15 of the Moon Agreement, for example, are considered to be stronger than the provisions of the Outer Space Treaty. See Nicolas Mateesco Matte, 'The Draft Treaty of the Moon, Eight Years Later' (1978) III AnnAir&SpaceL 511, 516.

[71] See Lawrence D Roberts, 'A Lost Connection: Geostationary Satellite Networks and the International Telecommunication Union' (2000) 15(3) BerkeleyTechLJ 1095; See also Steven A Levy, 'Institutional Perspectives on the Allocation of Space Orbital Resources: The ITU, Common User Satellite Systems and Beyond' (1984) 16(2) CaseWResJIntlL 171.

projects, partnering of Cold War adversaries[72] in the International Space Station (ISS) project is a classic instance of space cooperation operating in practical terms. Despite the recent threat by Russia to exit ISS by 2024 in the wake of the Russia–Ukraine War,[73] the spirit of cooperation in ISS is worthy of appreciation.

7 Rescue and Return Norms

The space activities during the initial stages have been predominantly exploratory in nature. Such exploration was considered to be beneficial for every State, including the developing States. With the entry of human beings into outer space, the human entrants were regarded as representatives of mankind in outer space. This is reflected in declaring the astronauts to be envoys of mankind in outer space. This was first recognised in the UN General Assembly Resolution 1962 and subsequently concretised under Article V of the Outer Space Treaty. As a natural corollary to the elevated status of the astronauts as representatives of mankind, the Outer Space Treaty imposed an obligation of emergency assistance to astronauts by all States Parties. It also mandates States to disclose publicly any dangerous phenomena that they discover in outer space or on celestial bodies so as to prevent any hazard to the health or life of astronauts.

The above mandate of the Outer Space Treaty has been carried forward by the Rescue Agreement. In 1967, both the United States and the Soviet Union had setbacks in their human space missions due to the death of astronauts in the Apollo 1 and Soyuz 1 disasters.[74] This prompted a quick response to put in place detailed norms on the rescue and return of astronauts and space personnel under the Rescue Agreement. The duty to notify, the duty to search and rescue personnel and the duty to return them to the launching authority are laid down concretely in the Rescue Agreement. Separate norms on recovery and return of space objects are provided in addition to the rescue and return of personnel. While the duty to search and rescue personnel is absolute, the recovery and return of space objects are contingent upon the request of the launching authority. Moreover, the cost of the latter must be borne by the launching authority.[75] However, no such stipulation

[72] Governments of Russia and the United States collaborated with those of Canada, Japan and the member States of the European Space Agency to establish ISS. See Agreement among the Government of Canada, Governments of Member States of the European Space Agency, the Government of Japan, the Government of the Russian Federation, and the Government of the United States of America Concerning Cooperation on the Civil International Space Station (adopted 29 January 1998, entered into force 27 March 2001) TIAS 12927 (1998).

[73] Pjotr Sauer, 'Russia Says it Will Quit International Space Station after 2024' *The Guardian* (26 July 2022) <www.theguardian.com/science/2022/jul/26/russia-opt-out-international-space-station-2024-build-own-outpost> accessed 25 July 2023.

[74] See 'A Look Back at Space Disasters in the History of Mankind' *The Economic Times* (29 January 2017) <https://economictimes.indiatimes.com/news/science/a-lookback-at-space-disasters-in-the-history-of-mankind/articleshow/56838897.cms> accessed 26 July 2023.

[75] See Rescue Agreement, art 5(5).

regarding shouldering the burden of cost by the launching authority is made in relation to the search and rescue of the personnel. This distinction stems from the principle of astronauts as envoys of mankind, which is logically extended to space personnel but obviously not to space objects.

8 Responsibility and Liability Aspects

Reference to the norms on State responsibility and international liability for damage caused by space activities is seen initially in UN General Assembly Resolution 1962. Articles VI and VII of the Outer Space Treaty have carried them into the domain of hard law. As States were the only players in the domain of space activities during the negotiation of the UN space treaties, the Outer Space Treaty attributes State responsibility and international liability for space activities only to States. In shouldering international responsibility, States are also obligated to ensure that the provisions of the Outer Space Treaty are complied with by both governmental and non-governmental entities in carrying on space activities.[76] Attribution of State responsibility for private space activities also necessitates authorisation and continuing supervision of such activities by the appropriate State. In case of any space activity by an international organisation, Article VI of the Outer Space Treaty attributes responsibility collectively to such organisation as well as individually to all participating States in such organisation.

Since the payment of monetary compensation is directly related to the aspect of international liability, more detailed provisions on international liability for damage caused by space activities can be seen under Article VII of the Outer Space Treaty, read together with the Liability Convention. Article VII of the Outer Space Treaty channelises the liability on the launching State/s; however, it fails to distinguish between absolute liability and fault-based liability in different instances of damage caused by space activities. The Liability Convention takes up this task of distinction and also clarifies the application of joint and several liability[77] when more than one launching State is involved in causing damage. It also envisages the peaceful settlement of compensation claims through diplomatic channels.[78] Upon failure to reach an agreement on payment of compensation through diplomatic negotiations, the Liability Convention provides for the establishment of a Claims Commission to settle the matter amicably.[79]

[76] See Bin Cheng, 'Article VI of the 1967 Space Treaty Revisited: "International Responsibility", "National Activities", and "The Appropriate State"' (1998) 26(1) JSpaceL 7, 13.

[77] See Trevor Kehrer, 'Closing the Liability Loophole: The Liability Convention and the Future of Conflict in Space' (2019) 20(1) ChiJIntlL 178, 198–200.

[78] See Liability Convention, art IX.

[79] See ibid art XIV.

9 Environmental Protection

The entry into outer space took place during a time when there was not much international awareness of the need for environmental protection. The Stockholm Declaration, the most notable international instrument relating to the protection of the global environment and sustainability, came into existence only in 1972.[80] Accordingly, initial developments in international space law did not take into account the environmental implications of space activities. This is clearly evident in the approach of UN General Assembly Resolution 1962. This Resolution only mentioned the need to have due regard to the corresponding interests of all States in carrying on space activities. A weak narration on the need for international consultation is also found to ensure the avoidance of any harmful interference with the activities of other States.[81] However, there was no mention of the need to protect the environment of either outer space or the Earth.

When the Outer Space Treaty was drafted, a sentence on environmental protection was inserted unceremoniously in Article IX, between the sentences on due regard and international consultation, mentioned above. This addition reads as follows:

> States Parties to the Treaty shall pursue studies of outer space, including the moon and other celestial bodies, and conduct exploration of them so as to avoid their harmful contamination and also adverse changes in the environment of the Earth resulting from the introduction of extraterrestrial matter and, where necessary, shall adopt appropriate measures for this purpose.

Deplorably, neither 'harmful contamination' nor 'adverse changes' is defined under the Outer Space Treaty. The vagueness of this provision has resulted in several misinterpretations and space environmental damage is continuing unabated.[82] An increasing amount of space debris along with chemical, biological and radiological contamination is posing a serious threat of reversing the benefits of space exploration in the near future. Efforts to develop comprehensive international regimes on debris mitigation as well as planetary protection have failed, and they have only remained in the form of non-binding policies and guidelines.[83] Even though the Moon Agreement has incorporated better provisions to protect

[80] Minna Palmroth and others, 'Toward Sustainable Use of Space: Economic, Technological, and Legal Perspectives' (2021) 57 SpacePoly 1.

[81] See UNGA Res 1962 (XVIII) (13 December 1963) para 6.

[82] See Kevin J Gaston and others, 'Environmental Impacts of Increasing Numbers of Artificial Space Objects'(2023) 21(6) FrontEco&Envir 289.

[83] See Sandeepa Bhat B., 'Application of Environmental Law Principles for the Protection of the Outer Space Environment: A Feasibility Study' (2014) XXXIX AnnAir&SpaceL 323, 338–40.

the celestial environment,[84] it has very limited application for the reasons already outlined.

10 Registration

The last significant aspect of international space law to be discussed in connection with the United Nations' contribution is the registration of space objects. Reference to UN General Assembly Resolution 1962 shows that the aspects of jurisdiction and control are associated with the State of registration of space objects.[85] This has been carried forward under Article VIII of the Outer Space Treaty. However, the Outer Space Treaty did not stipulate the mechanism for the registration of space objects. With the expansion of space activities, the need for registration and transparency became increasingly apparent for a variety of reasons. These reasons include determination of jurisdiction and control, identifying the launching authority for the return of personnel or space objects under the Rescue Agreement, identifying the launching State for fixation of liability for damage caused and determination of applicable laws to deal with various issues arising out of activities in outer space.

In order to cater to the above requirements, the Registration Convention established a system of national and international registration of space objects by the launching State. The international registration at the UN registry is instrumental in maintaining transparency. However, it has limited success, as Article IV of the Registration Convention mandates States to provide information about only the 'general function of the space object' and not about 'all functions of the space object'. Hence, States deploying dual-use satellites have only provided information on their commercial uses and have kept the military uses confidential.[86] Moreover, there are many instances of States either not registering their space objects or delaying their registration.[87] Since the Registration Convention does not stipulate any consequence for failure to register, States have often failed to take it seriously.

11 Efforts to Cover Private Space Financing

Any discussion on international space law remains incomplete unless we cover the developments in private space financing. Although efforts to deal with legal aspects of private space financing are not made at the UN level, but rather at the level of

[84] Moon Agreement, art 7.

[85] See UNGA Res 1962 (XVIII) (13 December 1963) para 7.

[86] Yun Zhao, 'Revisiting the 1975 Registration Convention: Time for Revision?' (2004) 11 AustJIntlL 106, 119–20.

[87] See Ram S Jakhu and others, 'Critical Issues Related to Registration of Space Objects and Transparency of Space Activities' (2018) 143 ActaAstro 406, 410–11.

the International Institute for the Unification of Private Law (UNIDROIT),[88] their significance in the current era of space commercialisation is enormous. Since space activities require a huge amount of investment, the private space player cannot expect the space projects to be funded entirely within the same jurisdiction: cross-border investment becomes essential. However, the application of differing national laws to contracts involving space financing undermines the predictability of the return on investment for creditors.[89] This has disincentivised potential investors in the space sector.

It has become apparent that investor confidence in the space sector can be fostered only with a transparent regime of financing. The problem of unpredictable applicable financial laws is found pressing not just in the space sector but also in other high-value equipment, especially those which move from one jurisdiction to another. Hence, UNIDROIT has attempted to establish an international regime to harmonise national financing laws relating to mobile equipment.[90] The regime consists of the Cape Town Convention[91] together with area-specific protocols that are adopted in the domain of different types of mobile equipment.[92] The UNIDROIT system provides for the creation and registration of international interests, the rule of priority, default remedies and insolvency remedies for creditors to ensure the existence of a creditor-friendly atmosphere in financing mobile equipment. Unfortunately, it is yet to enter into force in the space sector.

12 Conclusion

It is evident that international space law has increased by leaps and bounds over the years. As an offshoot of international law, international space law has established itself as an independent branch within a relatively short period. However, this is not to say that international law is of no relevance in dealing with the space sector. As discussed above, international law still forms the basis and fills in the gaps of international space law wherever required. The essence of international space law lies in the assertion of the common rights of all in outer space and celestial bodies. Since space and its resources are the bounties of nature, respecting the rights of

[88] See Paul B Larsen and Juergen A Heilbock, 'UNIDROIT Project on Security Interests: How the Project Affects Space Objects' (1999) 64(3) JAirL&Com 703.

[89] 'Cape Town Convention on Financing of High-Value, Mobile Equipment' (2004) 98(4) AJIL 852, 852.

[90] See Martin J Stanford and Alexandre de Fontmichel, 'Overview of the Current Situation Regarding the Preliminary Draft Space Property Protocol and its Examination by the COPUOS' (2001) 6(1) UnifLRev 60.

[91] Convention on International Interests in Mobile Equipment (adopted 16 November 2001, entered into force 1 April 2004) 2307 UNTS 285.

[92] In the space sector, the Protocol to the Convention on International Interests in Mobile Equipment on Matters Specific to Space Assets 2012 has been drafted.

all while using them attains great significance. Hence, international cooperation, prohibiting individual claims, peaceful uses and environmental protection have become central themes of international space law. Modern-day developments in space commercialisation have necessitated the evolution of new principles and norms, which are undertaken at both international and national levels.

2
THE NEVER-ENDING PROBLEM OF DEMARCATION: ADDRESSING THE AIR/SPACE BOUNDARY ISSUE IN INTERNATIONAL AND DOMESTIC LAW

Christopher J Newman[*]

1 Introduction

Attempts to identify a universally accepted and applicable boundary between air and space have been a topic of contention among legal scholars, policymakers, and scientists since the early years of spaceflight.[1] The quest for an explicit demarcation between these realms and differentiating between air and space would, at first blush, appear to be a relatively straightforward task of identifying an appropriate altitude. The reality, however, has shown it to be a much more nuanced discussion between a broad range of stakeholders incorporating the function and nature of both the craft involved and the overall mission, as well as a spatialist element relating to the location and altitude of the craft.[2] Various solutions have been posited, all of which acknowledge the issues at stake but each of which has so far been unsuccessful in providing a definitive, universally accepted resolution.[3] This chapter will provide a significant contribution to understanding the complexities inherent

* Thanks are due to Dr Lauren Napier of Northumbria Law School for her assistance on the writing of this chapter. All errors and omissions are the responsibility of the author.

[1] Indeed some of the foundational discussions come before the launch of Sputnik, specifically JC Cooper, 'High Altitude Flight and National Sovereignty' (1951) 4 ILQ 411–18 and Andrew G Haley, 'Space Law and Metalaw: Jurisdiction Defined?' (1957) 24 JAirL&Com 286–303. See also FB Schick, 'Space Law and Space Politics' (1961) 10 ILQ 681–706.

[2] For the origins and discussions on functionalist and spatialist approaches, see Bin Cheng, 'Legal Regime of Air Space and Outer Space: The Boundary Problem, Functionalism versus Spatialism: The Major Premises' (1980) 5 AnnAir&SpaceL 323, 335–38.

[3] For considered discussions on the extant position, see Marietta Benko and Engelbert Plescher, *Space Law: Reconsidering the Definition/Delimitation Question and the Passage of Spacecraft through Foreign Airspace* (Eleven Publishing 2013) and Paul Dempsey and Maria Manoli, 'Suborbital Flights and the Delimitation of Air Space vis-à-vis Outer Space: Functionalism, Spatialism and State Sovereignty' (2017) 42 AnnAir&SpaceL 209–51.

Christopher J Newman, *The Never-Ending Problem of Demarcation* In: *International Space Law in the New Space Era.* Edited by: Sandeepa Bhat B., Dilip Ukey, and Adithya Variath, Oxford University Press. © Christopher J Newman 2024.
DOI: 10.1093/9780198909415.003.0002

in this seemingly never-ending problem of demarcation and will examine options for moving forward on this issue.

The air/space boundary problem is of significance because there are numerous practical issues at play, beyond simply being abstract academic discourse. On an elemental level, the demarcation of the air/space boundary represents the point of transition between two fundamentally conflicting legal regimes. Airspace is under the exclusive sovereign control of subjacent states. Outer space, on the other hand, is not subject to national appropriation by claims of sovereignty. Accordingly, the question of the demarcation between air and space has legal, technological, and geopolitical ramifications. Matters such as sovereign control, jurisdiction, liability, and insurance all hinge on a clear understanding of which legal regime is operating. Consequently, finding a definitive demarcation point has far-reaching implications that resonate in both theoretical and practical contexts.

This chapter aims to present a rigorous analysis of the multifaceted factors contributing to the air/space boundary issue. There will be an exploration of the differing attempts to establish a demarcation line, ranging from altitude-based, spatialist definitions to more functionalist approaches that examine the nature and function of the craft. The extant position, such as it is, in international law, and agreements such as the Outer Space Treaty 1967,[4] will also be considered, although it should be noted at the outset that the 1967 Treaty and its progenitor treaties are silent on the physical location of the air/space boundary. Moreover, in an original contribution to the discussion, this chapter will examine the challenges posed by technological advancements that blur the lines between air and space, including the rise of suborbital flights, and the development of High-Altitude Pseudo Satellites/High-Altitude Platform Stations (HAPS).[5] It will be suggested that these new applications necessitate a revisiting of the functionalist approaches and a revivification of regulation within the specific altitudes being proposed.

In respect of methodology, the very nature of the topic means that there will be a mixed approach to the legal analysis contained herein. There will be consideration of such doctrinal law as exists in the area, and this includes international treaties and other binding agreements, as well as national legislative attempts to demark the air/space boundary. The chapter will also engage with theoretical perspectives surrounding boundaries, highlighting the inherent difficulties in drawing a clear line between two complex and interconnected realms. To understand the broader picture of how the question of demarcation could be addressed, the operation of so-called 'soft law'—non-legally binding agreements and international standards—will be examined as well.

[4] Treaty on Principles Governing the Activities of States in the Exploration and Use of Outer Space, Including the Moon and Other Celestial Bodies (adopted 27 January 1967, entered into force 10 October 1967) 610 UNTS 205 (Outer Space Treaty).

[5] For background discussion, see 'HAPS—High-Altitude Platform Systems' (*ITU*) <www.itu.int/en/mediacentre/backgrounders/Pages/High-altitude-platform-systems.aspx> accessed 2 October 2023.

It is important to note that the central thesis of this work is very much that sovereignty and demarcation of outer space are, fundamentally, political questions that need solutions by states. When those political questions are answered they will provide the contours of the legal regime that operates in airspace and outer space. Much of the discussion regarding the demarcation and delineation of outer space and airspace has its basis in the science and engineering of aerospace and astronautical studies. This work does not seek to evaluate demarcation on a scientific basis and does not claim any scientific credentials to do this. Where scientific studies are used, they come with this caveat and are deployed to provide additional context.

The work will draw from interdisciplinary research and examine competing attempts to resolve the demarcation issue. The chapter will start with an examination of the definitions that do exist in international law, and the sources of those definitions, both of airspace and outer space, from a purely spatialist perspective. There will then be a critique of the attempts to identify the start of outer space, taking account of both functionalism and spatialism as possible approaches. The chapter will conclude by proposing a hybrid system that utilises both functionalism and specialism within existing legal instruments. By taking account of the latest technological developments, such as the nascent space tourism industry[6] and—crucially—the development of HAPS, this chapter will provide an original and contemporary insight into resolving the never-ending question of where airspace ends and where outer space begins.

2 Introducing the Spatialist Approach

States exercise sovereignty in a 'three-dimensional quasi-cone'.[7] If land boundaries and territorial sea provide the horizontal dimension, then airspace, at its most fundamental level, is the vertical area above the Earth, the skies, and the atmosphere, which are subject to state control. Delineation of airspace involves dividing the portion of the atmosphere above countries into different flight information regions (FIR) based on various factors such as altitude, purpose, and control. Each FIR is managed by an aviation authority. It is governments and international bodies, therefore, that have assumed responsibility for regulating airspace to ensure safe and efficient air traffic management.[8]

The delineation of airspace is essential for providing standards and regulations for aircraft operations, airspace control, and coordination among different aviation

[6] See Christopher J Newman, 'Near Space—the Air-Space Boundary Question, Astronauts and Space Tourism' (2021) 3(29) ROOM:SpaceJAsgardia 80–85.

[7] Francis Lyall and Paul Larsen, *Space Law: A Treatise* (2nd edn, Routledge 2018) 135.

[8] For further details, see 'Understanding Airspace' (*National Air Traffic Services)* <www.nats.aero/airspace/introduction/> accessed 2 October 2023.

stakeholders. Different classes of airspace have different rules and restrictions. There is also airspace above non-sovereign areas on the Earth, such as the high seas, and delineation of airspace can also include the delineation of airspace and outer space. As a matter of air law and air traffic management, the airspace above the high seas may be subject to air traffic control and administration. For the purpose of this discussion, the focus will be on the issues related to airspace as it relates to state sovereignty.[9]

Demarcation, on the other hand, refers to the setting or marking of boundaries or limits between two distinct areas or regions. In the context of airspace and outer space, demarcation would be the establishment of a clear boundary between the highest point of airspace (typically at a specific altitude) and the beginning of outer space. To understand the difficulties in demarcation, it is necessary to first consider *where* a boundary might exist between the two competing legal systems of air and space law. Whilst there have been several ways suggested, the essence of these can be aggregated into two fundamental positions.

The first of these approaches to demarcation can be described as the *spatialist* approach. This is perhaps the most intuitive, as it looks at the physical location occupied by a craft and determines the legal regime in relation to that position.[10] Yet the idea of borders, which tends to work well in the terrestrial context, does not work quite so efficiently when considering higher altitudes in the vertical airspace above the subjacent states. While altitude may provide a fixed point, an agreement is not universal as to where that fixed point should be,[11] even though there is a general understanding of where the boundaries lie.[12]

3 The Delineation of Airspace

Before looking at the upper boundaries between airspace and outer space, it is useful to briefly consider the way in which airspace is delineated and organised. This is important for two reasons. First, airspace is a significantly more mature regime than outer space, being developed through international agreements dating back to 1919.[13] The second reason is that civilian aviation is concerned with the

[9] See Lyall and Larsen (n 7) 136 and Outer Space Treaty (n 4) for further details.

[10] Ram S Jakhu, Tommaso Sgobba, and Paul Stephen Dempsey (eds), *The Need for an Integrated Regulatory Regime for Aviation and Space* (Springer-Verlag 2011) 50–61.

[11] For a comprehensive overview on the origins of the attempts to create a legal boundary for airspace and outer space, see Lyall and Larsen (n 7) 135–62.

[12] Hao Liu and Fabio Tronchetti, 'The Exclusive Utilization Space: A New Approach to the Management and Utilization of Near Space' (2019) 40(3) UPaJIntlL 537, 548.

[13] The 1910 Conference on the regulation of Air Navigation saw the first proposals for international coordination of airspace and aircraft registration. This, coupled with rapid developments in aerospace technology and the events of the First World War, led to the International Convention on Aerial Navigation, Paris (adopted 13 October 1919, entered into force 11 July 1922) 11 LNTS 173 (Paris Convention). Art 1 of the 1919 Convention recognised that 'every Power has complete and exclusive authority over the airspace above its territory'. For details as to how the language of this provision created

safety of the vast number of passengers who use aircraft every day. The regime contrasts starkly with space travel, where the overwhelming majority of missions involve robots, not humans. Similarly, every major country will have at least one airport whereas the number of launch nations is only 13,[14] despite nearly 65 years since the launch of the first artificial satellite. How aviation law manages significant numbers of aircraft safely and predictably should attract the envious gaze of students of space law.[15]

The basis of modern public international aviation law can be found in the 1944 Convention on International Civil Aviation (known colloquially as the Chicago Convention).[16] Article 1 of the Chicago Convention affirms earlier aviation agreements and provides that '[e]very state has complete and exclusive sovereignty over the airspace above its territory'. Furthermore, as can be found in Article 3, this applies to civil aircraft.[17] Additionally, under the terms of the Chicago Convention, the International Civil Aviation Organization (ICAO) was formed. This is a specialised agency of the UN that coordinates the planning and development of international air transport to ensure safe and orderly growth in civilian aviation.[18]

The legal regime for airspace, therefore, is controlled by states and a domain which is used by aircraft.[19] ICAO defines an aircraft as 'any machine that can derive support in the atmosphere from the reactions of the air other than the reactions of the air against the Earth's surface'.[20] Annex 7 to the Chicago Convention proceeds to define a whole range of craft that is a subset of aircraft, including power-driven and non-power-driven aircraft, lighter than air and heavier than air craft. This

international custom by means of a treaty provision (art 1 does not limit the recognition to contracting parties), see Lyall and Larsen (n 7) 140. For a more general appraisal of the development of air law during the interwar years, see Malgorzata Polkowska, 'The Development of Air Law: From the Paris Conference of 1910 to the Chicago Convention of 1944' (2008) 33 AnnAir&SpaceL 59.

[14] Thirteen countries and the European Space Agency. See 'Countries Capable of Launching Space Rockets' (*statista.com*) <www.statista.com/chart/27792/countries-capable-of-launching-space-rockets/> accessed 2 October 2023.

[15] International air law is governed by a large number of multilateral agreements. See Dempsey and Manoli (n 3) 202 for further details.

[16] Convention on International Civil Aviation (adopted 7 December 1944, entered into force 4 April 1947) 15 UNTS 295 (Chicago Convention).

[17] ibid art 3(a), which provides that the Convention shall apply only to civil aircraft and is not applicable to state aircraft (defined in art 3(b)). State aircraft of one state are not permitted to fly over the territory of another state without authorisation, whilst art 5 provides the details of when flyover of civilian aircraft will be permitted.

[18] Contained within the Annexes to the Chicago Convention (n 16) (and adopted by the Council of ICAO) are technical rules and procedures, known as Standards and Recommended Practices (SARPs). These ensure uniformity and are designed to improve the rules governing air travel. SARPs are not a substantive part of the Convention and are not binding upon states. In practice, however, SARPs are a significant part of the overall air governance framework.

[19] Matthew T King, 'Sovereignty's Gray Area: The Delimitation of Air and Space in the Context of Aerospace Vehicles and the Use of Force' (2016) 81(3) J AirL&Com 377, 402–03, and 408.

[20] Chicago Convention (n 16) Annex 7(1) Definitions.

broad definition covers a wide variety of powered craft (naturally including such things as aeroplanes and helicopters) and includes balloons and gliders.

Through the wide range of international agreements and ICAO coordination, the control of aircraft in the airspace is broken down globally into Flight Information Regions (FIR). These FIR are then further delineated by controlled and uncontrolled regions of airspace. Within these two categories, there are four subsets: controlled, uncontrolled, special use, and other airspace.[21] For this discussion, and in trying to identify the upper limits of airspace, it is controlled airspace, managed by national civil aviation authorities, by means of air traffic control, which is of most interest. The most significant area of controlled airspace, according to the classification adopted by the United States (US) Federal Aviation Authority (FAA), is Class A airspace.[22] This is the airspace where commercial passenger planes operate, starting from an altitude of 18,000 ft (5.5 km). The core of civilian operations now takes place around 38,000 ft (11.6 km)[23] to 40,000 ft (12.2 km) at FL400,[24] and Class A airspace ends at 60,000 ft (18.3 km or FL600).[25] If the aircraft is operating above FL600 then it is still within Class E controlled airspace, but this has no upper altitude.[26]

The conclusions that can be drawn from this assessment of airspace assist greatly in identifying the issues that cause such problems in the demarcation of an airspace/space spatial boundary. First, as will be shown, the regime governing airspace is significantly more mature than that which governs space. This is wholly understandable. The fruits of aerospace technology are available for mass consumption

[21] See *NATS* (n 8).

[22] Title 14 Code of Federal Regulations §71.33 defines Class A airspace areas as:

(a) That airspace of the United States, including that airspace overlying the waters within 12 nautical miles of the coast of the 48 contiguous States, from 18,000 feet MSL to and including FL600 excluding the states of Alaska and Hawaii.

(b) That airspace of the State of Alaska, including that airspace overlying the waters within 12 nautical miles of the coast, from 18,000 feet MSL to and including FL600 but not including the airspace less than 1,500 feet above the surface of the earth and the Alaska Peninsula west of longitude 160°00′00″ West.

(c) The airspace areas listed as offshore airspace areas in subpart A of FAA Order JO 7400.11G (incorporated by reference, see § 71.1) that are designated in international airspace within areas of domestic radio navigational signal or ATC radar coverage, and within which domestic ATC procedures are applied.

[23] Liu and Tronchetti (n 12) 540.

[24] Flight Level refers to a level of constant atmospheric pressure related to a reference datum of 29.92 inches of mercury. Each is stated in three digits that represent hundreds of feet. For example, flight level (FL) 250 represents a barometric altimeter indication of 25,000 feet. Federal Aviation Administration (FAA), 'Pilot/Controller Glossary', 'Aeronautical Information Manual' (*FAA*, 29 March 2018) <www.faa.gov/air_traffic/publications/media/pcg_basic_chgs%201_2_9-13-18.pdf> accessed 2 October 2023.

[25] Most European civil aviation authorities have regulated airspace below FL660 or 66,000 ft (20.11 km), see *NATS* (n 8). For the purpose of this discussion, FL600 will be the recognised limit of commercially controlled airspace.

[26] Title 14 Code of Federal Regulations Part 71 subpart E(a) states that (inter alia) Class E airspace consists of the airspace of the United States, including that airspace overlying the waters within 12 nautical miles of the coast of the 48 contiguous states and Alaska, extending upward from 14,500 feet mean sea level (MSL) up to, but not including 18,000 feet MSL, and the airspace above FL600.

by a wide number of nations. Millions of people use planes, and the safe use of these craft depends on an efficient and coherent system of regulation and governance. This is just not the case for space technology, where a relatively small number of states can access space and an even smaller number can be said to be dominant in the use of space.

Similarly, the delineation of airspace broadly corresponds with understood terrestrial land and maritime borders. Territorial sovereignty can be asserted by subjacent states and must be respected by other states.[27] While FL600 represents the limits of commercially controlled airspace, it does not represent the upper limits of state sovereign airspace. Experimental aircraft have flown beyond FL600, and a US U2 aircraft operated by the Central Intelligence Agency (CIA) to fly over the old Soviet Union without authorisation was shot down at 70,500 ft (21.5 km) in the ultimate expression of protection of that sovereignty. That the US did not protest the shooting down means it can safely be said that vertical sovereignty extends *at least* to 21.5 km.[28] The absolute nature of sovereign airspace also means that the areas of operation for commercial aircraft are well-established routes provided for by a whole range of multilateral agreements.[29]

All of these agreements, and indeed the ICAO, have focused upon the delineation of airspace to manage that airspace peacefully and safely. Neither the Chicago Convention nor any subsequent agreement explicitly provides the upper limit of sovereign airspace.[30] Equally, there is no explicit mention of how the regime managing airspace works *with respect* to the adjacent regime of space law. It will, therefore, be necessary to examine the operative elements of international space law to identify not only the areas of responsibility but also the areas of conflict and uncertainty that could arise between these two systems of governance.

4 The Law of Outer Space and the Spatialist Approach to Demarcation of Outer Space

The legal basis for human exploration of outer space can be found in the Outer Space Treaty of 1967. Article I of the Treaty provides signatory states with the rights to use, explore, and conduct scientific exploration of outer space on the basis of equality, in accordance with international law, and with free access to all areas of celestial bodies. The Treaty then places agreed limitations on these rights to prevent unfettered exploration and exploitation. Yet despite identifying the scope of the Treaty as covering 'outer space, including the Moon and other celestial bodies',

[27] King (n 19) 418.
[28] Bin Cheng, *Studies in International Space Law* (OUP 1997) 105.
[29] King (n 19) 418.
[30] ibid 406.

it does not contain any precise definition of 'outer space' or 'celestial bodies'. As one writer notes, 'it is astonishing that there is no commonly accepted definition of the outer space and the celestial bodies'.[31]

The Outer Space Treaty creates a legal regime that is not only separate from air law but also diametrically opposed to it. Article I of the Treaty arguably eliminates the notion of sovereignty in outer space.[32] Article II goes a stage further and explicitly forbids national appropriation of outer space by states by claim of sovereignty, by means of use or occupation, or by any other means.[33] The Treaty affirms that outer space is not to be regarded as a lawless frontier and affirms that the rule of law exists by means of Article III, which provides that states shall carry on the use of outer space in accordance with international law. The general rules of international law apply to relations between states wherever they occur, be it on land, sea, in airspace, or outer space.[34] Alas, this does not assist in the quest for a demarcated border given the silence of air and space law on the altitude where air law stops and gives way to the law of outer space.

This provides a dilemmatic choice. At some stage, the sanctity of national sovereign airspace must yield to the realities of the space environment. As well as the prohibitions on appropriation and sovereignty explicitly made in the Treaty, Lyall and Larsen illustrate the difficulty of the 'lighthouse effect'. If state sovereignty were unlimited and extended indefinitely upwards (creating a 'cone of sovereignty') the Moon and other celestial bodies would, every so often, come within that cone producing irrational and 'absurd' flickers of sovereignty that appeared and disappeared with the rotation of the Earth.[35]

Additionally, the physical realities of the orbital environment mean that satellites that are in Low Earth Orbit (LEO) orbit the Earth at a speed over 28,000 kph. This means they will fly over some countries at such a speed as to make any sort of enforcement extremely difficult. But more than that, given the number of space objects in orbit, the inference is that overflight by satellites is permitted, and no agreement is required. There needs to be, therefore, a point where the sovereignty recognised by air law gives way to the sovereign-free regime of the Outer Space Treaty.

The atmosphere of the Earth does not simply cease to exist at a given altitude. Instead, it gets less dense as the altitude increases.[36] There is no single, identifiable

[31] Stephan Hobe, 'Article I' in Stephan Hobe, Bernhard Schmidt-Tedd, and Kai-Uwe Schrogl (eds), *Cologne Commentary on Space Law*, vol I (Carl Heymanns Verlag 2009) 32.

[32] Ram Jakhu, 'Legal Issues Relating to the Global Public Interest in Outer Space' (2006) 32 JSpaceL 31, 39.

[33] Outer Space Treaty (n 4) art II.

[34] Olivier Ribbelink, 'Article III' in Stephan Hobe, Bernhard Schmidt-Tedd, and Kai-Uwe Schrogl (eds), *Cologne Commentary on Space Law*, vol I (Carl Heymanns Verlag 2009) 65.

[35] Lyall and Larsen (n 7) 144.

[36] Jonathan C McDowell, 'The Edge of Space: Revisiting the Kármán Line' (2018) 151 ActaAstro 668, 669–70.

point where the physical environment changes from being 'Earth' to 'space'. Instead, theorists have sought to identify an altitude where the characteristics of the atmosphere change sufficiently to be recognised as the point where aeronautics gives way to astronautics.[37] Aerospace engineer and physicist Theodore von Kármán examined the nature of aerodynamic lift and calculated the altitude at which (crudely put) the thinness of the air would prevent a craft from relying on lift from the air and would have to travel at a velocity so fast, that it would cease relying on lift produced. Instead, it would have to rely on centrifugal force and become an orbiting craft instead of an aircraft supported by lift from the air.

At present, therefore, 100 km (62 miles) is the altitude that has the broadest acceptance amongst the international community[38] as the boundary where space begins. Yet whilst broadly accepted, this 100 km boundary has been subject to criticism by numerous academics, scientists, and aerospace engineers. It is not the purpose of this discussion to engage with the scientific and technical dimensions of the air/space boundary. There have been discussions both in the United Nations Committee of the Peaceful Uses of Outer Space (COPUOS) and more broadly to identify the spatial boundary since the dawn of human activity in outer space.[39] As with any human construct, the spatial boundary between Earth and outer space depends on the twin tracks of establishing a scientific rationale for the boundary and that boundary gaining sufficient acceptance within the international community for it to be broadly accepted.

Writing in 1961, Andrew Haley suggested that the proposal by the *Fédération aéronautique internationale* (FAI) for a 100 km line as a boundary 'coincides with the Kármán line theory'.[40] Yet, despite the widespread recognition of the 100 km line as representing the altitude at which space begins, this is not a universally accepted position. Numerous academics and engineers have argued that 100 km is too high. Hobe proposes that the true location of the von Kármán line is around 84 km. McDowell, in verifying the true von Kármán line as being at 84 km (52 miles), has gone on to posit that there is a moderately well-defined boundary of space at a boundary of 80 km (50 miles) which roughly corresponds to the mesopause.[41] This is the 'outermost physical boundary of the Earth's atmosphere where meteors typically burn up'. This also corresponds to the US Armed Forces' recognition of the height needed to gain astronaut wings.

With an altitude of 110 km being generally regarded as outer space, it is clear that the status of the zone between 80 and 110 km can be described as being 'highly

[37] S Sanz Fernández de Córdoba, 'The 100 km Boundary for Astronautics' (*Fédération aéronautique internationale*, 21 June 2004) <www.fai.org/page/icare-boundary> accessed 2 October 2023.

[38] Stephan Hobe, 'Legal Aspects of Space Tourism' (2007) 86(2) NebLRev 439, 442.

[39] UNCOPUOS, 'Historical Summary on the Consideration of the Question on the Definition and Delimitation of Outer Space' UN Doc A/AC.105/769/Add.1 (3 February 2020) 1–3.

[40] Thomas Gangale, 'The Non Kármán Line: An Urban Legend of the Space Age' (2017) 41(2) JSpaceL 151, 159.

[41] McDowell (n 36) 676.

controversial'.[42] Indeed, since 1967 the issue of the definition and delimitation of outer space has been held to be an issue of sufficient importance within the field of space law to be a standing agenda item at COPUOS.[43] Yet despite this self-proclaimed urgency, consensus continued to evade the Committee on this issue.[44] Some states within COPUOS recognised that the delimitation of outer space could start at 100 km above sea level, the distance now synonymous with the von Kármán line. Other states felt that the current framework had presented no practical difficulties and had served everyone well, and activities in outer space were flourishing. They advocated that things should continue as they were until there was a demonstrated need and a practical basis for developing such a definition or delimitation.[45]

Different national legislators have sought to address the demarcation by including a boundary above which a licence for outer space activity is required.[46] Indeed, Australia, in 2006 specifically stated that there is no actual definition of 'outer space' in domestic Australian law and Australia recognises that there is no internationally accepted definition or delimitation of the term. Identifying the 100 km altitude in the Act was not an attempt to define or delimit 'outer space' but merely a practical clarification of where the 1998 Act applies.[47] Kazakhstan[48] and Denmark,[49] following suit, *did* use domestic legislation to define the demarcation point between outer space and airspace. In respect of broader international organisations, the International Telecommunications Union (ITU) has referred to the 100 km line as being usually assumed to be the boundary between Earth's atmosphere and outer space [50]

With recent advances in the space tourism industry, specifically within the US, these discussions are moving from a scholarly to a more practical focus.[51] This change came about following the successful launch, in July 2022, of Virgin

[42] Hobe (n 31) 31.

[43] UNCOPUOS (n 39) 4–8.

[44] UNCOPUOS works by consensus and the strenuous objection of one or more states is sufficient to obstruct progress on an issue. See Olavo de Oliveira Bittencourt Neto, 'Delimitation of Outer Space and Earth Orbits' in Yanal Abul Failat and Anél Ferreira-Snyman (eds), *Outer Space Law: Legal Policy and Practice* (Globe Law and Business 2022) 49, 54.

[45] UNCOPUOS (n 39) 3.

[46] For example, the Space Activities Act 1998 in Australia as amended by Space (Launches and Returns) Act 2018 (Cth), s 8 defines a space object as (a) an object the whole or a part of which is to go into or come back from an area beyond the distance of 100 km above mean sea level; or (b) any part of such an object, even if the part is to go only some of the way towards or back from an area beyond the distance of 100 km above mean sea level.

[47] UNCOPUOS, 'National Legislation and Practise Relating to Definition and Delimitation of Outer Space' UN Doc A/AC.105/865/Add.1 (20 March 2006).

[48] As a counter to the Australian approach, the Kazakhstan Outer Space Activities Act, adopted on 6 January 2012, provides that 'outer space is defined as space extending beyond airspace at a height exceeding 100 kilometres above sea level', art 1, para 6.

[49] Outer Space Act, s 4.4 (Denmark) defines 'outer space' as 'Space above the altitude of 100km above sea level'.

[50] ITU Res 763 (WRC-15).

[51] 'We need to adopt a definition of space, even if it is not a legal one. I do not argue that there needs to be a single definition of "space" that applies in all contexts. Physicists, lawyers, and historians may need a boundary of space for different purposes and to address different questions; the "edge of space" might be defined differently in different international fora. Nevertheless, it is useful for those definitions to be

Galactic's SpaceShipTwo and the suborbital tourism vehicle, the VSS Unity, followed nine days later by Blue Origin's New Shepard. The VSS Unity achieved an apogee of 86 km, while the New Shepard craft reached just over 107 km.[52] The success of space tourism as an industry depends on the passengers going 'into space'. The demarcation between airspace and outer space now has a commercial edge to add to the theoretical discussions. The start of non-governmental, recreational spaceflight represents a renewed challenge for those looking at the legal boundaries of space and Earth.

5 Objections to the Spatialist Approach

Clearly, the lack of consensus as to where to draw the line of demarcation is the fundamental problem with the spatialist approach: 100 km may be broadly accepted, but mere broad acceptance is not enough to satisfy the setting of a boundary between two fundamentally different legal regimes. At present there is both enough dissent and insufficient evidence of established state practice amongst the major space nations to see 100 km become customary international law. This lack of certainty and broad acceptance alone creates a fundamental problem with the spatialist position. There are, however, other issues with the spatialist position that need exploring.

The first is the political dimension of sovereignty in respect of how much airspace above their territory states will be willing to concede to have a scientifically valid boundary demarcating the start of outer space. Having accepted that there is a dispute over the 100 km demarcation line in international diplomatic circles, it is hard to see how states will respond to an attempt to lower that line for the sake of achieving a scientifically more plausible boundary whilst ceding possibly 20 km of vertical sovereign territory. 'Territory is an existential matter for states and threats to it are not taken lightly.'[53] Despite the deliberations in COPUOS stressing that any demarcation line would not necessarily create an upper ceiling for airspace, it is hard to see how this would practically be the case. The development of hypersonic glide vehicle technology may well mean that states are ever more conscious about the sanctity of the upper limits of airspace[54] and reluctant to define the upper limits of sovereignty.

based on a common and accurate understanding for the physical conditions at the air-space boundary', McDowell (n 36) 669.

[52] Newman (n 6) 85.

[53] King (n 19) 395.

[54] Iain Boyd, 'China's Hypersonic Missiles Threaten US Power in the Pacific' (*The Conversation*, 24 May 2023) <https://theconversation.com/chinas-hypersonic-missiles-threaten-us-power-in-the-pacific-an-aerospace-engineer-explains-how-the-weapons-work-and-the-unique-threats-they-pose-206271> accessed 2 October 2023.

By adopting an altitudinal boundary there are also issues regarding the physical transit through the boundary. The first concerns those craft which skirt the boundary of outer space, (such as the aforementioned VSS Unity, which reached an altitude of 86 km, satisfying the McDowell original 'Kármán' line definition of space).[55] Crafts such as these spend most of their mission life in the zone beneath any physical line of demarcation for space and even the Blue Origin craft, which exceeded the 100 km barrier, was above this line for only a few minutes.[56] As Dempsey and Manoli identify, if a legal question arose during a flight, near the point of demarcation between airspace and outer space, it might be extremely difficult to determine on which side of the line the event occurred.[57] Furthermore, a continuing act might mean the legal question occurred in both legal systems, one which is subject to the sovereignty of a country and one which at the moment has no well-established dispute resolution mechanisms. It is not hard to see how this could lead to significant confusion and uncertainty.

6 Functionalist Approach to Jurisdiction

Looking at the location of the vehicle and deciding the applicable regime based on its altitude appears to ignore the purpose and nature of both the craft and the mission. The Chicago Convention explicitly provides that the regime of air law that it outlines applies to aircraft.[58] Conversely, the Outer Space Treaty provisions apply to 'space objects'.[59] Such questions do not require an examination of the location. Instead, they refer to the *function* of the craft in question. Such an approach points to the second possible method of deciding which legal regime is appropriate.

This approach to identifying the legal status of a craft from its characteristics and role is known as the *functionalist approach*. Instead of examining the altitude at which a craft is operating it concentrates upon 'the nature or purpose of a given activity'.[60] Strict functionalism eschews altitude and instead looks at the status, mechanical and technological characteristics, and possibly even its destination.

[55] Newman (n 6) 81.

[56] The entire mission of New Shepard lasted only 11 minutes; for full details, see 'New Shepard' (*Blue Origin*) <www.blueorigin.com/new-shepard/> accessed 2 October 2023.

[57] Dempsey and Manoli (n 3) 222.

[58] See above, but specifically art 3(a) of the Chicago Convention (n 16) provides that the Convention shall apply only to civil aircraft and is not applicable to state aircraft (defined in art 3(b)).

[59] 'Objects launched into space' are mentioned in arts VII and VIII of the Outer Space Treaty (n 4). The term space object is defined by reference to itself in the Convention on International Liability for Damage Caused by Space Objects (adopted 29 March 1972, entered into force 1 September 1972) 961 UNTS 187 (Liability Convention) and the Convention on Registration of Objects Launched into Outer Space (adopted 14 January 1975, entered into force 15 September 1976) 1023 UNTS 15 (Registration Convention) as 'including component parts of a space object as well as its launch vehicle and parts thereof'.

[60] Stephan Hobe, Geradine Meishan Goh, and Julia Neumann, 'Space Tourism Activities—Emerging Challenges to Air and Space Law' (2007) 33 JSpaceL 359, 362.

If the vehicle can 'derive support in the atmosphere from the reactions of the air other than the reactions of the air against the Earth's surface',[61] then, at first blush, it would appear to be an aircraft and subject to air law. On the other hand, if the vehicle was designed to operate in outer space, needs rocket thrust to move it through airspace, and 'can move in outer space without support from air and with a power source not dependent on external oxygen',[62] then it almost certainly will be a space object.

7 Hybrid 'Space Planes', HAPS, and Functionalism

One of the core reasons that nations have not been in a hurry to make progress on the air/space boundary is that, simply put, there have been no issues with the current ambiguity. Air travel and air law have developed as they have needed to do, and there has been no requirement to address the boundary as most space objects simply transit through the boundary to be operational in outer space. Where there is a failed launch, the authorities have liaised successfully to prevent civil aviation accidents, and there is already a clear understanding that the safety of civil aviation is paramount.[63] This is the case under present technical conditions. However, the question arises as to how to manage once technology allows planes to operate higher in the atmosphere and also how to deal with space planes that exhibit characteristics of both aircraft and spacecraft.

The idea of a spacecraft that can also travel through airspace is not a new one. The Space Shuttle 'during its launch and ascent was supported by rockets and during the orbital flight, then an aircraft during descent and landing'.[64] The shuttle, being classed as a 'state aircraft', would not have been subject to the terms of the Chicago Convention, and functionalists would point to the design and purpose of the Shuttle being a spacecraft that uses aeronautics only for a short period when it is coming in to land. The VSS Unity poses a more interesting challenge to a functionalist. It is clearly designed and intended to take people into 'space'. Yet it uses a carrier aircraft to raise its altitude to FL500; from there, it separates and uses a rocket engine to take it up to its perigee. It spends only a few minutes in space, then uses aerodynamics to return to an airfield. Despite the mission focus being on spending time in space, (a) the craft does not achieve orbit; (b) almost all the time is spent below 80 km; and (c) for the majority of the time, the craft is either attached

[61] Chicago Convention (n 16) Annex 7(1) Definitions.

[62] Dempsey and Manoli (n 3) 212.

[63] William Ailor and Paul Wilde, Requirements for Warning Aircraft of Reentering Debris," 3rd International Association for the Advancement of Space Safety Conference, Rome, Italy, October 21–23, 2008 quoted in William Ailor, *Large constellation disposal hazards* (Aerospace Corporation 2020) available online at <https://aerospace.org/sites/default/files/2020-01/Ailor_LgConstDisposal_20200113.pdf>

[64] Dempsey and Manoli (n 3) 213.

to an aircraft or behaves like an aircraft.[65] The legal status of the craft, therefore, is somewhat opaque. Given that the craft is only in space for a very short time, it would seem sensible to classify it under air law as most of the apparatus is predicated on using the atmosphere for lift (the carrier aircraft and the return journey).

In trying to unpick the legal status of each craft, a functionalist approach might be of assistance.[66] Clearly the 'launch phase', where the carrier aircraft and the VSS Unity are attached, would be subject to air law. The carrier aircraft would continue to be subject to air law for the duration of its flight. Once the VSS Unity separates from the carrier aircraft, using its rocket engine to achieve apogee, its purpose would be to reach outer space. Consequently, at separation from the carrier aircraft, the VSS Unity could become a 'space object'. In line with the Registration Convention,[67] 'space law should continue to apply even upon re-entry, descent and landing phases of the spacecraft'.[68] It is suggested that this approach works, save for one core question: if the VSS Unity is designed only to go to 80 km, can it be truly said to be a 'space object'? The functionalist approach can seemingly never escape that fundamental spatialist question.

The question of the status of such hybrid craft is significant because it gives rise to a whole range of legally significant issues. The first is liability. Passenger liability on aircraft is extremely well defined by international agreement,[69] with a system of unlimited liability for air carriers in the case of the death or injury of a passenger, and a limited liability framework in case of delay.[70] International space law, on the other hand, has a treaty framework outlining liability on a state-to-state basis for damage caused by space objects,[71] but it is (understandably) in no way as developed as air law. Indeed, there is no international agreement that provides for passenger liability. It is therefore likely, as has emerged in the US[72] and United Kingdom (UK),[73] that national legislation will step in to provide for suborbital space tourism passenger protection.[74] Given the fact that such tourism, for the foreseeable future, will almost certainly occur over a country's sovereign airspace, this is a sensible development. As the market matures, it may be that the international community will coalesce and be able to enter into a binding agreement. Presently, the space tourism

[65] This is, of course, the National Aeronautics and Space Administration (NASA)/United States Air Force (USAF) demarcation as to where space begins. See Newman (n 6) 83.

[66] Hobe (n 60) 363.

[67] Registration Convention (n 59).

[68] Hobe (n 60) 364

[69] Convention for the Unification of Certain Rules for International Carriage by Air (adopted 28 May 1999, entered into force 4 November 2003) 1999 WL 33292734, arts 17–37 (Montreal Convention).

[70] Hobe (n 60) 367.

[71] Liability Convention (n 59).

[72] US Commercial Space Launch Amendments Act, Pub L 108-492, 23 December 2004, US Commercial Space Launch Competitiveness Act, Pub L 114-90, 25 November 2015.

[73] The UK Space Industry Act 2018 and Space Industry Regulations 2021.

[74] Hobe (n 60) 369.

market is not sufficiently developed—either technically or commercially—to support such negotiations.

The question of hybrid craft extends not only to space tourism but also to the equally nascent space launch market. This involves the use of an aircraft (usually an adapted 747) to take a launcher up to an altitude of around 35,000 ft (10.5 km). The launcher then separates, and a rocket engine ignites, taking it to an altitude of between 500 and 1000 km. Again, according to the approach suggested by Hobe, where the rocket and the aircraft are connected, they are a single aircraft. When the rocket separates, as it is using a rocket engine, not lift from the air, and as the mission is to enter outer space and deposit satellites, it would be a space object and then subject to both national and international space law. This means that there would be some overlap in the licensing regimes. In the UK, for example, the regulator for both air and outer space activity is the Civil Aviation Authority,[75] which would need to license both the air and space elements. In jurisdictions where the air and space regulators are separate entities, this might lead to some confusion in licensing authority. Nonetheless, at present the air launch market is sufficiently small[76] to allow for a more bespoke approach within national legislation.

This hybrid functionalist approach has been criticised due to the legal uncertainty as to which regime is applicable at a given moment.[77] One of the key moments of vulnerability in air launch and indeed in the space tourism mission is the separation of the carrier aircraft from the space object.[78] If there is a critical failure at that point, deciding which regime is operative could be challenging. For the moment, national legislation will be the preferred way for states to ensure that investigatory and regulatory functions for these missions are clear and unambiguous. Only when the space tourism and air launch economies develop to an appropriate scale would international engagement be desirable.

Both the air launch and space tourism industries present some problems for the demarcation of space. But, as has been demonstrated, these problems are not insurmountable. Nor can the demarcation issue be said definitively to be a chilling factor in these nascent industries. One area, however, where boundary and demarcation may well be of interest is in the realms of High-Altitude Pseudo Satellites or High-Altitude Platforms (HAPS). As the title suggests, these are aerial platforms which can emulate the station-keeping facilities of a satellite at a sufficient altitude

[75] Space Industry Regulations 2021, Pt 2, reg 3.

[76] The collapse of air launch provider Virgin Orbit in April 2023 illustrates how vulnerable the air launch market is at the moment compared with other launch providers. See Sharmila Kuthunur, 'Virgin Orbit Shuts Down After Selling Key Assets to 3 Aerospace Companies' (*Space.com*, 24 May 2023) <www.space.com/virgin-orbit-ceases-operations-sells-assets> accessed 2 October 2023.

[77] Michael Chatzipanagiotis, 'Beware of the Hybrids: Legal Issues of Air Launches' (2019) 70 ProceIntlAstroCongress 12.

[78] Stephen Clark, 'Pilot Dies in Crash of Virgin Galactic Rocket Plane' (*Spaceflightnow*, 31 October 2014) <https://spaceflightnow.com/2014/10/31/virgin-galactics-spaceshiptwo-suffers-anomaly-during-test-flight/> accessed 2 October 2023.

to cover a wide area without either the cost of launch or interfering with controlled airspace.[79] There is a significant amount of development work being undertaken into a craft that can operate for extended periods in the 20–35 km zone.[80]

Whilst regulated airspace falls into the upper limits of this zone, the 20 km limit also provides the current technological threshold for some of the services provided by the air navigation service providers, as well as for traditional air traffic management (ATM) systems.[81] Should HAPS continue to develop as commercially viable options both in regional telecommunications and also as tools for earth observation, more regulation for high altitude operations will be needed. Currently, this is done in New Zealand[82] by means of dedicated national legislation. The European Concept for Higher Airspace Operations (ECHO) group has also been engaged in examining operations that are carried out at altitudes above the flight levels that are controlled but which are not yet in outer space.[83]

The approach adopted by New Zealand is particularly instructive as to how national legislation could be used to circumvent the difficulties posed by the demarcation of outer space. The 2017 Act defines 'high altitude' as an altitude above FL600 and 'above the highest upper limit of controlled airspace' (around 2.9 km). The Act defines the point at which 'high altitude' activities begin[84] but remains silent as to the upper boundaries of New Zealand's sovereign airspace. This has been criticised as 'merely shifting the problems concerning the appropriate legal regime',[85] whereby some missions above controlled airspace may qualify as both high-altitude activities and space activities, whereas others may only be high-altitude activities without any clear dividing line between the two and the respective legal regimes.[86]

The counter to this argument is that such a distinction is as much a matter of fact as it is a matter of law. Elements of functionalist approaches—such as the destination and mission parameters of the activity—will be covered by the licensing regime.[87] Nonetheless, despite the legislation in place in New Zealand, Australia,

[79] Jesus Gonzalo and others, 'On the Capabilities and Limitations of High Altitude Pseudo Satellites' (2018) 98 ProgAerospaceSci 37, 37.

[80] ibid 38.

[81] Frontex, European Border and Coast Guard Agency, 'HAPS Market Report: Research Study on High Altitude Pseudo Satellites' (*Frontex*, 2023) <https://frontex.europa.eu/assets/EUresearchprojects/2023/FX_HAPS_WP1_-_Market_Report_V4.pdf> accessed 2 October 2023.

[82] Outer Space and High-Altitude Activities Act 2017; Public Act 2017 No 29, date of assent 10 July 2017 (hereinafter referred to as the NZ Space Act 2017).

[83] See 'Regulatory Perspectives Emerging for Higher Airspace Users' (*Eurocontrol*) <www.eurocontrol.int/article/regulatory-perspectives-emerging-higher-airspace-users> accessed 2 October 2023.

[84] It is significant that the legislation does not choose to muddy the waters by mentioning the 'orbital' and 'sub orbital' distinction.

[85] Frans G von der Dunk, 'Kiwis in Space: New Zealand's "Outer Space and High-Altitude Activities Act"' (2017) 60 ProceColloqLOuterSpace 453, 455.

[86] ibid 456.

[87] NZ Space Act 2017 (n 82) Part 2 subpart 2 for payload licences and subpart 6 for High Altitude Licences.

and the UK, amongst others, it is clear that both international and domestic legal systems are not yet equipped to deal effectively with the regulation of HAPS that operate for a prolonged period above controlled airspace. Initiatives such as ECHO are exploring multilateral solutions. It is quite clear that until a clearer picture of how the technology might develop and how the economics of HAPS manifest themselves, a neat delineation of airspace up to and including outer space will remain elusive.

8 The Way Forward

States are understandably reluctant to limit their own sovereignty, given that the technology is in the early stages of development. The argument used by states within COPUOS, that there have not been any issues in respect of the boundary to require urgent action in this area, remains a suitable pretext for 'kicking the can down the road'. One might argue that, given the problems facing the outer space environment from orbital debris, and the strains being placed on the radio frequency (RF) spectrum, COPUOS does indeed have more pressing matters to deal with than addressing a problem that has not yet caused any serious legal issues.

There have been numerous suggestions as to how to deal with the impasse on demarcation. Some have suggested a new international order, delimiting the atmosphere from a boundary drawn at 120–160 km down to controlled airspace. Pelton has led the way in calling for a new regime to govern the atmosphere of the Earth.[88] Dempsey and Manoli[89] similarly suggest establishing an Outer Space Zone (above 120 km) subject to space law, a Near Space Zone, between 50 and 120 km,[90] and an Air Space Zone below 50 km retaining the current air law regime. Others have suggested the creation of an international body in charge of safety to ensure that safety issues, standardisations, and enforcement come from a single entity to avoid regulatory confusion.[91] Similarly, there are arguments requiring that the current airspace boundary be reduced to reflect the scientific reality of satellite behaviour more accurately.[92] These are extremely well-argued and conceptually sound ways forward. Unfortunately, despite the merits of such reform, ideas

[88] Joseph N Pelton, 'The Need for a New Integrated Global Regulatory Regime for Air and Space: The Case for Regulation of the Protozone' (3rd Manfred Lachs Conference, Montreal, March 2015) <www.mcgill.ca/iasl/files/iasl/mlc4_presentation_j_pelton.pptx> accessed 2 October 2023.

[89] Dempsey and Manoli (n 3) 235–37.

[90] 'A mix of the UNCLOS Contiguous Zone and Exclusive Economic Zone, open to innocent passage by all, with aerial safety and navigation rules established (presumably by ICAO), 153 enforced by States in Flight Information Regions, and whose aerospace and launch vehicles are subject to Air Law. It could also be restricted against overflights by foreign State aircraft absent permission of the underlying State.' ibid 236.

[91] Michael Chatzipanagiotis and George Kyriakopoulos, 'One-Stop Shop Safety Regulation: Should We Do it and How?' (2019) 6 JSpaceSafetyEngineer 212–18.

[92] McDowell (n 36) as discussed above.

of revivifying and renovating the current set of international agreements in this area are presently unworkable due to the current geopolitical situation. Given the current, fractured state of international diplomacy, trying to establish any sort of consensus, especially where States must limit their national sovereignty, looks to be an almost impossible task.

There is a paradox inherent in the central thesis of this discussion. There has been a never-ending problem with respect to demarcation—as the latest COPUOS pronouncements will testify. Yet this problem has not caused any significant issues. States have been able to defend their vertical sovereignty. Space activity has developed in LEO with the rise of large constellations and genuine reusability in launch vehicles. The problems with the space tourism industry have little to do with the demarcation of outer space and much more to do with lingering concerns about safety and the still considerable price of a ticket.[93] HAPS development does present some concerns with respect to the legal environment, but that is largely a matter of regulating the 20–50 km zone rather than the demarcation of outer space.

If there is to be some sort of resolution to the demarcation issue, it is likely that it will start with more countries using national space legislation to recognise the 100 km altitude as the spatial upper limit of airspace for the purpose of marking a divide between the airspace and space law legal regimes. Several countries have already started doing this,[94] and as their number increases, together with recognition by non-governmental entities, this could form enough of a critical mass to 'crystallise the law'.[95] The establishment of a distinct boundary could be easily ratified within COPUOS, but, again, the politics of this might make it impractical for the foreseeable future. This is not terminal for widespread acceptance of an air/space boundary, especially if suborbital, high-altitude, and space tourism activities retain their current national flavour.

The acceptance of such a spatialist border could be accompanied by a functionalist approach to the question of overflight by space objects. It is probably safe to say that the principle of overflight by satellites is accepted as a part of customary international law, and it is certainly embedded within Articles I and II of the Outer Space Treaty. The question of overflight and transit through airspace for space objects would need to be addressed, as has been identified by Bittencourt Neto.[96] Any state whose airspace is traversed by a space object of another state should receive notification and have the right to object for reasons of national security. The state whose territory is being overflown should not impose onerous or unreasonable requirements and should not routinely refuse such requests, being mindful of free access to space as provided for in Article I of the Outer Space Treaty. This principle

[93] Jonathan Amos, 'Virgin Galactic Grounded over Branson Spaceflight Mishap' *BBC News* (2 September 2021) <www.bbc.co.uk/news/science-environment-58421796> accessed 2 October 2023.

[94] See above (n 48).

[95] Lyall and Larsen (n 7) 149.

[96] Bittencourt Neto (n 44) 62.

does not need embedding in another international treaty; instead, it could be incorporated within the deliberations of ICAO and reflect the reality of airspace transit permissions that exist within commercial and controlled airspace.[97] Similarly, it would allow the international community to establish norms of behaviour with respect to re-entry and the management of debris that falls to Earth.

9 Conclusion

Space law and air law are both products of the twentieth century. While air law developed due to the ubiquity of aviation in a whole range of countries, space law remained the preserve of a few space-faring nations. This is why there is such an imbalance in the maturity of the regimes. Security and strategic considerations continue to bedevil the question of the demarcation of outer space, but the reality of the situation is that the boundary to outer space is already developing at 100 km. Anything below this level is unlikely to be accepted by states fearing an erosion of sovereignty and concerned about the development of possible threats from above-controlled airspace. Anything above this level is likely too high to reflect the reality of where space begins. A fusion of accepting this spatialist boundary and a recognition that incorporating functionalist methods of identifying a space vehicle is clearly the short-term resolution to the never-ending problem of demarcation.

[97] Cf Bittencourt Neto (n 44) 62, who advocates a series of internationally binding rules to cover all eventualities. It is suggested, at this stage, that it is not necessary to do this. Instead, whilst piecemeal and bespoke arrangements are not as comprehensive, they do allow for a regime to be 'stress tested' and any issues identified before the international community signs on.

3

THE CONTRIBUTION OF THE UNITED NATIONS TO THE DEVELOPMENT OF INTERNATIONAL SPACE LAW

Steven Freeland and Anne-Sophie Martin

1 Introduction: Genesis of International Space Law

The corpus of international law that regulates the conduct of space activities was born in the context of the Cold War, during a period of important technological, strategic and ideological antagonism between the United States and the Union of Soviet Socialist Republics (USSR).[1] Yet, despite the then prevailing geopolitical situation, several important major legal principles were developed, primarily through the United Nations, in the period following the launch by the Soviet Union, on 4 October 1957, of Sputnik 1, the first artificial satellite in orbit.[2]

At the time of the Sputnik mission, there were no specific rules regulating space activities conducted by States.[3] The same year, faced with the increasing interest in space of the international community, and the imperative to clarify important issues, given the tense geopolitical climate, the United Nations adopted a resolution[4] on the question of the use of outer space for peaceful purposes, recalling that 'the sending of objects through outer space shall be exclusively for peaceful and scientific purposes'. This was intended to assuage growing concerns in the international community that, in the context of the Cold War, outer space might become yet another field for intense rivalries between the superpowers.

The United States launched its first orbital satellite, Explorer 1, on 31 January 1958.[5] Shortly afterwards, the United Nations General Assembly (UNGA) adopted Resolution 1348 (XIII), which led to the creation of the 'United Nations Committee

[1] Carl Q Christol, *The Modern International Law of Outer Space* (Pergamon 1982); Philippe Achilleas, 'Le droit international de l'espace' in Philippe Achilleas (ed), *Droit de l'espace* (Larcier 2009) 13–16.

[2] 'Sputnik and the Dawn of the Space Age' (*NASA History Division*) <https://history.nasa.gov/sputnik.html> accessed 6 July 2023.

[3] Clarence Wilfred Jenks, *Space Law* (Stevens 1965); Manfred Lachs, 'The International Law of Outer Space' (1964) 113 RdC 1–115; Mireille Couston, *Droit spatial* (Ellipses 2014) 12ff.

[4] UNGA Res 1148 (XII) (14 November 1957).

[5] 'Explorer 1' (*NASA Jet Propulsion Laboratory*) <www.jpl.nasa.gov/missions/explorer-1> accessed 6 July 2023.

Steven Freeland and Anne-Sophie Martin, *The Contribution of the United Nations to the Development of International Space Law* In: *International Space Law in the New Space Era*. Edited by: Sandeepa Bhat B., Dilip Ukey, and Adithya Variath, Oxford University Press. © Steven Freeland and Anne-Sophie Martin 2024. DOI: 10.1093/9780198909415.003.0003

for the Peaceful Uses of Outer Space' (UNCOPUOS), as an ad hoc Committee composed of 18 Member States.[6] UNCOPUOS was established to 'consider *inter alia* the activities and resources of the United Nations, the specialized agencies and other international bodies relating to the peaceful uses of outer space, organizational arrangements to facilitate international cooperation in this field within the framework of the United Nations, and legal problems which might arise in programmes to explore outer space'.[7]

In 1959, the UNGA converted the UNCOPUOS, which had 24 members at the time, into a permanent subsidiary body, pursuant to Articles 7(2) and 22 of the United Nations Charter,[8] and reaffirmed its mandate in Resolution 1472 (XIV).[9] The UNCOPUOS was established as an auxiliary organ of the UNGA and not as an independent international organization founded on a legally binding treaty, as is the case for some other specialized agencies of the United Nations.[10]

Since then, the UNCOPUOS has served as the principal UN agency for international cooperation in the peaceful exploration and use of outer space. UNCOPUOS represents a multilateral forum for contact between governmental and non-governmental organizations concerned with outer space activities. Since 1958, the Committee has served as the main body responsible for the development of international space law,[11] illustrating how 'the United Nations has been involved in space activities since the beginning of the first "space race" '.[12]

While the Soviet Union (initially) maintained its lead over its 'rivals', notably by successfully sending into space the first human, Yuri Gagarin, on 12 April 1961, United States President Kennedy delivered a historic speech on 25 May 1961, claiming that an American would tread the lunar floor before the end of that decade.[13] The lure of space represented an increasingly visible national priority of, in particular, the two then-prevailing superpowers. At a time when the Soviet Union and the United States were embarking on a race for the Moon, things also began to accelerate at the United Nations in relation to space law, with the Member

[6] UNGA Res 1348 (XIII) (13 December 1958).

[7] Vladimir Kopal, 'Origins of Space Law and the Role of the United Nations' in Christian Brünner and Alexander Soucek (eds), *Outer Space in Society, Politics and Law* (Springer 2011) 221–33.

[8] Charter of the United Nations (adopted 26 June 1945, entered into force 24 October 1945) 1 UNTS XVI, art 7(2) provides that 'such subsidiary organs as may be found necessary may be established in accordance with the present Charter'; art 22 provides that 'the General Assembly may establish such subsidiary organs as it deems necessary for the performance of its functions'.

[9] UNGA Res 1472 (XIV) (12 December 1959).

[10] Sergio Marchisio, 'The Evolutionary Stages of the Legal Subcommittee of the United Nations Committee on the Peaceful Uses of Outer Space (COPUOS)' (2005) 31 JSpaceL 219–42.

[11] Francis Lyall and Paul B Larsen, *Space Law A Treatise* (Routledge 2018) 13–14.

[12] Steven Freeland, 'What Sort of Space "Race" Should We Be Pursuing?' (*Australian Institute of International Affairs*, 15 December 2021) <www.internationalaffairs.org.au/australianoutlook/what-sort-space-race-should-we-be-pursuing/> accessed 6 July 2023.

[13] 'Address to Joint Session of Congress May 25, 1961, JFK's Moon Shot Speech to Congress' (*John F. Kennedy, Presidential Library and Museum*) <www.jfklibrary.org/learn/about-jfk/historic-speeches/address-to-joint-session-of-congress-may-25-1961> accessed 30 July 2023.

States resolving to secure a framework for space activities by adopting binding rules before the first scheduled (hu)manned flights to the Moon. It was seen as important by all countries and accepted by the United States and the Soviet Union, to put in place a set of international standards that would allow for the exploration and responsible use of space in the interests of all States.

The landing of the Apollo 11 mission on 20 July 1969 put an end to the first major phase of competition between these two rival space powers. The 1970s were marked by the development of space applications, and the need to regulate them, in particular in the fields of telecommunications, television and remote sensing by satellite. The 1980s opened the phase of commercialization and privatization of space activities, which had been very limited until then, with the commercial space economy beginning to accelerate in the mid–late 1990s.[14]

During these developments, the international legal framework for outer space activities has largely been developed through the consensus process within the UNCOPUOS.[15] The UNCOPUOS is composed of two Subcommittees, created in 1961, 'the Scientific and Technical Subcommittee' (STSC) and 'the Legal Subcommittee' (LSC), both of which, like UNCOPUOS sitting in plenary, focus on civil and commercial activities. The STSC meets every year and the mandate of the Subcommittee is 'to discuss questions related to the scientific and technical aspects of space activities,'[16].

The LSC also meets every year to 'discuss legal questions related to the exploration and use of outer space'[17]. The Committee, in particular through the LSC,

[14] Matt Weinzierl and Mehak Sarang, 'The Commercial Space Age Is Here' *Harvard Business Review* (12 February 2021) <https://hbr.org/2021/02/the-commercial-space-age-is-here> accessed 6 July 2023; Andrea Sommariva, 'The Evolution of Space Economy: The Role of the Private Sector and the Challenges for Europe' (*ISPI*, 11 December 2020) <www.ispionline.it/en/pubblicazione/evolution-space-economy-role-private-sector-and-challenges-europe-28604> accessed 6 July 2023; Eng Teong 'Commercialization of Space Activities—The Laws and Implications' (2017) 82 JAirL&Com 1, 145–68; Jai Galliott (ed), *Commercial Space Exploration—Ethics, Policy and Governance* (Routledge 2015); Stephan Hobe, 'The Impact of New Developments on International Space Law (New Actors, Commercialization, Privatization, Increase in Number of "Space-faring Nations")' (2010) 15 UnifLRev 869; 'Value of Space Economy Reaches $464 Billion in 2022 Despite New Unforeseen Investment Concerns' (*Euroconsult*) <www.euroconsult-ec.com/press-release/value-of-space-economy-reaches-424-billion-in-2022-despite-new-unforeseen-investment-concerns-2/#:~:text=Value%20of%20Space%20Economy%20reaches,new%20unforeseen%20investment%20concerns%20%2D%20Euroconsult> accessed 6 July 2023; see also 'Space Economy Report' (9th edn, Euroconsult 2022).

[15] Ram S Jakhu, 'Sixty Years of Development of International Space Law' in Stephan Hobe (ed), *Air Law, Space Law, Cyber Law—the Institute of Air and Space Law at Age 90* (Carl Heymanns Verlag 2016) 76–84.

[16] *e.g.* space weather, global health, near-Earth objects (planetary defence), the use of space technology for socio-economic development or disaster management support, global navigation satellite systems, remote sensing by satellite, space debris, the use of nuclear power sources, and the long-term sustainability of outer space activities.

[17] *e.g.* status and application of the five United Nations treaties on outer space, the definition and delimitation of outer space, national space legislation, legal mechanisms relating to space debris mitigation and remediation measures, legal issues relating to the exploration, exploitation and utilization of space resources, suborbital flights and human transportation, non-legally binding instruments, legal aspects of space traffic management, small-satellites, and international mechanisms for cooperation in the peaceful exploration and use of outer space.

has made a significant contribution to the development of international space law with the adoption of five treaties, five sets of Principles, as well as a broad range of other 'soft law' instruments.[18] The Committee reports to the Fourth Committee of the General Assembly (Special Political and Decolonization Committee), which adopts an annual resolution on international cooperation in the peaceful uses of outer space.[19]

Since its creation, the membership of the UNCOPUOS has become more inclusive, making it one of the largest committees in the United Nations.[20] In addition to States, a number of international organizations, including both intergovernmental and non-governmental organizations, have obtained Permanent Observer status within the UNCOPUOS and its Subcommittees. As noted, UNCOPUOS deals with civil and commercial activities of space activities.[21] The Committee discusses questions relating to current and future activities in space, as noted above.

Supported by its Secretariat, the United Nations Office for Outer Space Affairs (UNOOSA), UNCOPUOS continues to serve as a forum for international deliberation for maintaining outer space for peaceful purposes. As of January 2023, there were 102 Member States of the UNCOPUOS and a further 49 Permanent Observers[22] comprising both non-governmental organizations and international intergovernmental organizations.

[18] Stephan Hobe, 'Space Law—an Analysis of its Development and its Future' in Christian Brünner and Alexander Soucek (eds), *Outer Space in Society, Politics and Law* (Springer 2011) 476–90; see also Irmgard Marboe (ed), *Soft Law in Outer Space* (Böhlau 2012).

[19] Tanja Masson-Zwaan and Mahulena Hofmann, *Introduction to Space Law* (Wolters Kluwer 2019) 10–11.

[20] 'Committee on the Peaceful Uses of Outer Space: Membership Evolution' (*United Nations Office for Outer Space Affairs*) <www.unoosa.org/oosa/en/ourwork/copuos/members/evolution.html> accessed 6 July 2023; UNGA Res 76/76 (15 December 2021).

[21] Issues related to the 'military' and 'weapons-related' aspects associated with space activities fall more within the mandate of other United Nations agencies such as the Geneva-based Conference on Disarmament and the First Committee of the General Assembly. However, given the increasingly dual-use nature of space activities, we have observed that, over recent years, the discussions, particularly within the LSC, now involve some elements related to what might be regarded as 'non-peaceful' space activities, although this development has met with some resistance particularly among several of the major space-related powers; see UN Doc A/AC.105/1260 (19 April 2022) 51:

> the universalisation and improvement of the implementation of the rules set out in the United Nations treaties on outer space were essential first steps in ensuring compliance with the three main principles that must govern space activities: (a) the freedom of access to outer space for peaceful uses; (b) the preservation of the security and integrity of satellites in orbit; and (c) the defence and security interests of States in space.

Moreover, during its sessions the LSC addresses various topics that are highly relevant for 'security matters', such as space traffic management, small-satellites, mega-constellations and cybersecurity risks (ibid 193).

[22] 'Committee on the Peaceful Uses of Outer Space: Observer Organizations' (*United Nations Office for Outer Space Affairs*) <www.unoosa.org/oosa/en/ourwork/copuos/members/evolution.html> accessed 6 July 2023; see also 'Committee on the Peaceful Uses of Outer Space: Observer Organization' (*United Nations Office for Outer Space Affairs*) <www.unoosa.org/oosa/en/ourwork/copuos/members/copuos-observers.html> accessed 10 July 2023.

In addition, within the United Nations system, various other agencies deal with space as part of their competencies.[23] In particular, the Geneva-based International Telecommunication Union (ITU) has an important role to play in the regulation of spectrum/frequency and orbits, which are to be used in an equitable, rational and efficient manner as required under Article 44 of the ITU Constitution.[24] Indeed, its Constitution, Convention and Administrative Regulations (including the Radio Regulations, which regulate certain radiocommunication services and the utilization of radio frequencies) are fundamental for all uses of space.

It is also worth mentioning that the World Meteorological Organization (WMO), International Civil Aviation Organization (ICAO), International Maritime Organization (IMO), World Intellectual Property Organization (WIPO) and United Nations Education, Scientific and Cultural Organization (UNESCO) also have, within their respective mandates, an interest in 'space', though they do not directly contribute to the development of space law. Outside the framework of the United Nations, the International Institute for the Unification of Private Law (UNIDROIT) has contributed to the further development of possible future space-related regulations with the finalization of the Space Protocol,[25] but that instrument has not yet entered into force.[26]

Over the years, the UNCOPUOS has been the major intergovernmental forum through which States have been encouraged to 'act collectively to promote the peaceful exploration and use of outer space through a variety of mechanisms, including United Nations treaties and Principles on outer space, UNGA resolutions and other relevant instruments dealing with the safe, secure, stable and sustainable use of outer space'.

The international law regulating space activities also influences relevant national space laws, which in turn interact with the international system.[27] Space law is not a self-contained regime. It is a branch of international law.[28] According to

[23] Lyall and Larsen (n 11) 12.

[24] Constitution of the International Telecommunication Union (adopted 22 December 1992, entered into force 1 July 1994) 1825 UNTS 143.

[25] Protocol to the Convention on International Interests in Mobile Equipment on Matters Specific to Space Assets (adopted 9 March 2012) <www.unidroit.org/instruments/security-interests/space-protocol/> accessed 6 July 2023. As noted, this instrument is not in force.

[26] Daniel Porras, 'Entering into Force: Promoting UNIDROIT's Space Protocol Among Emerging Space Actors' in Souichirou Kozuka (ed), *Implementing the Cape Town Convention and the Domestic Laws on Secured Transactions* (Springer 2017) 369–72; Protocol to the Convention on International Interests in Mobile Equipment on Matters Specific to Space Assets (adopted 9 March 2012) <www.unidroit.org/instruments/security-interests/space-protocol/status/> accessed 6 July 2023; Ignacio Tirado and Bernhard Schmidt-Tedd, 'Status of and Way Forward for the UNIDROIT Space Protocol' (IISL and ECSL Space Law Symposium 2021, Sixtieth Session of the LSC of the UNCOPUOS) <www.unoosa.org/documents/pdf/copuos/lsc/2021/05.pdf> accessed 6 July 2023.

[27] Steven Freeland, Ram S Jakhu and David Kuan-Wei Chen, 'The Sources of International Space Law: Revisited' (2018) 67 GerJAir&SpaceL 606–67.

[28] Frans Von der Dunk, *Advanced Introduction to Space Law* (Edward Elgar 2020) 8–9; Thomas Neger and Edith Walter, 'Space Law—an Independent Branch of the Legal System' in Christian Brünner and Alexander Soucek (eds), *Outer Space in Society, Politics and Law* (Springer 2011) 234–45.

Article III of the 1967 Treaty on Principles Governing the Activities of States in the Exploration and Use of Outer Space, including the Moon and Other Celestial Bodies,[29] States shall 'carry on activities in the exploration and use of outer space, including the Moon and other celestial bodies, in accordance with international law, which includes the Charter of the United Nations'.

This chapter now moves to an analysis of the space law-making process within the UNCOPUOS by taking into particular account the major role played by the LSC in the adoption of binding treaties and other (soft law) instruments. It then considers the legal challenges that the Committee faces due to the development of new space activities and a growing number of actors, both private and public, involved in space programmes. Finally, it addresses another major aspect of space activities—the security dimension—and the role of the First Committee of the UNGA.

2 The 'Phases' of Law-Making within the UNCOPUOS Legal Subcommittee

2.1 Initial Phase: Adoption of Legally Binding Instruments

According to the common interpretation of Article 38(1) of the Statute of the International Court of Justice,[30] treaties represent a material 'source' of international law.[31] Up until 1979, the regulatory frameworks established through the LSC/UNCOPUOS consensus-based process were characterized by the adoption of binding instruments in the form of five United Nations Space Treaties: i) the 1967 Treaty on Principles Governing the Activities of States in the Exploration and Use of Outer Space, including the Moon and Other Celestial Bodies (Outer Space Treaty); (ii) the 1968 Agreement on the Rescue of Astronauts, the Return of Astronauts and the Return of Objects Launched into Outer Space[32] (Rescue and Return Agreement); (iii) the 1972 Convention on International Liability for Damage Caused by Space Objects[33] (Liability Convention); (iv) the 1975 Convention on Registration of Objects Launched into Outer Space[34] (Registration

[29] Treaty on Principles Governing the Activities of States in the Exploration and Use of Outer Space, Including the Moon and Other Celestial Bodies (adopted 27 January 1967, entered into force 10 October 1967) 610 UNTS 205 (Outer Space Treaty).

[30] Statute of the International Court of Justice (adopted on 24 June 1945, entered into force 24 October 1945) XV UNCIO 355.

[31] Lyall and Larsen (n 11) 36.

[32] Agreement on the Rescue of Astronauts, the Return of Astronauts and the Return of Objects Launched into Outer Space (adopted 22 April 1968, entered into force 3 December 1968) 672 UNTS 119 (Rescue Agreement).

[33] Convention on International Liability for Damage Caused by Space Objects (adopted 29 March 1972, entered into force 1 September 1972) 961 UNTS 187 (Liability Convention).

[34] Convention on Registration of Objects Launched into Outer Space (adopted 14 January 1975, entered into force 15 September 1976) 1023 UNTS 15 (Registration Convention).

Convention); and (v) the 1979 Agreement Governing the Activities of States on the Moon and Other Celestial Bodies[35] (Moon Agreement).

When the law-making 'process' began within the UNCOPUOS shortly after its establishment, there was no binding instrument in place to regulate the exploration and use of outer space. However, the launch of Sputnik 1 in 1957 had already resulted in the 'almost instantaneous' recognition of two customary rules: the freedom of space and the non-appropriation of outer space. Indeed, with regard to the former, Judge Manfred Lachs of the International Court of Justice (ICJ) observed, shortly after the Outer Space Treaty had been finalized, that 'the first instruments that men sent into outer space traversed the air space of States and circled above them in outer space, yet the launching States sought no permission, nor did the other States protest. This is how the freedom of movement into outer space, and in it, came to be established and recognised as law within a remarkably short period of time.'[36]

International customary law relating to space activities was born, as is evidenced by the consistent practice of space-faring States at and from that pivotal time.[37] In this context, the Committee agreed on the terms of what became the 1963 Declaration of Legal Principles, which was subsequently adopted as Resolution 1962 (XVIII) by the UNGA. This Resolution is considered as an important declaratory tool of the major principles (at least with respect to certain provisions) by then already representing customary international law,[38] as part of the process of evolving international law, and guiding States' behaviour in conducting space activities.[39]

Yet, despite the fact that the 1963 Declaration of Principles was a very significant instrument in the evolution of the formal rules governing the exploration and use of outer space, it was intentionally drafted as a *non-binding* set of principles.[40] Their non-binding nature was further emphasized by the very next resolution passed on the same day by the UNGA. That Resolution, dealing with the 'International co-operation in the peaceful uses of outer space' recommended to the Member States that 'consideration should be given to incorporating in international agreement

[35] Agreement Governing the Activities of States on the Moon and Other Celestial Bodies (adopted 18 December 1979, entered into force 11 July 1984) 1363 UNTS 3 (Moon Agreement).

[36] *North Sea Continental Shelf Cases (Federal Republic of Germany v Denmark and Federal Republic of Germany v The Netherlands)* (Judgment) [1969] ICJ Rep 3, 230 (Dissenting Opinion of Lachs J).

[37] Sergio Marchisio, *The Law of Outer Space Activities* (Edizioni Nuova Cultura 2022) 22–24; Sergio Marchisio, *International Law* (Giappichelli 2021) 48–49; Bin Cheng, *Studies in International Space Law* (OUP 1997) 191–96.

[38] Steven Freeland and Ram S Jakhu, 'The Relationship between the Outer Space Treaty and Customary International Law' (2016) 23 ProceColloqLOuterSpace 183–95.

[39] Gaetano Arangio-Ruiz, 'The Normative Role of the General Assembly of the United Nations and the Declaration of Principles of Friendly Relations' (1972) 137 RdC 419–572.

[40] See UNGA Res 1962 (XVIII) (13 December 1963); 'Space Law: The Law of Outer Space' (*Georgetown Law Library*) <https://guides.ll.georgetown.edu/c.php?g=1037047&p=7518561> accessed 6 July 2023.

form, in the future as appropriate, legal principles governing the activities of States in the exploration and use of outer space'.[41]

The adoption of the 1963 Declaration was a major first step towards the finalization of a multilateral treaty, the Outer Space Treaty, clarifying and consolidating the rules to be applied to outer space activities, as well as lessening the risk of unilateral State practice. In 1967, a decade after Sputnik 1, the world community came together at the height of Cold War brinksmanship to adopt the Outer Space Treaty. There are currently 112 countries that are parties to the Treaty, which underlines 'the common interest of all humanity to explore and use outer space for peaceful purposes'.[42] The Treaty declares that 'space, including the Moon and celestial bodies, are free to be explored and used by all States on a basis of equality and in accordance with international law'.[43]

Departing from the structure of international law, 'the Outer Space Treaty codifies the most significant principle of law for enhancing the common interest of all in space', in order to thwart potential colonization ambitions in space. By declaring that outer space is not subject to national appropriation,[44] the Outer Space Treaty established a foundational governance system based on mutual understanding and friendly relations. The principle of non-appropriation relates to outer space, the Moon and other celestial bodies as a whole.[45]

As noted, the Outer Space Treaty also codifies in Article I some core principles of international space law which have, over the years, achieved widespread support, including the freedom to engage in the exploration and use of outer space, which represents the province of (hu)mankind; freedom of scientific investigation in outer space; and international cooperation in such investigation. The Treaty is framed on the basis that outer space falls within the category of *res communes omnium*[46]—that is, while prohibiting national appropriation, it recognizes the right to undertake space activities with due diligence, by considering the corresponding interests of other States. That said, this freedom is not completely unfettered, as States must comply with each of the fundamental principles within the Treaty as they exercise their rights.

[41] UNGA Res 1963 (XVIII) (I) (13 December 1963) 1 (emphasis added).

[42] Outer Space Treaty (n 27) Preamble para 2.

[43] ibid art I.

[44] ibid art II.

[45] Steven Freeland and Ram S Jakhu, 'Article II' in Stephan Hobe, Bernhard Schmidt-Tedd and Kai-Uwe Schrogl (eds), *The Cologne Commentary on Space Law*, vol I (Carl Heymanns Verlag 2009) 44–63.

[46] 'Res Communis' (*Oxford Reference*) <www.oxfordreference.com/view/10.1093/oi/authority.20110803100408305> accessed 7 July 2023 (defining *res communis* as '[a]n area of territory that is not subject to legal title of any State'); Martin Svec, 'Outer Space, an Area Recognised as Res Communis Omnium: Limits of National Space Mining Law' (2022) 60 SpacePoly 101473 <https://doi.org/10.1016/j.spacepol.2021.101473> accessed 6 July 2023; Sergio Marchisio, *International Law* (Giappichelli 2021) 188; Charlotte Ku, 'The Concept of Res Communis in International Law' (1990) 12(4) HistEurIdeas 459–77.

The Outer Space Treaty sets forth a particular form of international responsibility for space activities. Article VI provides that 'State Parties to the Treaty shall bear international responsibility for national activities in outer space . . . whether such activities are carried on by governmental agencies or by non-governmental entities and for assuring that national activities are carried out in conformity with the provisions of the treaty'.[47] This principle encompasses a *sui generis* form of responsibility for 'national activities in outer space', over and above the general international law relating to State responsibility[48] within the traditional meaning, which will also be applicable in relevant circumstances.

As mentioned, the legal framework codified in the Outer Space Treaty has been supplemented by four other treaties, the Registration Convention, the Liability Convention, the Rescue and Return Agreement and the Moon Agreement, all negotiated and agreed upon by a consensus within the LSC.[49] Under this legal framework, space activities carried out by States, international organizations and private entities have prospered with the advent of 'NewSpace',[50] enabling the development of new space technology and innovative applications used in our day-to-day life on Earth.

Some of these treaties, however, have not achieved the same level of broad acceptance (at least in a practical sense) among space-faring countries. Thus, one of the LSC's aims is to increase the global acceptance of these treaties, encouraging States to ratify or adhere to them as they represent a significant and beneficial tool for the conduct of space activities.[51] In the same vein, the LSC also fosters States that have already adopted these instruments to look at effective ways to implement them within their respective domestic laws.

[47] See Michael Gerhard, 'Article VI' in Stephan Hobe, Bernhard Schmidt-Tedd and Kai-Uwe Schrogl (eds), *The Cologne Commentary on Space Law*, vol I (Carl Heymanns Verlag 2009) 103–25.

[48] See *Military and Paramilitary Activities in and against Nicaragua (Nicaragua v United States of America)* (Merits) [1986] ICJ Rep 14, 113 and 115; *Case concerning the Gabcikovo-Nagymaros Project (Hungary v Slovakia)* [1997] ICJ Rep 7, 50, 79 and 83; *Application of the Convention on the Prevention and Punishment of the Crime of Genocide (Bosnia and Herzegovina v Serbia and Montenegro)* [2007] ICJ Rep 43, 377–458; International Law Commission, 'Draft Articles on Responsibility of States for Internationally Wrongful Acts' (November 2001) UN Doc A/56/10.

[49] vladimir Kopal, 'Introduction to the United Nations Treaties and Principles on Outer Space' (2003 Workshop on Capacity Building in Space Law, ST/SPACE/14(2003)) 11–25 <www.unoosa.org/pdf/publications/st_space_14E.pdf> accessed 6 July 2023.

[50] 'What is NewSpace' (*NewSpace2060*) <https://newspace2060.com/2019/01/28/what-is-newspace/> accessed 6 July 2023. (According to NewSpace2060, NewSpace is 'the emergence of the private spaceflight industry. New space ventures are increasingly created, such as private launch companies, small satellite constellations, or sub-orbital tourism, as well as more specific efforts to reinvent the traditional space industry supply chain'.)

[51] UN Doc A/AC.105/1243 (24 June 2021) 57 ('The delegation expressing that view also expressed the view that the five treaties contributed to the safe and peaceful conduct of space activities and were for the benefit and in the interests of all countries'); see also Annex I, Report of the Chair of the Working Group on the Status and Application of the Five United Nations Treaties on Outer Space (24 June 2021): 'The delegation expressing that view also expressed the view that the five treaties contributed to the safe and peaceful conduct of space activities and were for the benefit and in the interests of all countries'; UN Doc A/AC.105/C.2/117 (26 January 2022) 15–22.

As of 1 January 2022,[52] in addition to the 112 States Parties to the Outer Space Treaty, there were 99 States Parties to the Rescue and Return Agreement, 72 States Parties to the Registration Convention, 98 States Parties to the Liability Convention and (only) 18 States Parties to the Moon Agreement.

2.2 Second Phase: Adoption of the UNGA Principles

Whilst there are many reasons for the low(er) level of ratification of certain of the United Nations space treaties,[53] it was evident that the discussions regarding the last of those treaties, the Moon Agreement, unearthed an 'ideological divide', largely between the major industrialized/space-faring States at the time and other countries. This stemmed in part from their respective interpretations of the 'Common Heritage of (hu)Mankind' principle[54] and mirrored similar disagreements that followed the incorporation of this concept into the United Nations Convention on Law of the Sea (UNCLOS)[55] shortly thereafter.[56]

The consequence of these disagreements was that, from a *realpolitik* perspective, it was not possible thereafter for a further binding space instrument to be negotiated and agreed upon through the consensus decision-making processes of the UNCOPUOS, at least for the foreseeable future. This continues to be the case up to this day, particularly given the increasingly complex and multipolar contemporary global geopolitical paradigm.

Yet, as became clear over the subsequent years, the continuous and rapid development of space-related technology meant that an even greater array of activities in outer space were possible and, indeed, were being undertaken, not only by States

[52] UN Doc A/AC.105/C.2/2022/CRP.10 (28 March 2022).

[53] See also Ram Jakhu, Stephan Hobe and Steven Freeland, 'The Appropriateness of the Moon Agreement for Lunar Exploration and Use' (United Nations Global Lunar Conference, Beijing, 31 May–3 June 2010).

[54] Art XI (1) of the Moon Agreement (n 33) provides that '[t]he moon and its natural resources are the common heritage of mankind'.

[55] United Nations Convention on the Law of the Sea (adopted 10 December 1982, entered into force 16 November 1994) 1833 UNTS 397 (UNCLOS).

[56] Art 136 of UNCLOS (ibid) provides that '[t]he Area and its resources are the common heritage of mankind'; the Area is defined in art 1(1)(1) of UNCLOS as follows:

> 'Area' means the seabed and ocean floor and subsoil thereof, beyond the limits of national jurisdiction.

When the terms of UNCLOS were finalized, there was considerable controversy regarding the structure and operation of the regime for the deep sea bed to be established under that treaty, so much so that the industrialized countries declined to sign/ratify it in its original form. As a result, it became necessary to 'supplement' the regime through the finalization of an additional agreement in 1994—the New York Agreement Relating to the Implementation of Part XI of the United Nations Convention on the Law of the Sea (adopted 28 July 1994, entered into force 28 July 1996) 33 ILM 1309 (New York Agreement)—following which virtually all of the industrialized countries (apart from the United States) have subsequently become States Parties to UNCLOS.

themselves but also increasingly by the private sector.[57] The need for, and reliance upon, space technology came to become a standard feature of the development of virtually all industrialized (as well as many non-industrialized) countries. With this demand also came the development of significant commercial 'industries' in space, which gave rise to the ongoing development of additional technology and thus the further potential for the use of outer space.

Thus, over time, there has emerged an increasing tendency for the international community to resort to and rely upon what has typically been referred to as 'soft law' instruments—initially in the form of UNGA resolutions—to set out various additional principles relating to the exploration and use of outer space.[58]

As a result, after the adoption of the last of the United Nations space treaties in 1979, the work of the LSC continued from there with the finalization of draft sets of Principles that were subsequently adopted by the UNGA between 1982 and 1996. By their very nature, of course, they are non-binding instruments. Following the first phase of United Nations 'law-making' for space, involving fundamental principles codified as binding international legal obligations, for various geopolitical and other reasons, in this phase non-binding instruments thus emerged as an appropriate method to deal with certain fields of space activities in the absence of a willingness of the international community to negotiate new binding instruments.[59] This new phase of 'law-making' (although not in the strict sense) began with the adoption of declarations of principles, not binding per se,[60] to 'regulate specific kinds of activities, for instance, the use of artificial satellites for international direct television broadcasting, remote sensing, and the use of nuclear power sources for non-propulsive purposes in outer space'.

Four Declarations of Principles were negotiated by the LSC and then endorsed by the General Assembly of the United Nations, namely the 1982 Principles Governing the Use by States of Artificial Earth Satellites for International Direct Television Broadcasting (Broadcasting Principles),[61] the 1986 Principles Relating to Remote Sensing of the Earth from Outer Space (Remote Sensing Principles),[62] the 1992 Principles Relevant to the Use of Nuclear Power Sources in Outer Space (Nuclear Power Sources Principles),[63] and the 1996 Declaration on International

[57] Victoria Samson, 'The Complicating Role of the Private Sector in Space' (*Bulletin of the Atomic Scientists*, 17 January 2022) <https://thebulletin.org/premium/2022-01/the-complicating-role-of-the-private-sector-in-space/> accessed 6 July 2023.

[58] Tanja Masson-Zwaan and Mahulena Hofmann, *Introduction to Space Law* (Wolters Kluwer 2019) 40–47.

[59] Stephen Gorove, *Developments in Space Law. Issues and Policies* (Martinus Nijhoff 1991) 293–302.

[60] Vladimir Kopal, 'The Role of the United Nations Declarations of Principles in the Progressive Development of Space Law' (1988) 16 JSpaceL 5–20.

[61] Principles Governing the Use by States of Artificial Earth Satellites for International Direct Television Broadcasting, UNGA Res 37/92 (10 December 1982) (Broadcasting Principles).

[62] Principles Relating to Remote Sensing of the Earth from Outer Space, UNGA Res 41/65 (3 December 1986) (Remote Sensing Principles).

[63] Principles Relevant to the Use of Nuclear Power Sources in Outer Space, UNGA Res 47/68 (14 December 1992) (Nuclear Power Sources Principles).

Cooperation in the Exploration and Use of Outer Space for the Benefit and in the Interest of All States, Taking into Particular Account the Needs of Developing Countries (Benefits Declaration).[64]

Concerning the legal status of these various Principles, even if they were intended per se simply as recommendations when they were adopted, they have certainly not been completely divorced from the process of developing and reinforcing customary international rules. The critical element is represented by the practice of States in following at least certain aspects of the Principles. However, some of them now seem more substantially set within international space law—for instance, the free observation of the Earth from outer space—while others appear to be less established, with some still in the process of (possibly) gaining more universal legal applicability in the future.

The 1982 Broadcasting Principles[65] have international political, economic, social and cultural implications, particularly given that the question of influencing the content of direct television and other forms of information sources is still delicate in some societies.[66] These Principles raise issues relevant to national sovereignty, non-intervention, cultural independence and the flow of information.

A lack of consensus on the relevant issues relating to direct satellite television broadcasting undermined the possibility for the Resolution to have any significant normative value as an expression of the acceptance of these Principles as law. Indeed, it was a long process of debate, starting in the 1960s, which came to an end in the autumn of 1982, when the Resolution was adopted not by consensus but, unusually and exceptionally, by a majority vote.[67]

The Broadcasting Principles are an example of a soft law instrument whose principal soft law provisions of a substantive nature, incorporating a 'prior consent' requirement with respect to any proposed relevant activities,[68] were never accepted as binding—or even applicable—and never intended (at least by the major space-faring States) to be legally binding in the future.[69] Whilst some argue that 'the very existence of soft law instruments allows States to reach "agreement" in

[64] Declaration on International Cooperation in the Exploration and Use of Outer Space for the Benefit and in the Interest of All States, Taking into Particular Account the Needs of Developing Countries, UNGA Res 51/122 (13 December 1996) (Benefits Declaration).

[65] Carl Q Christol, 'Prospects for an International Regime for Direct Television Broadcasting' (1985) 34 ICLQ 142–58.

[66] Ram S Jakhu, 'Direct Broadcasting via Satellite and New Information Order' (1981) 8 SyracuseJIntlL&Com 375–90.

[67] Howard C Anawalt, 'Direct Television Broadcasting and the Quest for Communication Equality' (1984) 5 MichJIntlL 361–77.

[68] See Broadcasting Principles (n 59) Principles 13 and 14.

[69] As noted, the Broadcasting Principles (ibid) were exceptional in the evolution of both hard law and soft law instruments developed through the UNCOPUOS process, since they were not agreed by consensus. Instead, they were adopted by a vote at the UNGA—103 in favour, 13 against and 13 abstentions. They were not supported by the major space-faring States, which included virtually all of those that, at the time would be, or would likely to be, involved in direct television broadcasting via satellite.

circumstances where a treaty is simply not (politically) possible',[70] this is not always so and certainly was not the case with respect to the Broadcasting Principles. Rather, it was 'pushed through' the UNCOPUOS process by weight of numbers rather than the usual consensus mode. This was, however, a one-off occasion, with consensus otherwise being the norm in terms of decision-making within the United Nations governance framework. By comparison, the Remote Sensing Principles represent a successful consensus-led outcome in which an acceptable compromise was negotiated between the interests of the sensing States and the needs of the sensed States, representing at the time the majority of the developing countries.[71]

When they were adopted, the Remote Sensing Principles did not constrain programmes that had been already operated by States for many years.[72] In contrast, they reflected the fact that sensing States were convinced that their activity did not necessitate any consent, including no prior approval from sensed States. However, the sensed States were declared to have access to the primary data and the processed data concerning the territory under its jurisdiction on a non-discriminatory basis and on reasonable cost terms.[73] Thus, at least to some degree, the Remote Sensing Principles encoded pre-existing well-established practices of States before 1986, and the UNGA Resolution perhaps simply gave greater legitimacy and authenticity to, and endorsed, occurring practices.

Furthermore, later national practices have reiterated the key major elements of the 1986 legal framework. Nevertheless, due to what most regard as the inherently dual-use nature of space technologies, especially remote sensing activities that can be used for civil, commercial and military purposes at the same time, States have developed national programmes and adopted specific rules to regulate the use of data coming from these systems, by establishing data resolution limitations and access restrictions for the protection of national security interests.[74] As already

[70] For a criticism of this argument, see Jan Klabbers, 'The Undesirability of Soft Law' (1998) 67 ActScandJurisGent 381.

[71] Fabio Tronchetti, 'Legal Aspects of Satellite Remote Sensing' in Frans von der Dunk and Fabio Tronchetti (eds), *Handbook of Space Law* (Edward Elgar 2015) 501–53.

[72] Sergio Marchisio, 'The Evolutionary Stages of the Legal Subcommittee of the United Nations Committee on the Peaceful Uses of Outer Space (COPUOS)' (2005) 31 JSpaceL 219–42.

[73] Remote Sensing Principles (n 60) Principle XII.

[74] Anne-Sophie Martin, 'The 1986 UN Principles on Remote Sensing Dealing with the Dual-Use Nature of Space Imagery' (2021) 63 ProceColloqLOuterSpace 224–35; see also arts 50, 51, 52 and 53 of Regulation (EU) 2021/696 of the European Parliament and of the Council of 28 April 2021 establishing the Union Space Programme and the European Union Agency for the Space Programme and repealing Regulations (EU) No 912/2010, (EU) No 1285/2013 and (EU) No 377/2014 and Decision No 541/2014/EU. The United States operates a clear shutter control policy embedded in two different legislative instruments: (i) the Kyl-Bingaman Amendment to the 1997 National Defence Authorization Act which limits commercial satellite systems licensed by the Federal Government in providing imagery of Israel; (ii) the 2010 National and Commercial Space Programs Act and its programme LandSat. Title VII of the French Space Law of 2008, and arts 23 and 24, '*Données d'Origine Spatiale*', which provide specific provisions regarding remote sensing activity—in particular, each private '*exploitant primaire de données spatiales*' must declare its activity to the competent administrative authority. The Authority checks that the activities do not harm France's fundamental interests in defence, foreign policy and

stressed, these Principles took the form of a UNGA resolution and mainly stand as guidelines, given that they have largely not been enshrined in a formal treaty.[75] In addition, considering States' practice, some of the Principles seem more strongly embedded in customary law, whilst others tend to be more weakly established.[76]

The Nuclear Power Sources (NPS) Principles contain practical aspects concerning the use of this technology in space exploration,[77] including the operation of nuclear reactors on interplanetary missions and the storage of NPS objects in 'sufficiently high orbits' after the operational part of their mission.[78] The Principles also provide for a safety assessment and notification of re-entry.[79] Nevertheless, they only cover 'nuclear power sources devoted to the generation of electric power on board space objects for non-propulsive purposes, which have characteristics generally comparable to those of systems used and missions performed at the time of the adoption of the Principles'.[80] As such, they are not expressly applicable to situations (in the future) where it might be contemplated to use a nuclear power source for propulsive purposes.[81]

Subsequently, these Principles were further refined by the Safety Framework for Nuclear Power Source Applications in Outer Space, adopted by the COPUOS in 2009.[82] This has been drafted as a non-legally binding instrument, for utilization at a national level as a (voluntary) tool for national activities and initiatives. The Benefits Declaration recognizes the importance of 'international cooperation in the exploration and use of outer space for the benefit and the interests of all States'. It further elaborates the cooperation principle set out in Article I of the Outer Space Treaty. The Declaration underscores the fact that 'international cooperation

international obligations and, for these purposes, it can adopt restrictive measures. The German Act to Give Protection Against the Security Risk to the Federal Republic of Germany by the Dissemination of High-Grade Earth Remote Sensing Data, also known as Satellite Data Security Act (SatDSiG) of 2007 authorizes the German government to control the collection and dissemination of high-resolution satellite data in order to reduce any threat to national security, so as to ensure the peaceful coexistence of nations and Germany's international relations.

[75] This is to be contrasted with UNGA Res 1962 (XVIII) (13 December 1963).

[76] Sergio Marchisio, 'The 1986 United Nations Principles on Remote Sensing: A Critical Assessment' in *Studi di diritto internazionale, in onore di Gaetano Arangio-Ruiz* (Editoriale Scientifica 2004) 1311–40; Sergio Marchisio, 'Le régime de la télédétection' in Philippe Achilleas (ed), *Droit de l'espace* (Larcier 2009) 143–63.

[77] Anne-Sophie Martin and Steven Freeland, 'Back to the Moon and Beyond: Strengthening the Legal Framework for Protection of the Space Environment' (2021) 46(3) Air&SpaceL 441–44.

[78] Nuclear Power Sources Principles (n 61) Principles 3(1)(b) and (2)(a)(iii).

[79] ibid Principles 4 and 5.

[80] ibid Preamble.

[81] Nuclear Power in Space in Focus of IAEA and UN events (*World Nuclear News*, 21 February 2022) <https://world-nuclear-news.org/Articles/Nuclear-power-in-space-is-focus-of-IAEA-and-UN-eve> accessed 6 July 2023; 'Nuclear Technology Set to Propel and Power Future Space Missions, IAEA Panel Says' (*IAEA*, 18 February 2022) <www.iaea.org/newscenter/news/nuclear-technology-set-to-propel-and-power-future-space-missions-iaea-panel-says> accessed 6 July 2023; Steve Mirmina and David J Den Herder, 'Nuclear Power Sources and Future Space Exploration' (2005) 6(1) ChiJIntlL 149–75.

[82] UN Doc A/AC.105/943 (1 July 2009).

should promote the development of space science and technology, foster the development of relevant and appropriate space capabilities, and facilitate the exchange of expertise and technology among States'.

Today, international cooperation is a significant element in the development of complex space activities such as the launch or the ongoing operation of the International Space Station.[83] International cooperation reflects a consolidated practice in space activities. As noted, the adoption of non-legally binding instruments has increased in the past three decades due to the growing number of actors and the diversity of interests in space applications.[84] These instruments facilitate the drafting of new rules for reference and guidance; act as persuasive guidance of the behaviour of the parties, as they often may have a moral obligation not to violate these rules; and may also support the development of customary rules in the space field.[85]

Certainly, these soft law instruments have their advantages in many situations—as noted, they allow for the prescription of standards and guidelines, whilst enabling States (and international intergovernmental organizations) some room to manoeuvre in the context of their relationships vis-à-vis each other. As one of the authors has argued elsewhere, 'in the case of the legal regulation of outer space, this is almost an inevitable consequence of the fact that advances in space-related technology have continued to outweigh the capacity—and perhaps also the willingness—of the international community to come to a formalized agreement on any relevant legal rules'.[86]

2.3 Third Phase: Interpretation of Notions Contained in the Treaties through UNGA Resolutions

The third phase of United Nations regulatory development for space, commencing from the end of the 1990s, was characterized by an assessment of the value of, and adherence to, the United Nations space treaties, and an interpretation of concepts

[83] Steven Freeland, 'The US-Russian Space Station Mission Is a Study in Cooperation' *The Conversation* (1 May 2015) <https://theconversation.com/the-us-russian-space-station-mission-is-a-study-in-cooperation-41077> accessed 6 July 2023. Of course, even this fundamental notion of cooperation in space activities between Russia and, inter alia, the United States is challenged and threatened by geopolitical events on Earth, such as those we are (at the time of writing) witnessing as a result of the war in Ukraine; see Sam Tabahriti, 'The Head of Russia's Space Agency Says it Will End ISS Cooperation with NASA and the European Space Agency amid Western Sanctions' *Business Insider* (2 April 2022) <www.businessinsider.com/russia-dmitry-rogozin-roscosmos-international-space-station-nasa-2022-4?r=US&IR=T> accessed 6 July 2023.

[84] UN Doc A/AC.105/C.2/117 (26 January 2022) 63.

[85] UN Doc A/AC.105/C.2/2021/CRP.23 (7 June 2021).

[86] For a detailed discussion of the use of 'soft law' instruments in the international regulation of space activities, see Steven Freeland, 'For Better or For Worse? The Use of "Soft Law" within the International Legal Regulation of Outer Space' (2011) XXXVI AnnAir&SpaceL 409–45.

stemming from treaty law through UNGA resolutions.[87] This phase of the work of the UNCOPUOS/LSC is primarily focused on the appraisal of the existing space legal framework towards the formulation of further non-binding instruments that can be derived from the rights and obligations under the treaties currently in force. Some of the legal topics to be addressed through the LSC were approved after the Third United Nations Conference on the Exploration and Peaceful Uses of Outer Space (UNISPACE III) in 1999.[88]

In addition, the LSC has increasingly provided analysis and guidance on important questions with respect to the application of existing rules of international space law and their implementation.[89] In 2018, UNISPACE+50[90] offered a unique opportunity for UNCOPUOS Member States to 'reflect on more than 50 years of achievement in space exploration and use and to chart the future contribution of the Committee to the global governance of outer space activities'. This came at a time when more and more governmental agencies and non-governmental entities are venturing to explore and use space as a resource.[91]

UNISPACE+50 also provided Member States with the opportunity to look towards the future by strengthening the roles and activities of the Committee and its two Subcommittees, as well as UNOOSA, as its Secretariat, so as to ensure that they continue to operate as dedicated platforms for international cooperation in the peaceful uses of outer space, in the fulfilment and elaboration of their respective mandates.

Over the years, the LSC has evolved towards the review of several recurring relevant questions, including the status and application of the five United Nations treaties on outer space. Even though their implementation has highlighted several obstacles to their universality, in particular by non-space-faring States, nonetheless the number of ratifications and accessions has continued to grow quite consistently.[92]

[87] Stephan Hobe, *Space Law* (Nomos 2019) 42–43.

[88] 'Report of the Third United Nations Conference on the Exploration and Peaceful Uses of Outer Space' (Vienna, 19–30 July 1999) (18 October 1999) (UNISPACE III) UN Doc A/CONF.184/6.

[89] See the UNGA resolutions on the concept of launching State [UNGA Res 59/115 (10 December 2004)], the registration of space objects [UNGA Res 62/101 (17 December 2007)] and recommendations on national legislations [UNGA Res 68/74 (11 December 2013)].

[90] UNISPACE+50 celebrated the fiftieth anniversary of the first United Nations Conference on the Exploration and Peaceful Uses of Outer Space. It was held in Vienna (Austria) from 18 to 21 June 2018. It represented the fourth conference of the United Nations on the exploration and peaceful uses of outer space. The three previous global Conferences on the Exploration and Peaceful Uses of Outer Space—UNISPACE Conferences in 1968, 1982 and 1999—were organized to engage States and international organizations in furthering their cooperation in the peaceful uses of outer space; see 'UNISPACE+50' (*UNOOSA*) <www.unoosa.org/oosa/en/ourwork/unispaceplus50/index.html> accessed 6 July 2023.

[91] UNGA Res A/RES/73/6 (26 October 2018).

[92] In 2004, there were 98 State parties to the Outer Space Treaty (n 27), 88 to the Rescue Agreement (n 30), 82 to the Liability Convention (n 31), 45 to the Registration Convention (n 32) and 10 to the Moon Agreement (n 33), while, as noted in n 52, these have increased significantly, illustrating an ever-increasing number of States deciding that there is a benefit to formally adhering to the United Nations space treaties.

Moreover, the issue of the definition and delimitation of 'outer space' has been an Agenda item of the LSC almost from the start of its functioning. However, efforts to reach an agreement on a legally binding delimitation between airspace and outer space, or at least to adopt a common interpretation of these concepts, have not been successful. More recent attempts by the LSC to consider the legal aspects of sub-orbital flights have not (as yet) moved the issue closer toward any widely shared understanding of an appropriate legal framework for such activities.

Indeed, the question of the definition and delimitation of outer space is now only formally discussed every second year at LSC, which seems to indicate that this issue is still beyond agreement and that (at least for some States) it is not a pressing matter compared with other issues being considered by the LSC. Under the same item, the LSC has also engaged in discussions for many years on the legal status of the Geostationary Satellite Orbit (GSO). Whilst there is not always agreement on the direction of these debates due primarily to geopolitics, in 2000 an agreement was reached concerning the use of the geostationary orbit, as well as the ITU rules.[93]

An additional item on the agenda of the LSC that has been considered during this third phase of United Nations governance frameworks was the 'Application of the concept of the launching State'. This study aimed to identify and explain all aspects of the concept of 'launching State' as embodied in both the Liability Convention and the Registration Convention, also considering its application by States and international intergovernmental organizations, by taking into consideration new and emerging practices in space activities.

In 2004, after a discussion at UNCOPUOS, a resolution was adopted by the UNGA.[94] It suggests, inter alia, that States enact national legislation on

[93] UN Doc A/AC.105/738 (7 April 2000) Annex III. This involved the following notions:

> (a) In accordance with article 44 of the ITU Constitution, the satellite orbits and radio frequency spectrum are limited natural resources, which must be used rationally, efficiently, economically and equitably; (b) It is necessary to facilitate equitable access to the orbit/spectrum resources; (c) ITU has planned the use of certain frequency bands and services for the geostationary orbit; (d) In many frequency bands and services access to frequencies and satellite orbits, including the geostationary satellite orbit, takes place according to the principle of 'first come, first served'; (e) the current regulations on access to frequencies and satellite orbits in respect of bands and services may give rise to situations involving difficult processes of coordination among developed as well as developing countries.

The Legal Subcommittee therefore recommends that:

> Where coordination is required between countries with a view to the utilization of satellite orbits, including the geostationary satellite orbit, the countries concerned take into account the fact that access to that orbit must take place, inter alia, in an equitable manner and according to the ITU Radio Regulations. Consequently, in the case of comparable requests for access to the spectrum/orbit resource by a country already having access to the orbit/spectrum resource and a developing country or another country seeking it, the country already having such access should take all practicable steps to enable the developing country or other country to have equitable access to the requested orbit/spectrum resource.

See also UNGA Res A/RES/55/122 (27 February 2001) 4.

[94] UNGA Res 59/115 (10 December 2004).

authorization and supervision of space activities by private entities and agree on the conclusion of agreements related to joint launches. Since 2004, the LSC has also examined another matter linked to the application of the United Nations space treaties, namely the practice of States and international intergovernmental organizations in registering space objects. Experience arising mainly as a result of the practice of private operators and the commercialization of space missions has highlighted lacunae in the Registration Convention and disparities concerning the number of registered objects,[95] the territory of launch and the orbital parameters, as well as the function of a spacecraft.[96]

The debate culminated in the adoption of UNGA Resolution 62/101 of 17 December 2007 titled 'Recommendations on enhancing the practice of States and international intergovernmental organizations in registering space object',[97] which provides valuable insight on how to overcome current gaps and ensure a more harmonized application of the Registration Convention in the effective designation of space objects and launching State(s).

The LSC was also entrusted with the task of developing recommendations on national space legislation. As a result of its work, the UNGA adopted Resolution 68/74 of 11 December 2013,[98] which represents another very relevant and practical document advising States to consider a number of elements when adopting regulatory frameworks for their national space activities, considering, in particular, the specific needs and requirements of the particular State. In particular, States should determine: (i) the scope of activities to be covered by its national regulatory frameworks (for example, launch and on-orbit missions); (ii) international responsibility and national jurisdiction over space activities carried out from territory under its jurisdiction and control; (iii) the scope of a competent national authority granting authorization and licences; (iv) consistency with the relevant State's international obligations and other soft law instruments; (v) continuing supervision and monitoring by the appropriate State; (vi) the establishment and scope of its national registry; (vii) international liability, recourse, insurance requirements and indemnification procedures; and (viii) the relevant processes in the case of the transfer of ownership of a space object. As noted, the resolutions adopted by the UNGA might also serve as an interpretation of some of the existing international space law Principles of the five main United Nations space treaties.

[95] See 'United Nations Register of Objects Launched into Outer Space' (*UNOOSA*) <www.unoosa.org/oosa/en/spaceobjectregister/index.html> accessed 6 July 2023: 'to date over 89% of all satellites, probes, landers, crewed spacecraft and space station flight elements launched into Earth orbit or beyond have been registered with the Secretary-General.'

[96] Stephan Hobe and others, 'Current Issues in the Registration of Space Objects' in *Project 2001 Plus Global and European Challenges for Air and Space Law at the Edge of the 21st Century* (Proceedings of the Workshop, Cologne, 2005) 20–21.

[97] UNGA Res 62/101 (17 December 2007).

[98] UNGA Res 68/74 (11 December 2013).

3 New Legal Challenges to be Addressed within the UNCOPUOS

Moving to the present, work continues in an effort to further promote wider adherence and acceptance by States of the obligations contained in the United Nations space treaties, thus broadening international cooperation in the scientific and legal fields of the exploration and uses of outer space for peaceful purposes. UNISPACE+50 reaffirmed that 'the UNCOPUOS, its Subcommittees and UNOOSA should continue to coordinate efforts to strengthen the implementation of the United Nations treaties and principles on outer space'.[99]

Furthermore, the UNCOPUOS has adopted the Guidelines for the Mitigation of Space Debris of the COPUOS, endorsed by UNGA Resolution 62/217 of 22 December 2007, and the 21 Guidelines on the Long-Term Sustainability of Outer Space Activities, adopted by the STSC in June 2019.[100] The latter addresses 'the policy and regulatory framework for space activities; the safety of space operations; international cooperation, capacity-building and awareness; and scientific and technical research and development'. At its sixty-second session in 2019, the Committee decided to 'establish a new working group on the topic under the STSC'.[101] The Working Group aims 'to identify and study challenges, consider possible new guidelines for the long-term sustainability of space activities; share experiences, practices and lessons learned from the voluntary national implementation of the already adopted guidelines; and raise awareness in particular among emerging space nations'.[102]

At the moment, as noted, the international space community is, for a number of reasons, unwilling to adopt additional binding agreements, with geopolitical constraints as well as 'treaty fatigue' restricting any real attempts at multilateral treaty-making.[103] Non-legally binding instruments that can offer flexibility may often have political value and force.[104] Various aspects now codified in the existing treaties stemmed from several compromises rather than from a complete and seamless uniformity of views. Moreover, there may exist a certain element of risk in pushing for the development of new treaties or amendments to existing treaties, as these may instead re-open debate on the already agreed-upon issues.

The adoption of non-legally binding instruments could, in certain circumstances, actually help States to better abide by their international obligations. Soft law and specialized agencies have also helped the development of international

99 UNGA Res 73/6 (26 October 2018).

100 UN Doc A/74/20 (21 June 2019) 163 and Annex II.

101 ibid 165.

102 UN Doc A/AC.105/C.1/2022/CRP.13 (7 February 2022) 6.

103 Tanja Masson-Zwaan, 'Some Reflections on the Future of Space Law' in Stephan Hobe (ed), *Air Law, Space Law, Cyber Law—the Institute of Air and Space Law at Age 90* (Carl Heymanns Verlag 2016) 58.

104 Frans von der Dunk, 'Contradictio in Terminis or Realpolitik? A Qualified Plea for a Role of "Soft Law"' in Irmgard Marboe (ed), *Soft Law in Outer Space* (Böhlau 2012) 31–56.

space law to accommodate rapid technological changes and technical norms. Specialized agencies also play a vital role in setting regulatory standards and recommended practices. Some of them have practical effectiveness, others must be implemented by States through domestic Acts. The UNCOPUOS has for quite some time continued to consider its possible role in the elaboration of technical norms on space matters.[105]

In addition, the UNCOPUOS, together with its two Subcommittees, needs to consolidate its role by taking into account new challenges arising from emerging space activities, which will have to be appropriately considered from a legal and scientific point of view. One can think of issues such as the ongoing sustainability of outer space activities, space traffic management, the protection of the space environment, the exploration and use of space natural resources, the mitigation and removal of orbital debris, space tourism, regulation and registration of large constellations of small satellites, and the effects of satellite systems on astronomy, the latter of which was most recently included as a new Agenda item during the STSC 2022 session.[106]

Another example is that, in 2022, a consensus was reached within the LSC regarding the detailed Five Year Workplan for the next five years of the recently established (2021) Working Group on Legal Aspects of Space Resource Activities.[107]

4 Regulating the Security Dimension of Space Activities—The Role of the First Committee of the United Nations General Assembly

Space security is a significant element to consider in regulating activities conducted in outer space,[108] given new and emerging threats presented by space activities, such as ASAT tests,[109] spy satellite activities in orbit or cyberattacks on space infrastructure.[110] These are now part of the scope of activities undertaken by States,

[105] See Nandasiri Jasentulyana, 'Strengthening International Space Law: The Role of the United Nations' in Robert A Harris (ed), *International Organisations and Space Law: Their Role and Contributions* (ESA Publications Division 1999) 87–95.

[106] See UN Doc A/AC.105/1258 (23 February 2022) 267–68 (Item 18: 'General exchange of views on dark and quiet skies for science and society').

[107] UN Doc A/AC.105/C.2/2022/SRA/L.1 (5 April 2022).

[108] Rajeswari Pillai Rajagopalan, 'Changing Space Security Dynamics and Governance Debates' in Melissa de Zwart and Stacey Henderson (eds), *Commercial and Military Uses of Outer Space* (Springer 2021) 53–167.

[109] 'Fact Sheet: Vice President Harris Advances National Security Norms in Space' (*The White House*, 18 April 2022) <www.whitehouse.gov/briefing-room/statements-releases/2022/04/18/fact-sheet-vice-president-harris-advances-national-security-norms-in-space/> accessed 6 July 2023: 'Vice President Kamala Harris announced that the United States commits not to conduct destructive, direct-ascent anti-satellite (ASAT) missile testing, and that the United States seeks to establish this as a new international norm for responsible behavior in space.'

[110] Luke Shadbolt, 'Technical Study—Satellite Cyberattacks and Security' [2021] HDIGlobSpecialSE 3–13; Caroline Baylon, 'Challenges at the Intersection of Cyber Security and Space Security' (2014) Chatham House Research Paper 7–15.

and the destructive ASAT tests conducted by China in 2007, the United States in 2008, India in 2019 and Russia in 2021[111] are illustrative of this trend, as well as the dual-use nature of technology.[112]

Moreover, the current legal regime regarding the military use of outer space is seen by some elements as quite permissive, based on the reality that military activities have been carried out since the beginning of the 'space age'.[113] Indeed, the international legal framework reveals some limits in addressing new threats and new space technology in outer space and, to some, leaves a broad margin of appreciation in the interpretation of Article IV of the Outer Space Treaty, as well as flexibility in the way military activities in space are conducted. The UNCOPUOS has historically not dealt with military and security aspects,[114] notwithstanding their crucial significance to the way we are to regulate space activities moving forward.

Article IV(I) of the Outer Space Treaty provides for a 'de-nuclearized' governance regime but does not provide for a comprehensive demilitarization of space. States are not *expressly* prohibited from carrying out military activities or using conventional weapons in outer space. By contrast, Article IV(2) foresees a full demilitarization of the Moon and other celestial bodies. Article IV is the only article in the Outer Space Treaty that expressly addresses weapons and military uses of outer space,[115] although other provisions of the Treaty reinforce the general object and purpose of the Treaty that revolve around the peaceful uses of space.[116]

Article IV prohibits States from placing any weapons on the Moon or other celestial bodies (paragraph 2), but only expressly precludes States from placing certain weapons, such as weapons of mass destruction (WMD), into Earth orbit or elsewhere in space (paragraph 1).[117] This means that conventional weapons are not *expressly* banned from being placed into Earth orbit or into deep space. As a result, some argue that 'Article IV of the Outer Space Treaty presents only a partial limit with regard to the use of weapons in outer space and there are many aspects of the Outer Space Treaty that are relevant to, and indicative of the underlying

[111] Michelle Hanlon, 'The Rules of Space Haven't Been Updated in 50 Years, and the UN Says it's Time' *CNN* (3 January 2022) <https://edition.cnn.com/2022/01/03/world/space-law-united-nations-partner-scn/index.html> accessed 6 July 2023.

[112] Jakub Prazak, 'Dual-Use Conundrum: Towards the Weaponization of Outer Space?' (2021) 187 ActaAstro 397–405.

[113] Steven Freeland, 'The 2008 Russia/China Proposal for a Treaty to Ban Weapons in Space: A Missed Opportunity or an Opening Gambit' (2008) 51 ProceColloqSpaceL 261–71.

[114] See generally Annette Froehlich and Vincent Seffinga, *The United Nations and Space Security, Conflicting Mandates between UNCOPUOS and the CD* (Springer 2020).

[115] Kai-Uwe Schrogl and Julian Neumann, 'Article IV' in Stephan Hobe, Bernhard Schmidt-Tedd and Kai-Uwe Schrogl (eds), *The Cologne Commentary on Space Law*, vol I (Carl Heymanns Verlag 2009) 70–93.

[116] The imperative of peaceful purposes is specifically mentioned in the Preamble of the Treaty: 'Recognizing the common interest of all mankind in the progress of the exploration and use of outer space for peaceful purposes' (para 2); 'Desiring to contribute to broad international cooperation in the scientific as well as the legal aspects of the exploration and use of outer space for peaceful purposes' (para 4).

[117] UNGA Res 1884 (XVIII) (17 October 1963).

requirement that space is undertaken for peaceful purposes, although there have been diverse interpretations of the term "peaceful purposes"'.[118] Article 3 of the Moon Agreement stresses that 'the Moon (and other celestial bodies) must be used exclusively for peaceful purposes, and any threat or use of force or any other hostile act on the Moon is prohibited'. In this way, it reiterates the relevant parts of Article IV of the Outer Space Treaty but then goes even further.

In the continuity of the work within the UNCOPUOS, it is necessary to carefully consider the security aspects associated with space activities, and the risks of the 'weaponization' of space. The First Committee of the United Nations General Assembly[119] (Disarmament and International Security) plays an important role in terms of space security, as it deals with 'disarmament, global challenges and threats to peace that affect the international community, and seeks out solutions to challenges in the international security regime'. The mandate of the Committee is:

> to consider all disarmament and international security matters within the scope of the UN Charter or relating to the powers and functions of any other organ of the United Nations; the general principles of cooperation in the maintenance of international peace and security, as well as principles governing disarmament and the regulation of armaments; and promotion of cooperative arrangements and measures aimed at strengthening stability through reducing levels of armaments.

As noted above, the UNCOPUOS does *not* report to the First Committee, but rather to the Fourth Committee of the UNGA, which deals with 'Special Political and Decolonization'.[120] In this context, it is not uncommon for certain Member States of the UNCOPUOS to raise objections when they perceive that security-related discussions within the UNCOPUOS are beyond the mandate of the Committee and, as such, should be dealt with in other United Nations fora.[121] In this regard, the First Committee works in tandem with the United Nations Disarmament Commission and the Geneva-based Conference on Disarmament (CD).[122] Established in 1979, the Conference on Disarmament is one of the only standing multilateral disarmament negotiating bodies, with 65 Member States and 38 Observer States.[123]

[118] Stephan Hobe, *Space Law* (Nomos 2019) 76–77; Stephan Hobe, 'The Meaning of "Peaceful Purposes" in Article IV of the Outer Space Treaty' (2015) XL AnnAir&SpaceL 9-–24.

[119] 'Disarmament and International Security (First Committee)' (*United Nations*) <www.un.org/en/ga/first/> accessed 6 July 2023.

[120] 'Special Political and Decolonization (Fourth Committee)' (*United Nations*) <www.un.org/en/ga/fourth/> accessed 6 July 2023.

[121] UN Doc A/AC.105/1260 (19 April 2022) 110.

[122] 'Disarmament and International Security (First Committee)' (*United Nations*) <www.un.org/en/ga/first/> accessed 6 July 2023.

[123] 'Conference on Disarmament' (*United Nations Office for Disarmament Affairs*) <www.un.org/disarmament/conference-on-disarmament/> accessed 6 July 2023.

Security concerns within the context of space activities include the development of a specific regime related to the Prevention of an Arms Race in Outer Space (PAROS),[124] which has been a focal point for the international community within the United Nations and the CD since 1981.[125] Indeed, since 1980, the UNGA has passed a resolution on PAROS each year reiterating the necessity to maintain international peace and security in outer space.[126] PAROS is also on the list of competencies of the CD. However, the CD has not been able to provide any concrete agreement over the past 40 years, due to logjams, impasses and dysfunction in its decision-making processes.[127] Moreover, the legal obligations linked to PAROS are difficult to achieve[128] because both the UNCOPUOS and the CD have different objectives and operate under diverse legal approaches on the question of an 'arms race' in outer space.[129]

This is a geopolitically sensitive area since it addresses national security, interest and defence matters. The issues are framed by non-binding resolutions and 'best' practices. The most recent treaty proposal was presented jointly by Russia and China to the CD—the 'Draft Treaty on Prevention of the Placement of Weapons in Outer Space, the Threat or Use of Force against Outer Space Objects (PPWT)'.[130] Some States, notably the United States, have opposed some fundamental aspects arising from this draft treaty.[131] A complex and complicated matter that is often raised as impacting the possibility of negotiation of a binding treaty is the absence of concrete tools for the verification[132] of objects launched, which therefore

[124] UNGA Res 36/97(C) (9 December 1981); Steven Freeland, 'Peaceful Purposes? Governing the Military Uses of Outer Space' (2016) 18(1) EurJLRef 35–51; Ben Baseley-Walker, 'Outer Space, Geneva and the Conference on Disarmament: Future Directions' (2012) 28(1) SpacePoly 45ff.

[125] Péricles Gasparini Alves, *Prevention of an Arms Race in Outer Space: A Guide to the Discussions in the Conference on Disarmament* (United Nations Institute for Disarmament Research [UNIDIR] 1991) 1ff.

[126] Since the 1980s, every year the UNGA has passed a resolution on the prevention of an arms race in outer space (PAROS), which reminds the international community of 'the importance and urgency of preventing an arms race' and calling on States to 'refrain from actions contrary to that objective'; see eg UNGA Res 75/36 (7 December 2020); UNGA Res 76/230 (24 December 2021).

[127] Paul Meyer, 'Does the Conference of Disarmament Have a Future?' (2021) 4(2) JPeace&NuclDisarm 287–94.

[128] Setsuko Aoki, 'Law and Military Uses of Outer Space' in Ram S Jakhu and Paul Dempsey (eds), *Routledge Handbook of Space Law* (Routledge 2017) 207ff.

[129] Cassandra Steer and Matthew Hersch (eds), *War and Peace in Outer Space—Law, Policy and Ethics* (OUP 2021) 97ff; Anne-Sophie Martin, 'Forty Years of Negotiations on PAROS: Outcomes and New Challenges' (*SpaceWatch.Global Opinion*, November 2020) <https://spacewatch.global/2020/11/spacewatchgl-opinion-forty-years-of-negotiations-on-paros-outcomes-and-new-challenges/> accessed 6 July 2023; Sa'id Mosteshar, 'Space Law and Weapons in Space' in *Oxford Research Encyclopedia of Planetary Science* (OUP May 2019) 7 <https://oxfordre.com/planetaryscience/view/10.1093/acrefore/9780190647926.001.0001/acrefore-9780190647926-e-74?print=pdf> accessed 6 July 2023.

[130] The first version of the draft PPWT was submitted to the CD in 2008. The latest (revised) version is to be found in CD/1985 (12 June 2014) <https://undocs.org/CD/1985> accessed 6 July 2023.

[131] 'Statement by Ambassador Wood: The Threats Posed by Russia and China to Security of the Outer Space Environment' (*US Mission to International Organizations in Geneva*, 14 August 2019) <https://geneva.usmission.gov/2019/08/14/statement-by-ambassador-wood-the-threats-posed-by-russia-and-china-to-security-of-the-outer-space-environment/> accessed 6 July 2023.

[132] Daniel Porras, 'Eyes on the Sky—Rethinking Verification in Space' (UNIDIR, Space Dossier 4, 2019) <https://doi.org/10.37559/WMD/19/Space01> accessed 6 July 2023; Frank R Cleminson and

restricts the ability to definitively assess if those objects should be regarded as a space 'weapon'.[133] Indeed, there are significant challenges in agreeing to a multilateral understanding on the definition of a space weapon.

Both in the CD and the UNGA, a number of proposals have also been formulated for the development of non-legally binding instruments in the context of transparency and confidence-building measures (TCBMs).[134] On 2 December 2014, the UNGA adopted Resolution 69/32 titled 'No first placement of weapons in space'. It represents a relevant development in the area of space security and, notwithstanding the intensely political nature of the issues involved, indicates a growing recognition of the positive role that transparency and confidence-building measures play in the PAROS debate. These developments also indicate a willingness of the Member States of the United Nations to engage in measures which can promote a more secure space environment. The adoption of these instruments is of the utmost importance for making political commitments through unilateral declarations and/or bilateral agreements, in order to foster greater international cooperation in the field and promote the peaceful uses of outer space.

In 2017, UNGA Resolution 72/250 titled 'Further practical measures for the prevention of an arms race in outer space'[135] recalled the need to agree on and implement at the earliest opportunity a balanced and comprehensive programme of work in the CD, and requested the United Nations Secretary-General to 'establish a Group of Governmental Experts (GGE) to consider and make recommendations on substantial elements of an international legally binding instrument on the prevention of an arms race in outer space including *inter alia* on the prevention of the placement of weapons in outer space'. However, the two sessions held in 2018 and 2019 did not reach a consensus on a substantive report due to disagreements among the major space powers.[136] A new GGE, established by UNGA resolution 77/250, met in 2023 and then in 2024 to formulate recommendations by consensus in this field.[137]

Péricles Gasparini Alves, 'Space Weapons Verification: A Brief Appraisal' in Serge Sur (ed), *Verification of Disarmament or Limitation of Armaments: Instruments, Negotiations, Proposals* (UNIDIR 1992) 177. See also CD/1781 (22 May 2006); CD/1785 (21 June 2006).

[133] Jinyuan Su, 'Towards an Effective and Adequately Verifiable PPWT' (2010) 26(3) SpacePoly 152.

[134] Peter Martinez and others, 'Criteria for Developing and Testing Transparency and Confidence-Building Measures (TCBMs) for Outer Space Activities' (2014) 30(3) SpacePoly 91–97; Jana Robinson, 'Transparency and Confidence-Building Measures for Space Security' (2011) 27(1) SpacePoly 31–37. See also UNGA Res 76/55 (6 December 2021).

[135] UNGA Res 72/250 (24 December 2017); 'Group of Governmental Experts on Further Effective Measures for the Prevention of an Arms Race in Outer Space' (*United Nations Office for Disarmament Affairs*) <www.un.org/disarmament/topics/outerspace/paros-gge/> accessed 6 July 2023.

[136] 'Report by the Chair of the Group of Governmental Experts on Further Practical Measures for the Prevention of an Arms Race in Outer Space' (31 January 2019) <www.un.org/disarmament/wp-content/uploads/2019/02/oral-report-chair-gge-paros-2019-01-31.pdf> accessed 6 July 2023.

[137] UNGA Res 77/250 (30 December 2022).

In 2021, the First Committee of the United Nations approved five draft resolutions with the aim of preventing an arms race in outer space.[138] Discussions on space security have also been reinitiated with Resolution A/C.1/76/L.52 which has led to the establishment of a new Open-Ended Working Group (OEWG) on outer space security with the aim to make recommendations on substantial elements for the prevention of an arms race in outer space.[139] The OEWG met in Geneva in 2022 and 2023.[140] Participation is extended to intergovernmental institutions, commercial actors and civil society. The Working Group is empowered to review existing international agreements and normative frameworks that relate to threats in outer space arising from irresponsible behaviour. This is an important development in the field of space security—while the OEWG is being established through a UNGA resolution, it is located in Geneva in an effort to revitalize the disarmament debate and PAROS issues at the CD.[141]

5 A Way Forward—the Next 'Phase' of United Nations Contributions?

The United Nations, via its agencies, offers unique platforms for international cooperation in the exploration and use of outer space for 'peaceful purposes', for 'the global governance of outer space activities', for 'developing international space law', for 'fostering dialogue among spacefaring and emerging space nations', and for 'promoting the increased involvement of all countries in space activities'.

In particular, the Legal Subcommittee of the UNCOPUOS has created a (relatively) comprehensive, reliable and enduring scheme of international standards. It has been of fundamental importance for the development of space law, having inter alia agreed on treaties, as well as soft law principles, guidelines and best/current practices that supplement and interpret the treaties. Importantly, the UNCOPUOS enables an exchange of views on the myriad issues on its agenda, thus creating the

[138] 'Delegates Approve 5 Draft Resolutions, as First Committee Takes Action on Peaceful Use, Non-Weaponization of Outer Space, Chemical Weapons' (*United Nations Meetings Coverage and Press Releases*, 1 November 2021) <www.un.org/press/en/2021/gadis3676.doc.htm> accessed 6 July 2023.

[139] *See* Jessica West, 'The Open-Ended Working Group on Reducing Space Threats – Recap of the Fourth and Final Session August 2023', Ploughshares, January 2024.

[140] At the time of writing, the OEWG had concluded its fourth session in September 2023; see 'Open-ended Working Group on Information and Communication Technologies: Overview' (*United Nations Office for Disarmament Affairs*) <https://meetings.unoda.org/open-ended-working-group-on-information-and-communication-technologies-2021#:~:text=The%20OEWG%20is%20an%20inter,for%20any%20other%20UN%20process> accessed 6 July 2023; *see* Almudena Azcarate Ortega and Victoria Samson (eds), 'A Lexicon for Outer Space Security', UNIDIR, August 2023, <https://unidir.org/wp-content/uploads/2023/09/UNIDIR_Lexicon_for_Outer_Space_Security.pdf> accessed 4 March 2024

[141] Rajeswari Pillai Rajagopalan, 'Space Security Governance: Could a New Working Group Narrow the Divide?' *The Diplomat* (4 November 2021) <https://thediplomat.com/2021/11/space-security-governance-could-a-new-working-group-narrow-the-divide/> accessed 6 July 2023.

ground for a common understanding that can result in additional multilateral regulation.

The work of the UNCOPUOS is not over—indeed, if anything, its scope is expanding. It will continue in the coming years to debate further norms and standards related to new space activities through its two Subcommittees, which form the most suitable and universal environment to assess the current legal regime and consider further regulatory frameworks.

In addition, the First Committee of the UNGA and the CD play an important role for the development of (non-binding) instruments related to space security, especially on the issue of PAROS. It is important to find ways of overcoming the diplomatic impasses that restrict real progress on some of the important security-related matters discussed in those fora. Over the last few years, space activities have been witnessing significant changes in the structure, substance and intensity of outer space exploration. This is also reflected in the growing number of participants and new space activities, including the deployment of large constellations of small satellites, space resource activities and threats such as proliferating space debris and kinetic ASAT test activities.

The complexities make it even more important that international space law is supplemented by further governance instruments and implementation of national space legislation. In this regard, the relevant United Nations institutions described in this chapter must, as much as possible, be adaptable and responsive to the technological, geopolitical, economic, cultural, military and other aspects that are relevant in considering appropriate governance structures and instruments moving forward.

It is, therefore, necessary to make better use (and take advantage) of the UNCOPUOS, especially the Legal Subcommittee, with the adoption of more strategic decisions instead of debates on issues that have a long record of the exchange of views with little tangible action, and to foster greater interaction with other specialized agencies of the United Nations, including those outside the UNCOPUOS system, which have interests in space applications. Politically acceptable modalities need to be found to more comprehensively allow for the views of business and industry, academia and civil society to be part of the discussions, and to utilize participatory methods of consultation with delegations (for example, through more questionnaires during inter-sessional periods).

Overall, the United Nations has made a crucial and overarching contribution to the regulatory and governance regime for space activities. Its involvement has meant that space has 'worked' over the past seven decades, notwithstanding the technological developments and undoubted challenges. Yet, the contribution of the United Nations remains a continuous work-in-progress. Reform of existing UN agencies, both in terms of process and structure, may become increasingly appropriate, in order to (try to) keep up with a rapidly developing paradigm of space activities.

Reform requires a change of thinking and a recognition of the common interests that all stakeholders have in maintaining the safety, stability, security and sustainability of space and space activities. There are many important discussions to be had to ensure that the United Nations can, as effectively as possible—given the *realpolitik* and commercial realities—continue to make such positive and pivotal contributions. We look forward to those discussions and hope that they will result in ongoing dialogue at the multilateral level that best serves the interests of humanity.

4
OUTER SPACE TREATY AND FUNDAMENTAL PRINCIPLES

Setsuko Aoki

1 The Path to the Outer Space Treaty

1.1 The Establishment of the Forum to Discuss Peaceful Uses of Outer Space

No sooner had World War II ended than the United States of America (US) and the Union of Soviet Socialist Republics (USSR, or the Soviet Union) started research and development (R & D) of outer space. Both countries developed launch vehicles and satellites, mainly for military purposes at the early stage of space R & D. The US determined that the first rocket and satellite should be civilian and for purely scientific purposes to formulate a principle that outer space was free for the explanation and use for all States.[1] As the upper limit of airspace that was characterized with 'complete and exclusive sovereignty' was not provided in Article 1 of the Convention on International Civil Aviation,[2] as the reflection of customary international law, it was possible that some States might protest against the passage of a satellite above their airspace. Therefore, the US, which expected that it would be the first to launch a satellite, was cautious in making a state practice to establish a rule of international law for the freedom of exploration and use of outer space. It considered that military satellites could be launched after scientific satellites had established a right to pass over foreign States.[3]

Thus, the US found a good opportunity in the International Geophysical Year (IGY) to be held from 1 July 1957 to 31 December 1958 for its first satellite to be placed into orbit for purely scientific purposes. The US formally announced its planned satellite launch within the framework of the IGY on 29 July 1955,[4] which was followed by a similar but less formal statement by the USSR on 2 August 1955. It was the USSR that successfully launched the first artificial satellite, Sputnik 1, on

[1] Paul Stares, *The Militarization of Space: U.S. Policy 1945–1984* (Cornell University Press 1985) 33–35; Albert K Lai, *The Cold War, the Space Race, and the Law of Outer Space* (Routledge 2021) 14–15.

[2] Convention on International Civil Aviation (adopted 7 December 1944, entered into force 4 April 1947) 15 UNTS 295, art 1.

[3] Stares (n 1) 33–35; Lai (n 1) 15.

[4] Bin Cheng, *Studies in International Space Law* (Clarendon Press 1997) 3–10; Stares (n 1) 34–35.

Setsuko Aoki, *Outer Space Treaty and Fundamental Principles* In: *International Space Law in the New Space Era*.
Edited by: Sandeepa Bhat B., Dilip Ukey, and Adithya Variath, Oxford University Press.
DOI: 10.1093/9780198909415.003.0004

4 October 1957.[5] The US succeeded in placing its first satellite, Explorer 1, on 31 January 1958.[6]

The first United Nations General Assembly (UNGA) resolution referring to 'outer space' was adopted within two months of the successful launch of the first artificial satellite. UNGA Resolution (Res) 1148 (XII) stated that 'the sending of objects through outer space shall be exclusively for peaceful and scientific purposes'[7] to avoid the launch of nuclear-tipped ballistic missiles. This resolution, adopted in the First Committee of the UNGA dealing with disarmament and international security, was not necessarily made for pursuing peaceful uses of outer space. The difference between missiles and rockets was hardly recognized at that time, and UNGA Res 1148 (XII) was adopted to urge the negotiation of an inspection system to ensure that a nuclear warhead would not be put into a ballistic missile, as a first step to reach the eventual negotiation of a nuclear reduction agreement between the US and USSR. As any progress in nuclear disarmament was impossible at the height of the Cold War, and emerging space activities demonstrated rich potential far beyond the realm of disarmament, the US and USSR agreed to discuss question of the peaceful uses of outer space in a different organ, which resulted in UNGA Res 1348 (XIII)[8] and UNGA Res 1472 (XIV)[9] which established ad hoc and permanent Committee on the Peaceful Uses of Outer Space (COPUOS) respectively as a subsidiary organ of the UNGA.[10]

1.2 Rapid Development of Fundamental Principles through UNGA Resolutions

Fundamental principles of space activities had been developed rather rapidly through UNGA resolutions. The first COPUOS session held in 1961 produced UNGA Res 1721 A and B (XVI),[11] which included important recommendations determining the contours of international space law to be created. UNGA Res 1721 A (XVI) declared general principles such as outer space, including celestial bodies, being: (i) the object of international law; (ii) free for exploration and use by all States; and (iii) not subject to national appropriation.[12] UNGA Res 1721 B (XVI) called upon States launching objects into orbit or beyond to furnish information to the COPUOS through the UN Secretary-General (UNSG) for the registration of

[5] UN Doc A/AC.105/INF.2 (24 March 1962) 4.

[6] UN Doc A/AC.105/INF.1 (7 March 1962) 3.

[7] UNGA Res 1148 (XII) (14 November 1957) para 1(f).

[8] UNGA Res 1348 (XIII) (13 December 1958) Preamble paras 1(a), (b) and 2. Ad hoc COPUOS comprised 18 States.

[9] UNGA Res 1472 (XIV) (12 December 1959) Preamble para 1(a).

[10] Charter of the United Nations (adopted 26 June 1945, entered into force 24 October 1945) XV UNCIO 335, art 22 (UN Charter).

[11] UNGA Res 1721 (XVI) A and B (20 December 1961).

[12] UNGA Res 1721 (XVI) A (n 11) para 1(a), (b).

launchings and requested the UNSG to maintain a public registry of information on such objects.[13] This registration later developed into a unique system to exercise jurisdiction and control over objects in outer space without expressly conferring nationality.[14]

This was followed by two decisively important UNGA resolutions adopted in 1963: UNGA Res 1884 (XVIII)[15] and UNGA Res 1962 (XVIII).[16] The combination of these two resolutions constitutes a major part of the most important treaty to date on space activities: the Outer Space Treaty (OST).[17] UNGA Res 1884 (XVIII) was not made in the COPUOS but drafted in the UNGA First Committee as a disarmament issue,[18] which called on all States '[t]o refrain from placing in orbit around the earth any objects carrying nuclear weapons or any other kinds of weapons of mass destruction, installing such weapons on celestial bodies, or stationing such weapons in outer space in any other manner'.[19] This had been wholly incorporated in paragraph 1 of Article IV of the OST, while the text was slightly modified. The other resolution, UNGA Res 1962 (XVIII), titled 'Declaration of Legal Principles Governing the Activities of States in the Exploration and Use of Outer Space', included most recommendations in UNGA Res 1721 (XVI) A and B, and was incorporated in Articles I–III and V–IX of the OST with minimum modifications of the text. On the same day that UNGA Res 1962 (XVIII) was adopted, UNGA Res 1963 (XVIII), titled 'International Co-operation in the Peaceful Uses of Outer Space' was also adopted, recommending that UNGA Res 1962 (XVIII) be incorporated 'in international agreement form, in the future as appropriate'.[20] The same recommendation was repeated two years later in UNGA Res 2130 (XX).[21]

1.3 Two Superpowers' Decisive Power Enabling the Adoption of the Outer Space Treaty

On 7 May 1966, US President Lyndon B Johnson announced the intention to establish a space treaty to prevent any nation from claiming sovereignty over the Moon and other celestial bodies and to assure peaceful uses thereof, amidst the heated space race of lunar exploration between the US and USSR. After bilateral

[13] UNGA Res 1721 (XVI) B (n 11) paras 1–2.

[14] This is reflected in the Treaty on Principles Governing the Activities of States in the Exploration and Use of Outer Space, including the Moon and Other Celestial Bodies(adopted 19 December 1966, entered into force 10 October 1967) 610 UNTS 205, art VIII (OST).

[15] UNGA Res 1884 (XVIII) (17 October 1963); Cheng (n 4) 152–53.

[16] UNGA Res 1962 (XVIII) (13 December 1963); Cheng (n 4) 153–54.

[17] OST (n 14).

[18] It was originated by the proposals by the US and the USSR made in the Eighteen Nation Committee on Disarmament (ENDC) in 1962, which was later submitted to the First Committee of the UNGA.

[19] UNGA Res 1884 (n 15) para 2(a).

[20] UNGA Res 1963 (XVIII) (13 December 1963) para 1.

[21] UNGA Res 2130 (XX) (21 December 1965) para I.

consultation on the contents of the future treaty, held on 11 May, the USSR sent its draft treaty to the UNSG on 16 June 1966,[22] which 'largely reproduced the principles set forth in resolutions 1721, 1884, and 1962'.[23] On the same day, the US sent its draft treaty to the chair of the COPUOS. The scope of the US draft treaty was restricted to the Moon and other celestial bodies.[24] During the Fifth Session of the Legal Subcommittee (LSC) of the COPUOS held between 12 July and 4 August 1966, the US showed flexibility in agreeing with the Soviet's draft, partly because most States members of the COPUOS preferred the general treaty governing both outer void space and celestial bodies. The new elements proposed by the US, such as exclusively peaceful uses of and free access to all areas of celestial bodies, which was not provided in UNGA Res 1962 (XVIII), were also welcomed by most of the COPUOS States members.[25]

After this, it was not discussions in the COPUOS[26] but rather direct negotiations between the two superpowers which led to the successful agreement for the first space treaty. The revised draft treaties sent to the UNGA by the US and USSR on 17 September 1966 and 4 October 1966 respectively largely narrowed their differences. The two countries then continued negotiations outside the UN until an agreed text was sent to the UNGA on 15 December. That was initially adopted in the First Committee on 17 December 1966, then in the Plenary on 19 December 1966 as UNGA Res 2222 (XXI), which contained the Treaty on Principles Governing the Activities of States in the Exploration and Use of Outer Space, Including the Moon and Other Celestial Bodies (OST).[27] The OST was opened for signature on 27 January 1967 and entered into force on 10 October 1967, with the ratification of five States, including the United Kingdom (UK), the US, and the USSR, which were Depositary Governments of the OST.[28]

1.4 When Were the Fundamental Principles of Space Activities Generated?

The OST consists of UNGA Res 1962 (XVIII), corresponding to Articles I–III and V–IX;[29] UNGA Res 1884 (XVIII), incorporated in paragraph 1 of Article IV; and other provisions and elements adopted in the later negotiations of the COPUOS

[22] Cheng (n 4) 219–20.
[23] ibid 221.
[24] ibid 219–20.
[25] ibid 221–22.
[26] The Resumed Fifth Session of the COPUOS/LSC, held 12–16 September 1966, did not produce any additional compromise or agreement. UN Doc A/AC.105/C.2/SR.72 (19 October 1966); UN Doc A/AC.105/C.2/SR.73 (19 October 1966).
[27] UNGA Res 2222 (XXI) (19 December 1966); OST (n 14); Cheng (n 4) 223-25.
[28] OST (n 14) art XIV(2)–(3).
[29] Precisely speaking, the contents of arts I and IV of the OST had been enriched through the proposals and discussions in the COPUOS after 1963. Cheng (n 4) 219–24; see eg UN Doc A/AC.105/C.2/SR.57 (20 October 1966).

and UNGA. As a result, major parts of the basic principles of the OST had been solidly established by 1963, or within six years of the first successful launch of an artificial satellite. Therefore, it is sometimes asserted that the present space order had been already made by 'instant' customary international law before the OST was adopted.[30] Needless to say, caution is needed in finding rules of customary international law based on the close similarity of the text found in the UNGA resolutions and the later adopted treaty, as it is logically possible that each of the treaty provisions gradually developed into a rule of customary international law after its entry into force.

With this limitation in mind it may still be possible to assess that some of the norm-creating rules—for instance paragraph 1 of UNGA Res 1962 (XVIII) which provides that '[o]uter space and celestial bodies are free for exploration and use of all States'—had become rules of customary international law by 1963, due to the absence of objection to satellite launches since the US announcement of its launch plan in July 1955. In this regard, a survey of state practice shows that, by 1963, other than the US and USSR, only the UK and Canada had started operating satellites with the help of the US launch capability.[31] France became the third independent spacefaring State, with its own space launch vehicle and a satellite in 1965.[32] The situation around 1963 was therefore that only a few States were exploring and using outer space and the overwhelming majority of other States remained silent on the practice.

In general, the finding of a rule of customary international law on the basis of silence or acquiescence by many countries must be carefully evaluated.[33] Thus, the safe conclusion seems to be that it is not clear whether the first few years of space activities by a few nations could be enough to establish a rule of customary international law.[34] This exact point demonstrates the importance of the OST. As of January 2023, 112 States had ratified or acceded to the OST,[35] which includes almost all spacefaring States[36] and, as a result, whether a certain rule is customary international law or a treaty-based rule hardly makes any difference to the application of the OST.

[30] Cheng (n 4) 125–49.

[31] For instance, both the UK and Canada procured their first satellites, Ariel 1 and Alouette 1 in April and September 1962, respectively, with the help of the US launch capability. 'Ariel 1' (*US National Air and Space Museum*) <https://airandspace.si.edu/collection-objects/ariel-1-satellite/nasm_A19751410000> accessed 19 August 2022; 'Alouette 1' (*Canadian Space Agency*) <https://www.asc-csa.gc.ca/eng/satellites/alouette.asp> accessed 19 August 2022.

[32] France's first satellite, A-1 (later named Astérix), was launched by its Diamant rocket on 26 November 1965. UN Doc A/AC.105/INF.129 (11 March 1966) 1.

[33] See eg Malcolm N Shaw, *International Law* (8th edn, CUP 2017) 66–68.

[34] Cheng (n 4) 146–49.

[35] UN Doc A/AC.105/C.2/2023/CRP.3 (20 March 2023) 12.

[36] The Islamic Republic of Iran is not a Party to the OST, although it did sign. ibid 7.

2 The Exploration and Use of Outer Space for All Humankind

Article I of the OST establishes one of the most critical principles in the exploration and use of outer space. It provides:

> The exploration and use of outer space, including the Moon and other celestial bodies, shall be carried out for the benefit and in the interests of all countries, irrespective of their degree of economic or scientific development, and shall be the province of all mankind.
>
> Outer space, including the Moon and other celestial bodies, shall be free for exploration and use by all States without discrimination of any kind, on a basis of equality and in accordance with international law, and there shall be free access to all areas of celestial bodies.
>
> There shall be freedom of scientific investigation in outer space, including the Moon and other celestial bodies, and States shall facilitate and encourage international cooperation in such investigation.

There is no definition of 'outer space' nor is there a delimitation of outer space from airspace in the OST, while the area a satellite continues to orbit around the Earth is generally understood as a part of outer space.[37] Article I of the OST obligates States Parties to explore and use outer space for the common interests while at the same time it guarantees freedom of exploration and use as well as freedom of scientific investigation in outer space. The exploration is understood as 'the general finding out of something yet to be explored, which may or may not include scientific activity',[38] and 'use' means the second stage of activities after exploration for both economic and non-economic purposes. The term 'exploitation', which is not used in the OST, is regarded as a part of 'use' and which is conducted for economic gain.[39] 'Province of mankind' differs from the concept of the 'common heritage of mankind' found in the Moon Agreement[40] in that the former refers to the nature of activities, ,[41] while the latter provides the status of a certain area whose new concept was incorporated in a legally binding instrument for the first time in the Moon Agreement in 1979.

[37] The situation remains the same today. UN Doc A/AC.105/769/Add.1 (2 February 2020) reports a historical summary on the consideration of the question on the definition and delimitation of outer space.

[38] Stephan Hobe, Bernhard Schmidt-Tedd, and Kai-Uwe Schrogl (eds), *Cologne Commentary on Space Law*, vol I (Carl Heymanns Verlag 2009) 34 (Hobe and others).

[39] ibid 36. See also Stephan Hobe, *Space Law* (Nomos 2019) 74.

[40] Agreement Governing the Activities of States on the Moon and Other Celestial Bodies (adopted 5 December 1979, entered into force 11 July 1984) 1363 UNTS 3, art 11(1) (MA).

[41] Cheng (n 4) 476. On the concept of the 'province of mankind', see eg Ricky J Lee, *Law and Regulation of Commercial Mining of Minerals in Outer Space* (Springer 2012) 206–17.

How to reconcile the obligation to explore and use outer space for common interests and the right of the freedom for exploration and use of outer space had been the subject of heated discussion for some decades after the adoption of the OST. One position was that concrete benefits from space use such as telecommunications, meteorological information, and images of the Earth should be allocated to all countries, a view which tended to be supported by non-spacefaring States, especially in the 1970s. The other was that, to the extent that a certain space activity did not substantially harm the future activity of other States, the State concerned fulfilled the obligations for securing common interests, and no provision to other States of concrete benefits accruing to it was required. This was generally supported by spacefaring States.[42]

The divided interpretation was to a large extent resolved by the 1996 UNGA Res 51/122 titled 'Declaration on International Cooperation in the Exploration and Use of Outer Space for the Benefit and in the Interest of All States, Taking into Particular Account the Needs of Developing Countries' (Space Benefit Declaration).[43] This can be seen as guidance to interpret Article I of the OST.[44] The Space Benefit Declaration holds that 'States are free to determine all aspects of their participation in international cooperation' and underlines the importance of the maintenance of a mutually acceptable basis including contractual terms.[45] Recognizing the typical north–south problem in space activities, this Declaration aimed to arrive at a constructive solution not by forcing countries to do what might be too idealistic, but by bringing about mutually beneficial results through reasonable methods of international cooperation.[46]

It can be safely said that subsequent international cooperation has largely achieved a common interest of space exploration and use for all humankind. This seems to be reflected in the two important UNGA resolutions: one is UNGA Res 73/6 (2018) that commemorates the fiftieth anniversary of the OST; and the other is UNGA Res 76/3 (2021) titled '"Space2030" Agenda: Space as a Driver of Sustainable Development', which lauded the accomplishments of common interest to date enabled by international cooperation and urged States to further harness

[42] Melissa K Force, 'The Paradox of United States' Position on Regulation of Space Resource Extraction' in PJ Blount, Rafael Moro-Aguilar, and Kai-Uwe Schrogl (eds), *Proceedings of the International Institute of Space Law 2016* (Eleven 2017) 257–68.

[43] UN Doc A/Res/51/112 (13 December 1996) (Space Benefit Declaration).

[44] Stephan Hobe, Bernhard Schmidt-Tedd, and Kai-Uwe Schrogl (eds), *Cologne Commentary on Space Law*, vol III (Carl Heymanns Verlag 2015) 315–16.

[45] Space Benefit Declaration (n 43) para 2.

[46] Kai-Uwe Schrogl, 'Legal Aspects Related to the Application of the Principle that the Exploration and Utilization of Outer Space Should be Carried Out for the Benefits and in the Interest of All States Taking into Particular Account the Needs of Developing Countries' in Marietta Benkö and Kai-Uwe Schrogl (eds), *International Space Law in the Making: Current Issues in the UN Committee on the Peaceful Uses of Outer Space* (Editions Frontiers 1993) 195, 212–18.

the potential of space as a tool to solve global challenges and leverage space technologies to improve human security, safety, and quality of life worldwide.[47]

3 Non-Appropriation of Outer Space

3.1 Non-Appropriation of Outer Space by Both States and Private Persons

Article II of the OST provides that '[o]uter space, including the Moon and other celestial bodies, is not subject to national appropriation by claim of sovereignty, by means of use or occupation, or by any other means'. While this text is almost identical to paragraph 3 of UNGA Res 1962 (XVIII), entry into force of the OST was greatly appreciated in the space community because '[t]he world cannot be made to face the task of the correct legal interpretations of a slightly uncertain resolution' and because 'until such a treaty is signed, it cannot be stated with certainty whether or not the victor in the race to the moon may claim territorial sovereignty over the occupied area'.[48] Therefore, the first and greatest merit of this provision may be the clarification of the legal status of outer space as an area beyond national jurisdiction.

At first reading, the scope of 'national appropriation' may be ambiguous if it also includes private proprietary claims. The answer is, however, simple and clear from both the drafting history and logical treaty interpretation. During the drafting processes, it was clearly understood by delegations that both extending sovereignty to outer space and private ownership of any parts of outer space were prohibited. Outer space shall not be property for either States or non-governmental entities.[49] That is logical, because a private person cannot claim title to land without the endorsement of a State that has sovereignty (territorial jurisdiction) over such territories. It is true that a private person can possess or control some land in outer space if it has technological capability, but that would remain a mere fact, which could be transformed into ownership only when it is granted through relevant national laws with respect to the land concerned. As any State must not appropriate and exercise territorial jurisdiction in any part of outer space, the logical conclusion would be that a factual situation created by a person controlling some areas of outer space could not possibly be authorized as ownership.[50] This conclusion naturally concurs with Article I of the OST, which provides that 'there shall be free

[47] UN Doc A/RES/73/6 (31 October 2018); UN Doc A/RES/76/3 (28 October 2021).

[48] Gyula Gál, *Space Law* (Oceana Publications 1969) 195.

[49] Eugene Brooks, 'Control and Use of Planetary Resources' (1969) 11 ProceColloqLOuterSpace 339, 339–50; Cheng (n 4) 229–30; Hobe and others (n 38) 45–51.

[50] See eg Virgiliu Pop, 'Appropriation in Outer Space: The Relationship between Land Ownership and Sovereignty on the Celestial Bodies' (2000) 16 SpacePoly 275, 277–78; Shaw (n 33) 363–64.

access to all areas of celestial bodies'.[51] Although the interpretation of Article II alone leads to the prohibition of private acquisition of ownership of any areas of outer space, Article I reinforces the conclusion. Private appropriation of any parts of outer space is prohibited.[52]

Thus, any business to sell the soil of the Moon, Mars, or other celestial bodies cannot be legitimized.[53] But this conclusion does not solve the issue of actual possession by a private person. As of today, because no private person has possessed or controlled any part of celestial bodies with the intention to own, these actual issues have not surfaced, but once a private person is capable of possessing or controlling some areas of a celestial body, then even if ownership is not claimed, this would pose a question as to whether such is within the permissible scope of the exploration and use of outer space provided in Article I of the OST.

The principle of non-appropriation of outer space for both States and private persons would not be changed if a certain State is not a party to the OST, for non-appropriation of outer space has been established a rule of customary international law.[54]

3.2 Removal and Use of Minerals and Other Substances in Place on the Celestial Bodies for Scientific Purposes

The contemporary issue regarding Article II of the OST is as to whether the prohibition of national appropriation extends to natural resources—mineral resources and other abiotic substances such as water—on the celestial bodies. Theoretically, natural resources in outer space could include spacecraft orbits, radio spectrum, solar energy, etc., but in normal cases, discussions on space resources have been restricted to mineral resources and abiotic substances. This may be explained by the fact that the orbital resources have been allocated as 'limited natural resources'[55] by the International Telecommunication Union (ITU) for about half a century,[56] and energy from the sun seems inexhaustible.

[51] Art XII of the OST (n 14) restricts the access to '[a]ll stations, installations, equipment and space vehicles on the Moon and other celestial bodies' to representatives of other States Parties to the OST and on a basis of reciprocity. It should be noted that this provision was needed for safety reasons.

[52] Tanja Masson-Zwaan and Stephan Hobe (eds), *The Law of Outer Space: An Experience in Contemporary Law-Making, by Manfred Lachs, Reissued on the Occasion of the 50th Anniversary of the International Institute of Space Law* (Martinus Nijhoff 2010) 42–45.

[53] Francis Lyall and Paul Larsen, *Space Law: A Treatise* (2nd edn, Routledge 2018) 170–71.

[54] ibid 65; Hobe and others (n 38) 45–47.

[55] Constitution and of the International Telecommunication Union (adopted 22 December 1992, entered into force 1 July 1994) 1825 UNTS 143, art 44(2) (ITU Constitution).

[56] The ITU started allocating radio frequencies for space telecommunications in 1959. The 1973 amendment to the ITU Convention (Málaga Torremolinos Convention) inserted a 'limited natural resources' provision in art 33(2), which corresponds to art 44(2) of the ITU Constitution (n 55).

During the drafting phase of the OST, Belgium and France stated that not only land on the celestial bodies but also space resources were the targets of the prohibition of national appropriation.[57] As Article II of the OST did not refer to space resources, however, there had been different views on the implications since the adoption of the OST, if the Belgian-French proposal was rejected, or rather included but not explicitly specified.[58]

For the States Parties to the Moon Agreement, there is no issue of divergent views as it explicitly states that natural resources on any of the celestial bodies in the solar system are the common heritage of mankind, and natural resources in place shall not become property of any State or any private person.[59] It should however be noted that the Moon Agreement does not categorically prohibit property rights or ownership of natural resources on such celestial bodies. It is only 'natural resources in place' which shall not become property of any State or private person,[60] and the exploitation of the natural resources of celestial bodies in the solar system is legally permissible provided that it is conducted in accordance with the rules of an international regime to be established once such exploitation of natural resources—ie the non-scientific (including especially commercial) removal of space resources—is about to become feasible. An international regime will be established to satisfy conditions of the common heritage of mankind, the contours of which are not fully specified in the Moon Agreement.[61] While the Moon Agreement was adopted based on the consensus of the COPUOS, the fact that only 18 States have either ratified or acceded to the Agreement as of January 2023, and that none of the major spacefaring States are included therein would suggest that the value of the planned future regime of space resource exploitation under the Agreement is inevitably limited.

To date, the total volume of lunar rocks that the spacecraft of the US, USSR, and People's Republic of China (China) have brought to Earth amounts to around 384 kilogrammes.[62] Extremely small amounts of samples from asteroids and comets have also been brought back to Earth.[63] In those cases, the collection and removal of minerals in place have never been criticized nor become the issue of

[57] Fabio Tronchetti, 'Title IV—Space Resource Exploration and Utilization of the US Commercial Space Launch Competitiveness Act: A Legal and Political Assessment' (2016) 41 Air&SpaceL 123, 146 (fn 10).

[58] Cheng (n 4) 233; Tronchetti (n 57) 145–46.

[59] MA (n 40) arts 1, 11(1), (3).

[60] ibid art 11(3).

[61] ibid art 11(1)–(8).

[62] See eg 'Lunar Rocks and Soils from the Lunar Missions' (*NASA*) <https://curator.jsc.nasa.gov/lunar/index.cfm> accessed 1 August 2022. The breakdown is that the US brought back around 382 kg, the USSR around 0.326 kg, and China around 1.731 kg.

[63] Japan's Hayabusa 1 and 2 and the United States' Stardust took samples from asteroids and a comet. See eg 'Past Hayabusa' (*JAXA*) <www.isas.jaxa.jp/en/missions/spacecraft/past/hayabusa.html> accessed 1 August 2022; 'Stardust' (*NASA*) <https://solarsystem.nasa.gov/stardust/home/index.html> accessed 1 August 2022.

ownership of space resources of celestial bodies, perhaps because the results of the study have been publicized and samples shared with the international scientific community, as appropriate for the promotion of scientific investigation through international cooperation, which is exactly the purpose of Article I of the OST. Thus, it may be said that a rule of customary international law has been established on the removal of mineral and other substances from celestial bodies when such removal is conducted for the scientific purposes. In this regard, Article 6(2) of the Moon Agreement may be said to be the reflection of a rule of customary international law.[64]

It provides that:

> In carrying out scientific investigations and in furtherance of the provisions of this Agreement, the States Parties shall have the right to collect on and remove from the Moon samples of its mineral and other substances. Such samples shall remain at the disposal of those States Parties which caused them to be collected and may be used by them for scientific purposes. States Parties shall have regard to the desirability of making a portion of such samples available to other interested States Parties and the international scientific community for scientific investigation. States Parties nay in the course of scientific investigations also use mineral and other substances of the Moon in quantities appropriate for the support of their missions.

Hence, the issue to solve would be whether commercial exploitation, extraction, and free disposal of space resources would be a lawful activity for States which are not States Parties to the Moon Agreement. In case the answer is in the affirmative, the legal conditions for such activity should be studied, including as to whether a unilateral and free exploitation without any additional restrictions is permissible or a new multilateral framework should be created.

3.3 Contradicting Views on the 2016–2017 Sessions of the COPUOS/LSC

When the US passed the Space Resource Exploration and Utilization Act of 2015 as part of the US Commercial Space Launch Competitiveness Act,[65] the space community reacted strongly to its content, especially to the provision

[64] Force (n 42) 268; Stephan Hobe, Bernhard Schimidt-Tedd, and Kai-Uwe Schrogl (eds), *Cologne Commentary on Space Law*, vol II (Carl Heymanns Verlag 2013) 369–70.

[65] US Public Law 114-90 (25 November 2015). Relevant sections of Title IV of the US Commercial Space Launch Competitiveness Act are incorporated in 51 USC (National and Commercial Space Programs), ss 51301–51303.

that a US citizen is entitled to possess, own, transport, use, and sell a space resource obtained pursuant to the applicable US law in accordance with international law.[66] The US Act prompted heated discussion in the fifty-fifth session of the COPUOS/LSC held in April 2016. As Article II of the OST combined with Article I(2) does not explicitly prohibit the taking of resources from outer space, one interpretation was that exploiting and extracting space resources for commercial and non-commercial purposes would be lawful to the extent that such an activity is in accordance with rules of international law, including especially the UN treaties on outer space.[67] This reasoning is sometimes called the 'Lotus Principle', meaning that sovereign States may act in any way they wish as long as they do not contravene an explicit prohibition.[68] The other interpretation was that when taking space resources was not clearly prohibited by Articles I(2) and II of the OST, other Articles of the OST and UN treaties on outer space, as well as established rules in other fields of international law, should be considered to find *lex lata* on this subject. Exploring other fields of international law is required in particular due to Article III of the OST, which provides that States Parties shall carry on space activities in accordance with international law, including the Charter of the UN (UN Charter). This method of study is taken based on the reasoning of the 1996 Advisory Opinion of the International Court of Justice (ICJ) on the Legality of the Threat or Use of Nuclear Weapons, in which the ICJ was of the view that the non-existence of the established rule of international law explicitly prohibiting the use of nuclear weapons did not mean that the use of nuclear weapons had been recognized lawful, but that other fields of international law such as international humanitarian law and international environmental law should also be studied to find *lex lata* on the legality or illegality of the use of nuclear weapons in today's international society.[69]

Due to overwhelming interest in the legality of the recovery of space resources in outer space, a new agenda item was adopted for the 2017 session of the LSC session titled 'General exchange of views on potential legal models for activities in the exploration, exploitation and utilization of space resources'.[70]

[66] 51 USC (n 65) s 51303.

[67] Other than the OST, these are: Agreement on the Rescue of Astronauts, the Return of Astronauts and the Return of Objects Launched into Outer Space (adopted 19 December 1967, entered into force 3 December 1968) 672 UNTS 119 (Rescue Agreement); Convention on International Liability for Damage Caused by Space Objects (adopted 29 November 1971, entered into force 1 September 1972) 961 UNTS 187 (Liability Convention); Convention on Registration of Objects Launched into Outer Space (adopted 12 November 1974, entered into force 15 September 1976) 1023 UNTS 15 (Registration Convention); and MA (n 40).

[68] *The Case of the SS 'Lotus' (France v Turkey)* [1927] PCIJ Rep Series A No 10 18-20.

[69] *Legality of the Threat or Use of Nuclear Weapons* (Advisory Opinion) [1996] ICJ Rep 243–44.

[70] UN Doc A/AC.105/1113 (27 April 2016) para 251.

The long-standing US position that Article I of the OST recognizes the right of exploitation of space resources was explained in the COPUOS/LSC discussions in 2017, quoting the statement of US Secretary of State Cyrus Vance in 1979:

> [T]his 'non-appropriation' principle applies to the natural resources of celestial bodies only when such resources are 'in place.' The prohibition on national appropriation does not, however, limit 'ownership to be exercised by States or private entities over those natural resources which have been removed from their "place" on or below the surface of the moon or other celestial bodies.' Such removal, Secretary Vance further explained, is permitted by Article I of the Outer Space Treaty, which provides that 'outer space, including the moon and other celestial bodies, shall be free for exploration and use by all States'.[71]

In the same session, Belgium submitted a paper that stated that 'Article II OST must be read in combination with Article I, para. 2, of the same treaty, which connects the freedom of use with the non-discriminatory access for all nations, on an equal basis, to all parts of celestial bodies'.[72] Belgium was of the view that his would inevitably lead to the approach that an international legal framework should preferably be made to ensure equitable access to natural resources of extra-terrestrial origin and that the UNCOPUOS was the competent body to find such norms.[73]

In the COPUOS/LSC discussions since 2017 to date, no country has claimed the ownership of space resources in place. Neither has any country opposed the ownership of space resources after the extraction from the soil or subsoil of celestial bodies.[74] The difference of the views exists on legal conditions to exploit and extract space resources, including whether space resource exploitation must be conducted on the basis of international frameworks to be made .[75] This would be addressed in the COPUOS/LSC discussions. Following the decision to establish a five-year working group to consider legal aspects of space resources activities in 2021,[76] a detailed work plan and working methods were determined in 2022.[77] It is planned that recommended principles of space resources activities will be finalized in 2027, preferably to be adopted as a UNGA resolution.[78] Hence, legal conditions for the major part of space resource activities seem set to be clearer by the end of the 2020s. This might be understood the processes to find concrete and detailed

[71] Statement of the US Secretary of State Cyrus Vance in 1979 was reiterated by the US delegation on agenda item 14 of the COPUOS/LSC, 28 March 2017 <https://icms.unov.org/CarbonWeb/public/oosa/speakerslog/dc656751-ffce-4b22-977e-a0244f55010e> accessed 30 August 2022.

[72] UN Doc A/AC.105/C.2/2017/CRP.19 (28 March 2017) 2.

[73] ibid 4.

[74] UN Doc A/AC.105/1122 (18 April 2017) paras 251–66; UN Doc A/AC.105/1177 (30 April 2018) paras 229–65; UN Doc A/AC.105/1203 (18 April 2019) paras 239–67.

[75] ibid.

[76] UN Doc A/AC.105/1243 (24 June 2021) para 255.

[77] UN Doc A/AC.105/1260 (19 April 2022) Appendix 39–40.

[78] ibid 40.

rules on the exploration, exploitation and utilization of space resources from fundamental principles of 'non appropriation' of the OST through international cooperation.

4 Implications of Peaceful Uses of Outer Space

4.1 Permissible Scope of Military Use of Outer Void Space

'Peaceful uses of outer space' is certainly a fundamental principle of the OST, but the exact meaning of this term needs careful examination as it is not as simple as it seems at first glance.

Article IV of the OST provides:

> States Parties to the Treaty undertake not to place in orbit around the Earth any objects carrying nuclear weapons or any other kinds of weapons of mass destruction, install such weapons on celestial bodies, or station such weapons in outer space in any other manner.
>
> The Moon and other celestial bodies shall be used by all States Parties to the Treaty exclusively for peaceful purposes. The establishment of military bases, installations and fortifications, the testing of any type of weapons and the conduct of military manoeuvres on celestial bodies shall be forbidden. The use of military personnel for scientific research or for any other peaceful purposes shall not be prohibited. The use of any equipment or facility necessary for peaceful exploration of the Moon and other celestial bodies shall also not be prohibited.

Paragraph 1 of this Article is almost identical to UNGA Res 1884 (XVIII), paragraph 2(a), adopted in the First Committee of the UNGA in 1963. In outer void space, placing in orbit around the Earth objects carrying weapons of mass destruction (WMD) or otherwise stationing such weapons is prohibited. WMD were understood as 'atomic explosive weapons, radio-active material weapons, lethal chemical and biological weapons, and any weapons developed in the future' that have comparable destructive effects to such WMD that currently exist.[79] Naturally, the major target for the prohibition was atomic explosive weapons (nuclear weapons). It was agreed that an intercontinental ballistic missile (ICBM) carrying WMD was not prohibited as it did not orbit around the Earth, but navigated in a suborbital trajectory.[80] Similarly, the predominant interpretation was that the Fractional Orbital Bombardment System (FOBS) that had been developed by the USSR at the time of drafting the OST was not banned, as FOBS would de-orbit

[79] UN Doc S/C.3/32/Rev.1 (18 August 1948).

[80] US Senate Committee on Aeronautical and Space Sciences, *Outer Space Treaty Analysis and Background Data Staff Report* (1967) 26 (US Senate Staff Report).

before making an entire rotation of the Earth,[81] although some States, including Italy, were against that interpretation.[82] FOBS was prohibited by the US-USSR nuclear arms control agreement, SALT II,[83] in 1979. While SALT II never entered into force, both nations informally followed that prohibition and the dismantling of FOBS had been completed by May 1984.[84] The formal prohibition of FOBS was realized when the Treaty Between the US and the USSR on FurtherReduction and Limitation of Strategic Offensive Arms (START I) entered into force in 1994.[85] However, beyond the bilateral agreement, the prevailing interpretation remained that FOBS was not prohibited by Article IV(1) of the OST. Therefore, when China conducted a weapons test of a FOBS-type vehicle carrying a hypersonic glide vehicle (HGV) in October 2021, no reference to such a test being a possible violation of the OST was made, while serious concerns were expressed on its apparently destabilizing effect on space security.[86]

One peculiar point is that no mention was made of the nuclear weapons test in outer space in Article IV(1) of the OST. That part of nuclear disarmament had been addressed by the 1963 Treaty Banning Nuclear Weapon Tests in the Atmosphere, in Outer Space and Under Water (PTBT) adopted prior to the negotiation of the OST.[87] The PTBT was concluded because electromagnetic pulse (EMP) radiation generated by the nuclear tests conducted in outer space by the US and USSR damaged at least six satellites of the two States and the UK, and it was expected that appropriate monitoring systems of nuclear explosions in outer space, including the nuclear explosion detection satellites, would be soon operated.[88] Arms control in outer void space remains restricted to the prohibition of stationing WMD under the OST and the banning of nuclear explosions tests under the PTBT.

[81] Stares (n 1) 99 (Statement by Mr McNamara on 3 November 1967); Eric Stein, 'Legal Restraints in Modern Arms Control Agreements' (1972) 66 AJIL 255, 264.

[82] UN Doc A/7221 (10 September 1968).

[83] Treaty Between the United States of America and the Union of Soviet Socialist Republics on the Limitation of Strategic Offensive Arms (signed 18 June 1979) (SALT II). SALT II never entered into force.

[84] Braxton Brick Eisel, 'The FOBS of War' *Air Force Magazine* (1 June 2005) <www.airandspaceforces.com/article/0605fobs/> accessed 20 September 2022.

[85] Treaty Between the United States of America and the Union of Soviet Socialist Republics on Further Reduction and Limitation of Strategic Offensive Arms (signed 31 July 1991, entered into force 5 December 1994) art V(18)(c) (START I). START I was terminated on 5 December 2009.

[86] See eg Demetri Sevastopulo and Kathrin Hille, 'China Tests New Space Capability with Hypersonic Missile' *Financial Times* (London, 17 October 2021) <www.ft.com/content/ba0a3cde-719b-4040-93cb-a486e1f843fb> accessed 20 September 2022.

[87] Treaty Banning Nuclear Weapon Tests in the Atmosphere, in Outer Space and Under Water (adopted 5 August 1963, entered into force 10 October 1963) 480 UNTS 43 (PTBT). The fact that China, France, and the Democratic People's Republic of Korea are not Parties to the PTBT might affect the norm of the prohibition of the nuclear explosion test in outer space unless such prohibition has become a rule of customary international law. In case it has not become a customary international law rule, France would not be able to carry out a nuclear explosion test as it made a formal statement that it would move to underground nuclear testing in 1974. *Nuclear Tests Case (Australia v France)* [1974] ICJ Rep 265–67.

[88] James Clay Moltz, *The Politics of Space Security: Strategic Restraint and the Pursuit of National Interests* (Stanford University Press 2008) 119, 137.

The operation of military satellites which might be categorized as conventional weapons under the national laws of many countries is not prohibited.[89] Nor are the stationing of conventional weapons in outer void space and the testing of conventional weapons in orbits prohibited. Hence, today there is no legally binding rule to ban destructive anti-satellite (ASAT) tests using conventional weapons. Attention should be paid to the fact that the obligation to use outer void space 'exclusively for peaceful purposes' was not specified, in contrast to the first sentence of Article IV(2) which provides for arms control on the Moon and other celestial bodies. Although India proposed to also add that phrase in Article IV(1) during the drafting phase, and Argentina, Austria, Brazil, Canada, Iran, Japan, and the United Arab Republic supported the Indian proposal, it was not adopted.[90] But this does not necessarily mean that the obligation to explore and use outer space 'exclusively for peaceful purposes' does not exist, considering the provisions of other Articles of the OST, including especially Articles I and III. Some commentators have claimed that the obligation of 'exclusively for peaceful purposes' is also applicable to outer void space.[91]

In early days of space exploration and use, the US and the USSR had to establish the interpretation of 'peaceful', whether it equated with 'non-military' or 'non-aggressive'. Except in the short period through 1958, the US interpreted 'peaceful' as 'non-aggressive', which meant that military space activities were lawful unless they were explicitly forbidden by a rule of international law. This interpretation allows for military use of space falling short of the threat or use of force prohibited by Article 2(4) of the UN Charter. In contrast, the Soviet interpretation was 'peaceful' being 'non-military', thus no military activities in outer space being permissible.[92] The Soviet position in the early 1960s was clearly reflected in its proposal on the prohibition of 'the use of artificial satellites for the collection of intelligence information in the territory of foreign States' submitted to the First Session of the COPUOS/LSC held in 1962.[93] However, the successful operation of military satellites by the USSR led the two States to the understanding that 'peaceful' meant 'non-aggressive' although this was not publicly announced.[94] Worldwide state

[89] For instance, in case of the US, military satellites are categorized into Munitions List Category XV of the International Traffic in Armed Regulations under the Arms Export Control Act.

[90] UN Doc A/AC.105/C.2/SR/66 (21 October 1966) 5–6; UN Doc A/AC.105/C.2/SR/71 and Add.1 (21 October 1966) 10.

[91] Marko G Markoff, 'Disarmament and "Peaceful Purposes" Provisions in the 1967 Outer Space Treaty' (1976) 4(3) JSpaceL 11–12; Georgy P Zhukov, 'On the Question of Interpretation of the Term "Peaceful Use of Outer Space" Contained in the Space Treaty' (1969) 11 ProceColloqLOuterSpace 36, 36–38.

[92] Ivan A Vlasic, 'The Legal Aspects of Peaceful and Non-Peaceful Uses of Outer Space' in Bhupendra Jasani (ed), *Peaceful and Non-Peaceful Uses of Space: Problems of Definition for the Prevention of an Arms Race* (Taylor & Francis 1991) 37, 38–39; American Bar Foundation, *Committee on the Law of Space, Report to NASA* (1961) 25–26.

[93] UN Doc A/AC.105/L.2 (10 September 1962) para 8.

[94] See eg Gerald M Steinberg, *Satellite Reconnaissance: The Role of Informal Bargaining* (Praeger Publishing 1983) 59–63; D Brennan, 'Arms and Arms Control in Outer Space' in Lincoln P Bloomfield (ed), *Outer Space: Prospects for Man and Society* (Prentice-Hall 1962) 161.

practice to date seems to indicate that 'peaceful' means 'non-aggressive' pursuant to the methods of treaty interpretation set forth in Article 31(3)(b) of the Vienna Convention on the Law of Treaties (VCLT), which provides 'any subsequent practice in the application of the treaty which establishes the agreement of the parties regarding its interpretation'.[95]

As long as 'peaceful' means 'non-aggressive', with or without the obligation of space being explored or used 'exclusively for peaceful purposes', the permissible scope of military space activities in outer void space would be the same unless the adverb 'exclusively' changes the meaning of this phrase to 'non-military'. If the 'exclusively for peaceful purposes' obligation is implied in Article IV(1), States Parties to the OST may carry out military space activities except stationing WMD and resorting to the threat or use of force. In the same vein, without the imposition of 'exclusively for peaceful purposes' obligation, States Parties may not station WMD and may not resort to the threat or use of force in their military space activities due to Article 2(4) of the UN Charter. If 'exclusively' changes the meaning of the phrase from 'non-aggressive' to 'non-military', then there would be differences, as under the obligation of 'non-military' use the use of military satellites, and the testing of ballistic missiles, FOBS, and conventional weapons would not be permissible. But this interpretation cannot be taken, as it contradicts the text of Article IV(1) of the OST and state practice. In conclusion, in outer void space, (i) stationing WMD in orbits around the Earth or otherwise; (ii) conducting nuclear explosion tests; and (iii) resorting to the threat or use of force are prohibited.

4.2 Arms Control on the Moon and Other Celestial Bodies

Arms control regulations on celestial bodies differ from those in outer void space. The original US draft treaty in June 1966, which became the basis of Article IV(2), borrowed expressions largely from Article 1 of the 1959 Antarctic Treaty[96] that had successfully demilitarized Antarctica. Because of this, it is said that the US intention was to demilitarize the Moon and other celestial bodies modelled after the Antarctic Treaty.[97] However, the expression '*inter alia*' used in the Antarctic Treaty when enumerating concrete forbidden actions, which indicates that prohibited activities are not limited only to explicitly enumerated ones, was lacking in the US proposal.[98] This and the later modification in the Soviet proposal which led to the present Article IV(2) of the OST seem to show that a definite list of prohibited

[95] Vienna Convention on the Law of Treaties (adopted 22 May 1969, entered into force 27 January 1980) 1155 UNTS 331, art 31(3)(b).

[96] Antarctic Treaty (adopted 1 December 1959, entered into force 23 June 1961) 402 UNTS 71.

[97] US Senate Staff Report (n 80) 24–25; Cheng (n 4) 518–19; Alex Meyer, 'Interpretation of the Term "Peaceful" in the Light of the Space Treaty' (1969) 11 ProceColloqLOuterSpace 24, 24–29.

[98] Cheng (n 4) 247.

activities is the final result.[99] Thus, if 'peaceful' meaning 'non-aggressive' is applied, and if 'exclusively' is assessed as adding no legal implications for 'peaceful purposes', military activities other than those explicitly prohibited and falling short of the threat or use of force would be permissible, as is the case in outer void space, even though it is difficult to identify an example of an activity which would fit into that category today.

Today, there are two interpretations on the scope of arms control on the Moon and other celestial bodies. One view is that irrespective of the meaning attributed to the term 'peaceful purposes', adding the adverb 'exclusively' in combination with the drafting history of Article IV(2) leaves no doubt that the celestial bodies are demilitarized.[100] The other view is that 'exclusively' does not add any legal implication and, 'non-aggressive' use of outer space applies also to the Moon and other celestial bodies, which means military activities are possible on the celestial bodies as long as they do not correspond to the specifically prohibited activities and they fall short of the threat or use of force.[101]

The third and fourth sentences of Article IV(2) allow military personnel to conduct 'peaceful' research using 'any' equipment or facility for 'peaceful purposes', which is likely to include those converted from military equipment or facilities.[102]

In conclusion, there seems to be no clear answer as to whether only non-military use is permissible on the Moon and other celestial bodies. While the difference in the permissible scope of space activities remains only theoretical today, irrespective of the interpretation of "peaceful" and "exclusively", future exploration and use of the celestial bodies might pose a real problem.

5 Conclusion: OST—A Very Resourceful Treaty

It is true that the OST alone cannot fully address new issues, as it was drafted in 1966, decades before the flourishing privatization of outer space and the enormous increase in the problems of space debris. In addition, this would also be the case even in combination with UN treaties on outer space concluded later.[103] One

[99] ibid.

[100] ibid 248; Hobe and others (n 38) 82; Fabio Tronchetti, 'Legal Aspects of the Military Uses of Outer Space' in Frans von der Dunk and Fabio Tronchetti (eds), *Handbook of Space Law* (Edward Elgar 2015) 331, 340–41.

[101] Robert L Bridge, 'International Law and Military Activities in Outer Space' (1980) 13 AkronLRev 649, 659; Meyer (n 97) 29. Both authors however opined that it was very difficult to identify concrete possible military activity on celestial bodies due to the wide scope of prohibited activities.

[102] Hobe and others (n 38) 85.

[103] Rescue Agreement (n 67); Liability Convention (n 67); Registration Convention (n 67); and MA (n 40).

example would be the identifying of the launching States[104] that are jointly liable for damage caused by space objects. That is largely the result of increasing private procurement of the launch service from a foreign launch provider. As business contracts can be very complicated, the assessment of which State should be the State that 'procures the launching of an object in outer space'[105] cannot be necessarily clearly determined today.[106] And this can lead directly to the non-registration of a satellite operated by a private actor. As it is the registration of space objects that determines which State shall retain jurisdiction and control over such objects in outer space,[107] the uniquely and carefully designed responsibility, liability, and registration regime under the OST[108] certainly faces challenges in the twenty-first century.

The OST is however very resourceful and resilient, which enables the Treaty to keep up with the new era and new challenges. One example would be space debris mitigation. The phrase 'potentially harmful interference with activities of other States Parties' found in Article IX of the OST did not anticipate addressing space debris issues. It was mainly targeted at electronic jamming and other radio frequency interferences when the provision was negotiated,[109] but this phrase has been perfectly suitable in establishing necessary space debris mitigation measures.[110] This has been possible because the OST mostly provides principles on the basis of which many rules could be developed, with the help of other space-related treaties and also non-legally binding UN instruments relating to space activities. As the OST has been in the past, it will continue to be the most important treaty in space activities. The fundamental principles of the OST would be the freedom of exploration and use of outer space, taking into full account common interests and mutual benefits, non-appropriation and peaceful uses. All these principles are intertwined with one other and closely connected through the legal obligation of international cooperation, through which the first 60 years of space activities have successfully become the 'province of humankind'.

[104] OST does not use the term 'launching States', but the concept is provided in art VII of the OST with the identical text of the definitions of 'launching State' found in the Liability Convention and the Registration Convention. Liability Convention (n 67) art I(c); Registration Convention (n 67) art I(a).

[105] OST (n 14) art VII.

[106] See eg Setsuko Aoki, 'The Implications of the Registration of Space Objects in the 2020s: Possibility of the UK and Dutch Practice in Enhancing Registering Space Objects' in Comitato Promotore (ed), *Liber Amicorum Sergio Marchisio*, vol II (Editoriale Scientifica 2022) 943, 953–56.

[107] OST (n 14) art VIII.

[108] ibid arts VI–VIII.

[109] See eg US Senate Staff Report (n 80) 164.

[110] See eg Hobe and others (n 38) 176–79.

5

NORMS GOVERNING RESCUE AND RETURN

Adithya Variath

1 Introduction

The beginning of the space era, with the launch of the first artificial satellites orbiting Earth under an international scientific programme, the International Geophysical Year (IGY) (1957–1958), and the rapid advancement of rocket technology during that period made it necessary to create a special regime for outer space and celestial bodies.[1] The Antarctic Treaty, which was signed in Washington on 1 December 1959 and came into effect on 23 June 1961, set forth the legal guidelines for the scientific exploration of Antarctica based on knowledge gained during IGY, and served as a model for the regulation of space activities.[2] The Antarctic Treaty in many ways shaped the idea of sovereignty in international legal theory.[3] In contrast to that instrument, efforts to establish a global framework for regulating space operations were more reactive to the changing geopolitics and technological developments.[4]

[1] James M Wordie, 'International Geophysical Year' (1955) 2 JGlacio 482; FWG Baker, 'Chile, the International Geophysical Year, and the Antarctic' (2009) 45 Polar Record 85; 'International Geophysical Year, 1957–58: Antarctica, 1956' (1957) 8 Polar Record 357.

[2] The Antarctic Treaty (adopted 1 December 1969, entered into force 23 June 1963) 402 UNTS 71; Donald R Rothwell, 'Sovereignty and the Antarctic Treaty' (2010) 46 PolarRec 17; Gillian D Triggs (ed), *The Antarctic Treaty Regime: Law, Environment and Resources* (CUP 1987); Jean Georges Sauveplanne, 'Freedom and Sovereignty in Air and Outer Space' (1965) 12 NILR 228.

[3] Rothwell (n 2); Hans Zinnecker and others, 'Principles of the Antarctic Treaty' (2008) 33 EurAstronomSocyPublSeri 233; Rip Bulkeley, 'The Political Origins of the Antarctic Treaty' (2010) 46 PolarRec 9; Alessandro Antonello, 'The Antarctic Treaty as a Temporal Order' (2019) 55 PolarRec 323; Triggs (n 2) 49–50; Francisco Orrego Vicuna, 'The Antarctic Treaty System and the International Community' in Francisco Orrego Vicuna (ed), *Antarctic Mineral Exploitation: The Emerging Legal Framework* (CUP 1988) 458; Paul Arthur Berkman and Robert B Woyach, 'International Transitions in the Antarctic Treaty System' (1992) 28 PolarRec 244; Zou Keyuan, 'The Common Heritage of Mankind and the Antarctic Treaty System' (1991) 38 NILR 173; Christopher C Joyner, 'The Antarctic Treaty and the Law of the Sea: Fifty Years On' (2010) 46 PolarRec 14; Karen Scott, 'Institutional Developments within the Antarctic Treaty System' (2003) 52 ICLQ 473; Peter J Beck, 'Fifty Years On: Putting the Antarctic Treaty into the History Books' (2010) 46 PolarRec 4.

[4] MJ Peterson, 'The Use of Analogies in Developing Outer Space Law' (1997) 51 IntlOrg 245; D Goedhuis, 'Reflections on Some of the Main Problems Arising in the Future Development of Space Law' (1989) 36 NILR 247; 'Space Law: United Nations' (1977) 26(3) ICLQ 681; FB Schick, 'Space Law and Space Politics' (1961) 10 ICLQ 681; P Malanczuk, 'Space Law as a Branch of International Law' (1994) 25 NYIL 143; Jessica Schechinger, 'International Space Law' in André Nollkaemper and Ilias

Adithya Variath, *Norms Governing Rescue and Return* In: *International Space Law in the New Space Era*.
Edited by: Sandeepa Bhat B., Dilip Ukey, and Adithya Variath, Oxford University Press.
DOI: 10.1093/9780198909415.003.0005

The United Nations (UN) General Assembly set up a Special Committee on the Peaceful Uses of Outer Space (COPUOS) for that purpose, initially as an ad hoc committee in 1958 and then as an official organisation in 1959.[5] All cooperative space programmes supported by the UN and its Member States are now coordinated through the COPUOS. It established two subcommittees, one 'for legal matters' and the other 'for scientific and technical matters' and 'to examine specific suggestions for the growth of global collaboration in the fields of space research for peaceful purposes'. A comprehensive plan for this type of multilateral collaboration was unanimously endorsed by the UN General Assembly in Resolution 1721 (XVI) of 20 December 1961. The same resolution commended two fundamental principles to States for their guidance in space activities: 'the application of international law, including the United Nations Charter, to outer space and celestial bodies' and 'the freedom of all States to explore and use outer space and celestial bodies in accordance with international law without regard to national appropriation'.[6]

Vladimír Kopal notes that, '[w]hen the COPUOS and its Legal Subcommittee began their consideration of the topic, it was understood that the rule of law in outer space should be developed step by step in harmony with the actual needs of international cooperation in this new area of human activity and that all decisions in this regard should be adopted by consensus'.[7] A draft Declaration of the Basic Principles Governing the Activities of States Pertaining to the Exploration and Use of Outer Space proposed by the Union of Soviet Socialist Republics (USSR) was one of the documents presented to the Legal Subcommittee's first meeting in the spring of 1962.[8] The proposal included a set of fundamental rules that, in the

Plakokefalos (eds), *The Practice of Shared Responsibility in International Law* (CUP 2017) 451; Malcolm N Shaw, 'Air Law and Space Law' in *International Law* (5th edn, CUP 2003); D Goedhuis, 'Reflections on the Evolution of Space Law' (1966) 13 NILR 109.

[5] UNCOPUOS, 'International Cooperation in the Peaceful Uses of Outer Space' Report of the Legal Subcommittee on the Peaceful Uses of Outer Space on the Work of its Sixth Session (19 June–14 July 1967) UN Doc A/6804 Annex III; Philip C Jessup and Howard J Taubenfeld, 'The United Nations Ad Hoc Committee on the Peaceful Uses of Outer Space' (1959) 53 AJIL 877; UN Committee on the Peaceful Uses of Outer Space, Report of Legal Subcommittee (1962) 1 ILM 97–105; UNCOPUOS, 'Report of the Legal Subcommittee on the Work of its First Session (28 May–20 June 1962)' UN Doc A/AC.105/6 (9 July 1962); Agreement on Cooperation in the Exploration and Use of Outer Space for Peaceful Purposes (adopted and entered into force 24 May 1972) TIAS 7347.

[6] Treaty on Principles Governing the Activities of States in the Exploration and Use of Outer Space, Including the Moon and Other Celestial Bodies (adopted 27 January 1967, entered into force 10 October 1967) 610 UNTS 205 (Outer Space Treaty).

[7] Vladimir Kopal, 'Treaty on Principles Governing the Activities of States in the Exploration and Use of Outer Space, Including the Moon and Other Celestial Bodies' in RH Mankiewicz (ed), *Yearbook of Air and Space Law 1966* (McGill University Press 1968) 463.

[8] UNCOPUOS, 'Draft Declaration of the Basic Principles Governing the Activities of States Pertaining to the Exploration and Use of Outer Space proposed by the USSR' First Session of the Legal Subcommittee UNCOPUOS, UN Doc A/AC.105/C.2/L.1 (March 1962); see also 'Proposals for Treaty on Exploration of Outer Space' (1966) 5 ILM 1105; UNCOPUOS, 'Report of the Legal Subcommittee on the Work of its Second Session (16 April–3 May 1963) to the Committee on the Peaceful Uses of Outer Space' UN Doc A/AC.105/12 (6 May 1963).

USSR's opinion, were essential for any future or ongoing space-related operations. Some of the concepts in the draft Declaration, in particular, did not instantly win the endorsement of all the COPUOS Member States. Nevertheless, the notion gained traction, and a successful Declaration was prepared in 1963.

A collection of general principles that described the legal status of celestial bodies and outer space, as well as the parameters of what activities by States are legal in the space environment, were included in the 1963 Declaration. Additionally, it offered preliminary guidelines for addressing several previously well-known issues related to the space activities being conducted at the time, setting the stage for future space law regulation of space initiatives. Being a General Assembly resolution, it could not create legally enforceable standards of international law. Nevertheless, it was viewed as the foundation for a future legally binding treaty, even at the time of its passage. The space race between two major space nations to reach the Moon hastened the realisation of having a strong legal regime. The following proposals were made by them on 16 June 1966, following a brief diplomatic overture: 'a draft Treaty Governing the Exploration of the Moon and Other Celestial Bodies by the United States'[9] and 'a draft Treaty on Principles Governing the Activities of States in Exploration and Use of Outer Space, the Moon and Other Celestial Bodies by the USSR'.[10] Fortunately, the disparity in focus between the two projects could be overcome thanks to widespread acceptance of the topic's broader perspective, which was offset by a few minor concessions.

Beginning on 12 July 1966, the Legal Subcommittee had its fifth session in Geneva, where discussions on the agreed-upon topic proceeded in September of that same year in New York. The discussions covered two different types of subject: the first dealt with fundamental principles, the essence of which had already been outlined in the 1963 Declaration and could now be transferred into the draft Treaty with some additions and minor changes; the second elaborated the principle of international cooperation in space activities and dealt with some specific issues relating to its implementation. Preliminary discussions during the 1963 Declaration negotiations led to agreement on fundamental principles in the draft Treaty, while specific questions involving the second category led to controversial discussions.

The Preamble to the Outer Space Treaty spells out the desire 'to contribute to broad international cooperation in the scientific as well as the legal aspects of the exploration and use of outer space for peaceful purposes' and the belief 'that such cooperation will contribute to the development of mutual understanding and to the strengthening of friendly relations between States and peoples'.[11] It reflects the fact

[9] UNCOPUOS, 'Draft Treaty Governing the Exploration of the Moon and Other Celestial Bodies' UN Doc A/AC.105/32 (17 June 1966).

[10] UNCOPUOS, 'Draft Treaty on Principles Governing the Activities of States in the Exploration and Use of Outer Space, the Moon and Other Celestial Bodies' UN Doc A/6352 (16 June 1966).

[11] Outer Space Treaty (n 6).

that the Outer Space Treaty addressed scientific and technical needs, contributing to cold war détente and fostering cooperation.[12] The negotiations on the major space law instrument positively impacted UN space agreements. Shortly after the conclusion of the Outer Space Treaty, the Agreement on the Rescue and Return of Astronauts and the Return of Objects Launched into Outer Space was finalised. On 19 December 1967, the UN General Assembly, by a vote of 115–0, approved an 'Agreement on the Rescue of Astronauts, the Return of Astronauts, and the Return of Objects Launched in Outer Space', requested the Depositary Governments 'to open the Agreement for signature and ratification at the earliest possible date' and expressed its vision 'for the widest possible adherence to this Agreement'.

It was adopted by the General Assembly on 19 December 1967 (Resolution 2345 (XXII)), opened for signature on 22 April 1968, and entered into force on 3 December 1968.[13] The negotiating process was accelerated by some tragic accidents that caused the loss of lives of astronauts. The Agreement, elaborating on elements of Articles 5[14] and 8[15] of the Outer Space Treaty, provides that 'States shall take all possible steps to rescue and assist astronauts in distress and promptly return them to the launching State and that States shall, upon request, provide assistance to launching States in recovering space objects that return to Earth outside the territory of the Launching State'. As 'envoys of mankind', astronauts are held in high regard, and in space they are treated with the same humanitarian principles that ships' captains apply to distressed mariners on the high seas, providing

[12] Proposals for Treaty (n 8); Manfred Lachs, 'The Treaty on Principles of the Law of Outer Space, 1961–1992' (1992) 39 NILR 291; D Goedhuis, 'An Evaluation of the Leading Principles of the Treaty on Outer Space of 27th January 1967' (1968) 15 NILR 17.

[13] Agreement on the Rescue of Astronauts, the Return of Astronauts and the Return of Objects Launched into Outer Space (adopted 22 April 1968, entered into force 3 December 1968) 672 UNTS 119 (Rescue Agreement).

[14] Outer Space Treaty (n 6) art V:

> States Parties to the Treaty shall regard astronauts as envoys of mankind in outer space and shall render to them all possible assistance in the event of accident, distress, or emergency landing on the territory of another State Party or on the high seas. When astronauts make such a landing, they shall be safely and promptly returned to the State of registry of their space vehicle. In carrying on activities in outer space and on celestial bodies, the astronauts of one State Party shall render all possible assistance to the astronauts of other States Parties. States Parties to the Treaty shall immediately inform the other States Parties to the Treaty or the Secretary-General of the United Nations of any phenomena they discover in outer space, including the moon and other celestial bodies, which could constitute a danger to the life or health of astronauts.

[15] ibid art VIII:

> A State Party to the Treaty on whose registry an object launched into outer space is carried shall retain jurisdiction and control over such object, and over any personnel thereof, while in outer space or on a celestial body. Ownership of objects launched into outer space, including objects landed or constructed on a celestial body, and of their component parts, is not affected by their presence in outer space or on a celestial body or by their return to the Earth. Such objects or component parts found beyond the limits of the State Party to the Treaty on whose registry they are carried shall be returned to that State Party, which shall, upon request, furnish identifying data prior to their return.

unequivocal help as long as it does not put the ship or her crew in danger. Since the launching authority will benefit from the recovery and return, a benefit that in certain circumstances could amount to millions of dollars, these costs must be covered by the launching authority.

The General Assembly approved a decade-long effort to establish an international agreement on rescuing astronauts in distress, returning them, and restoring space objects, aiming to ensure the fulfilment of humanitarian and scientific objectives.[16] This chapter traces the development of the Treaty and explores how the Treaty will play a pivotal role in the coming age of space exploration.

2 Background and Setting the Context

With the launch of the unmanned Soviet Sputnik satellite in 1957, the international community became aware of the difficult task of regulating the conditions when a spacecraft lands on foreign soil and how it would remain the property of the state that launches it and should be returned. Lipson and Katzenbach, in their 'Report To the National Aeronautics and Space Administration on the Law of Outer Space' advocated that 'once a satellite is launched, it is beyond the control of human beings, and therefore its flight is analogous to the flight of a meteor. Because a meteor is the property of the nation in which it lands, a spent satellite would also be the property of the nation in which it lands, regardless of its point of origin.'[17] They suggested that a space object launched by humans was not similar to natural objects because of human control. The ownership of objects launched into outer space by states is established by Article VIII of the Outer Space Treaty.

Loftus Becker, Legal Adviser to the Department of State during the late 1950s, Congressman Kenneth B. Keating and the Counsel to the House Science and Astronautics Committee suggested that 'the return of a space object from foreign territory should be conditioned upon payment of compensation for any damages caused by the landing'.[18] Even the report of the 1958 Ad Hoc Committee on the Peaceful Uses of Outer Space established by the General Assembly noted 'the desirability of the conclusion of multilateral agreements concerning re-entry and landing, such agreements to contain suitable undertakings on cooperation and appropriate provisions on procedures. Among the subjects that might be covered by

[16] R Cargill Hall, 'Rescue and Return of Astronauts on Earth and in Outer Space' (1969) 63 AJIL 197; Piet-Hein Houben, 'A New Chapter of Space Law: The Agreement on the Rescue and Return of Astronauts and Space Objects' (1968) 15 NILR 121; Charles W Mathews, 'Astronautics and Guided Flight Section: The Space Shuttle and its Uses' (1972) 76 AeronautJ 19.

[17] Leon Lipson and Nicholas Deb Katzenbach, 'Report to the National Aeronautics and Space Administration on the Law of Outer Space' (Project Reporters American Bar Foundation, Chicago 1961).

[18] Paul Dembling, 'Aspects of the Law of Space Activities' (1961) 21(2) FedBarJ 235.

such agreements would be the return to the launching state of the vehicle itself and in the case of a manned vehicle provision for the speedy return of personnel'.[19]

After the Ad Hoc Committee submitted its 1959 report, the General Assembly established a permanent Committee on the Peaceful Uses of Outer Space, consisting of twenty-four members, and requested the Committee, inter alia, 'to study the nature of legal problems which may arise from the exploration of outer space'.[20] The Legal Subcommittee of the UN Committee on the Peaceful Uses of Outer Space laid the foundations for the Assistance and Return Agreement. The Legal Subcommittee held six sessions. With the exception of the sessions in 1964 and 1966, which were split into two halves, one session of around one full month has been held each year since it first convened in Geneva in June 1962.[21] President Kennedy and Chairman Khrushchev corresponded in 1962, just before the first meeting of the Legal Subcommittee, and each of them suggested a number of potential areas for US-USSR cooperation in the exploration and utilisation of space.[22] Chairman Khrushchev proposed an international agreement on the rescue and return of astronauts:

> At the present stage of man's inroads into outer space, it is very desirable to draft and conclude an international agreement providing for assistance in the search for and rescue of spaceships, sputniks and capsules that descend to the earth due to accident. Such an agreement seems even more necessary because the point in question here is the saving of the lives of cosmonauts, these intrepid explorers of universal space.[23]

Taking into account Chairman Khrushchev's suggestion, the Soviet Union introduced in the first session a detailed draft titled 'International Agreement on the Rescue of Astronauts and Spaceships Making Emergency Landings'.[24] The

[19] Ad Hoc Committee on the Peaceful Uses of the Outer Space, 'Report of the Legal Committee' UN Doc A/AC.98/2 (12 June 1959) 7.

[20] Jessup and Taubenfeld (n 5); International Cooperation (n 5); Report of Legal Subcommittee First Session (n 5).

[21] The first four sessions of the Legal Subcommittee are treated in detail in Paul G. Dembling and Daniel M. Arons, 'Space Law and the United Nations: The Work of the Legal Subcommittee of the United Nations Committee on the Peaceful Uses of Outer Space' (1966) 32 JAirL&Com 329. The Fifth Session is discussed in Paul G Dembling and Daniel M Arons, 'The United Nations Celestial Bodies Convention' (1966) 32 JAirL&Com 535; and in Paul G Dembling and Daniel M Arons, 'The Evolution of the Outer Space Treaty' (1967) 32 JAirL&Com 419.

[22] Agreement on Cooperative Activities in Outer Space: Soviet and American Institutions (adopted 1 August 1963) (1963) 2 ILM 902; Agreement on Cooperation (n 5); Klaus Knorr, 'On the International Implications of Outer Space' (1960) 12 WorPol 564.

[23] This letter is dated 20 March 1962 and printed in *The New York Times* of 22 March 1962. President Kennedy's letter dated 7 March 1962, appears in 'Public Papers of the Presidents of the United States: John F. Kennedy (1962)' <www.govinfo.gov/app/details/PPP-1962-book1> accessed 10 October 2023.

[24] UNCOPUOS, 'Draft International Agreement on the Rescue of Astronauts and Spaceships Making Emergency Landings' UN Doc A/AC.105/C.2/L.2 (1962).

Soviet draft included provisions covering notification of the launching state of an astronaut in distress, the rescue of an astronaut who has made an emergency landing on the territory of a foreign state, a similar rescue in the event of a landing on the high seas, the return of such astronauts to their own countries, and more, in addition to an article imposing a general duty on states to use every possible means at their disposal to rescue astronauts in distress.

In addition, the Soviet Union proposed a 'Declaration of Basic Principles Governing the Activities of States Relating to the Exploration and Use of Outer Space',[25] Article 9 of which provided for the return of space objects discovered outside the borders of the launching state, and the provision of all assistance to spaceships and their crews that might make emergency landings on the territory of a foreign state or on the high seas. At the same time, the United States submitted a draft resolution to the General Assembly that included three proposed 'principles' which broadly covered the same topics as the two Soviet drafts.[26]

The Soviet Union once again presented its draft Declaration of Basic Principles, which included an article on assistance and return, as well as its earlier draft Agreement, at the Second Session of the Legal Subcommittee, which met in New York in April 1963.[27] In addition, the United States presented a draft Declaration of Basic Principles that, in contrast to the Soviet draft Declaration, included provisions for the return of astronauts, aid for astronauts, and the return of spacecraft. A draft 'Code for International Cooperation in the Peaceful Uses of Outer Space' that included clauses on the rescue and return of astronauts as well as the return of spacecraft was also presented by the United Arab Republic.[28]

The need for a comprehensive international agreement on 'assistance and return' was questioned following the General Assembly's acceptance of the Outer Space Treaty, in light of the Treaty's inclusion of provisions establishing general duties in that regard. General Assembly resolution that commended the Outer Space Treaty for signature also reflected that many states believed such a comprehensive agreement on 'assistance and return' was required. In accordance with this directive, the Legal Subcommittee re-examined 'assistance and return' as one of the four topics on the agenda for its sixth session. Geneva hosted the Sixth Session from 19 June to 14 July 1967.[29] The treaty took its present shape during the three meetings held on 15 December 15. The draft agreement was endorsed by the Committee when it met the following day, on 16 December, and it was decided to send it to the General

25 UNCOPUOS Draft Declaration (n 8).

26 UNCOPUOS, 'Assistance to and Return of Space Vehicles and Personnel' UN Doc A/AC.105/C.2/L.3 (1962).

27 UNCOPUOS, 'Report of LSC Second Session' (n 8) Annex I G.

28 ibid.

29 UNCOPUOS, 'Legal Sub-Committee Special Session Summary Record of the Eighty-sixth Meeting' UN Doc A/AC.105/C.2/SR.86 (9 February 1968).

Assembly for review.[30] The pact was endorsed by the General Assembly on 19 December.

3 Analysing the Text of the Treaty

The Preamble to the Agreement on the Rescue of Astronauts, the Return of Astronauts and the Return of Objects Launched into Outer Space notes the provisions in the Outer Space Treaty which call for 'the rendering of all possible assistance to astronauts in the event of accident, distress or emergency landing, the prompt and safe return of astronauts, and the return of objects launched into outer space, desiring to develop and give further concrete expression to these duties, Wishing to promote international co-operation in the peaceful exploration and use of outer space, prompted by sentiments of humanity'.[31] The Treaty, which is a product of the six sessions of the Legal Subcommittee of the United Nations Committee on the Peaceful Uses, has around 95 ratifications and 24 signatures. Article 1 of the Treaty states that:

> Each Contracting Party which receives information or discovers that the personnel of a spacecraft have suffered an accident or are experiencing conditions of distress or have made an emergency or unintended landing in territory under its jurisdiction or on the high seas or in any other place not under the jurisdiction of any State shall immediately: (a) notify the launching authority or, if it cannot identify and immediately communicate with the launching authority, immediately make a public announcement by all appropriate means of communication at its disposal; (b) notify the Secretary-General of the United Nations, who should disseminate the information without delay by all appropriate means of communication at his disposal.[32]

This article forms a pivotal addition to the related provisions of Article V of the Outer Space Treaty.[33] Paul G. Dembling and Daniel M. Arons write that 'Article 1,

[30] UNGA, 'Report of the Committee on Peaceful Uses of Outer Space' UN Doc A/6804/Add.1 (16 December 1967).

[31] Rescue Agreement (n 13).

[32] ibid art 1.

[33] Outer Space Treaty (n 6) art V:

> States Parties to the Treaty shall regard astronauts as envoys of mankind in outer space and shall render to them all possible assistance in the event of accident, distress, or emergency landing on the territory of another State Party or on the high seas. When astronauts make such a landing, they shall be safely and promptly returned to the State of registry of their space vehicle. In carrying on activities in outer space and on celestial bodies, the astronauts of one State Party shall render all possible assistance to the astronauts of other States Parties. States Parties to the Treaty shall immediately inform the other States Parties to the Treaty or the Secretary-General of the United Nations of any phenomena they discover in outer space, including

on notification, was one of the least controversial, and reflects the essential humanitarian purpose of the Assistance and Return Agreement'.[34] Article V of the Outer Space Treaty is extensively supplemented by the provisions of this article. The word 'astronaut' is replaced with 'spacecraft personnel' in the Rescue Agreement, which seems to further restrict the scope of the commitment of the Outer Space Treaty. This distinction is significant because Article 10(1) of the Moon Agreement requires States to 'take all practicable measures to safeguard the life and health of persons on the Moon'.[35] As a result, everyone is subject to the obligation to rescue, not simply 'astronauts' or 'personnel' as specified in the Treaty and the Rescue Agreement, respectively. Since terms like 'astronaut' or 'personnel' will determine the applicability of such a convention to a situation that was not anticipated at the time of its formation, the application of the obligation to rescue and return astronauts during non-governmental spaceflights creates significant challenges.

Article 1 of the Assistance and Return Agreement expressly states that notification must take place prior to or concurrently with the obligation to provide assistance to astronauts under Articles 2 or 3, despite the fact that the obligation to notify could be inferred from the obligation to do so under the Outer Space Treaty. Additionally, Article 1 extends the obligation to inform to situations in which a Contracting Party discovers that the crew of a spacecraft has had a mishap or encountered difficulty in space or on a celestial body.

Additionally, Article 1 is a good example of the Treaty's expansive nature in numerous ways. The 'personnel' of a spacecraft is covered by Article 1 just as it is in Articles 2, 3, and 4. The term 'astronauts', which appears in Article V of the

> the moon and other celestial bodies, which could constitute a danger to the life or health of astronauts.

Paul G Dembling and Daniel M Arons, 'The Treaty on Rescue and Return of Astronauts and Space Objects' (1968) 9 William&MaryLRev 630:

> Throughout Article 1, the emphasis is on maximum insurance on providing the requisite notification in timely fashion to the launching authority or other Contracting Parties in order that assistance may be rendered in accordance with Articles 2 or 3. The word 'immediately' is used three times and 'without delay' once. Although it is contemplated that the Contracting Party would make an effort to identify and communicate with the launching authority before making a public announcement, any flexibility permitted by subparagraph (a) must be considered limited by the information which the Contracting Party has received regarding the condition of the astronauts. By requiring immediate notification to the Secretary-General, subparagraph (b) provides additional insurance that the information will reach the proper authorities as soon as possible.

[34] Dembling and Arons (n 33).

[35] Agreement Governing the Activities of States on the Moon and Other Celestial Bodies (adopted 18 December 1979, entered into force 11 July 1984) 1363 UNTS 3, art 10 (Moon Agreement):

> 1. States Parties shall adopt all practicable measures to safeguard the life and health of persons on the moon. For this purpose they shall regard any person on the moon as an astronaut within the meaning of article V of the Treaty on Principles Governing the Activities of States in the Exploration and Use of Outer Space, including the Moon and Other Celestial Bodies and as part of the personnel of a spacecraft within the meaning of the Agreement on the Rescue of Astronauts, the Return of Astronauts and the Return of Objects Launched into Outer Space.

Outer Space Treaty, could be read strictly to refer only to someone who pilots or operates a spaceship. The entire crew of a spacecraft, or even potential passengers, may be more accurately referred to as 'personnel' in this context. The obligation to notify only arises in instances of 'accident', 'distress', or 'emergency or unintended landing', which is more crucial to the Treaty's application. In every previous draft, the phrases 'accident', 'distress', and 'emergency' had been used. Although the term 'unintended landing' is unique to the Working Paper presented in the Legal Subcommittee's Special Session, it seems to be an accepted replacement for the earlier US suggestion to include 'mistake' as a distinct notion.

Article 2 provides that if a spacecraft's crew accidentally, distressfully, urgently, or unintentionally lands on a territory governed by a Contracting Party, that party must immediately take all reasonable measures to save them and provide them with all required aid. It must report the actions it is taking and their progress to both the launching authority and the Secretary-General of the United Nations. Article 3 states that those Contracting Parties who are in a position to do so shall, if necessary, extend assistance in search and rescue operations for such personnel in order to ensure their prompt rescue. This obligation arises in the event that information is received, or it is discovered, that the crew of a spacecraft has disembarked on the high seas or in any other location not subject to the jurisdiction of any State.

Article 4 states that the crew of a spacecraft must be safely and promptly returned to representatives of the launching authority if they accidentally land in territory under the jurisdiction of a Contracting Party due to an accident, distress, emergency, or unintended landing. They may also have been discovered on the high seas or in another area outside of any State's jurisdiction. Article 5(1) states: 'Each Contracting Party which receives information or discovers that a space object or its component parts has returned to Earth in territory under its jurisdiction or on the high seas or in any other place not under the jurisdiction of any State, shall notify the launching authority and the Secretary-General of the United Nations.'[36] Article 5(2) states: 'upon request from the launching authority and with assistance from that authority if requested, each Contracting Party having jurisdiction over the territory on which a space object or its component parts has been discovered shall take such actions as it considers practicable to recover the object or component parts.' Article 5 also emphasises the additional responsibility of States with regard to components.[37]

[36] Rescue Agreement (n 13) art 5.

[37] ibid:

> 3. Upon request of the launching authority, objects launched into outer space or their component parts found beyond the territorial limits of the launching authority shall be returned to or held at the disposal of representatives of the launching authority, which shall, upon request, furnish identifying data prior to their return. 4. Notwithstanding paragraphs 2 and 3 of this article, a Contracting Party which has reason to believe that a space object or its component parts discovered in territory under its jurisdiction, or recovered by it elsewhere, is of a hazardous or deleterious nature may so notify the launching authority, which shall immediately

Article 6 states:

> For the purposes of this Agreement, the term 'launching authority' shall refer to the State responsible for launching, or, where an international intergovernmental organization is responsible for launching, that organization, provided that organization declares its acceptance of the rights and obligations provided for in this Agreement and a majority of the States members of that organization are Contracting Parties to this Agreement and to the Treaty on Principles Governing the Activities of States in the Exploration and Use of Outer Space, including the Moon and Other Celestial Bodies.[38]

Regarding hazardous or toxic objects, the Contracting Party is not subject to any recovery or return obligations. It is an obligation of the launching authority to ensure that any potential harm or hazard associated with such objects is eliminated.

Article 7 opened the Treaty for signature to all States.[39] Article 8 states that amendments to this Agreement may be proposed by any State Party. The date on which a majority of the State Parties to the Agreement accept the amendments will determine the effect of the amendments for each State Party adopting them, and they will take effect for the other State Parties to the Agreement on the date on

take effective steps, under the direction and control of the said Contracting Party, to eliminate possible danger of harm. 5. Expenses incurred in fulfilling obligations to recover and return a space object or its component parts under paragraphs 2 and 3 of this article shall be borne by the launching authority.

[38] ibid art 6; Dembling and Arons (n 33):

> Article 6 is a decided improvement over the comparable provision in the Outer Space Treaty. Under Article XIII of the Outer Space Treaty, a Contracting Party has the option of dealing directly with the international organization or with one or more members of the organization which are parties to the Outer Space Treaty. Article XIII provided a compromise through which the Soviet Union could adhere to its position of not acknowledging that an international organization is able to acquire rights under the Treaty.

[39] Rescue Agreement (n 13) art 7:

> 1. This Agreement shall be open to all States for signature. Any State which does not sign this Agreement before its entry into force in accordance with paragraph 3 of this article may accede to it at any time. 2. This Agreement shall be subject to ratification by signatory States. Instruments of ratification and instruments of accession shall be deposited with the Governments of the United Kingdom of Great Britain and Northern Ireland, the Union of Soviet Socialist Republics and the United States of America, which are hereby designated the Depositary Governments. 3. This Agreement shall enter into force upon the deposit of instruments of ratification by five Governments including the Governments designated as Depositary Governments under this Agreement. 4. For States whose instruments of ratification or accession are deposited subsequent to the entry into force of this Agreement, it shall enter into force on the date of the deposit of their instruments of ratification or accession. 5. The Depositary Governments shall promptly inform all signatory and acceding States of the date of each signature, the date of deposit of each instrument of ratification of and accession to this Agreement, the date of its entry into force and other notices. 6. This Agreement shall be registered by the Depositary Governments pursuant to Article 102 of the Charter of the United Nations.

which they accept them. Article 9 states that '[a]ny State Party to the Agreement may give notice of its withdrawal from the Agreement one year after its entry into force by written notification to the Depositary Governments. Such withdrawal shall take effect one year from the date of receipt of this notification.'

The Rescue Agreement lacks clear enforcement procedures or sanctions for non-compliance. There are no particular procedures to deal with non-compliance or settle any conflicts, despite the fact that it defines the legal framework and obligations of States. The main objective of the Rescue Agreement is the rescue and return of astronauts. Although it mentions the return of space objects in passing, it does not address the expanding problems with space debris mitigation, removal, and accountability for space object damages. The Rescue Agreement will be difficult to implement effectively given the rapid development of space technology, particularly the rise of commercial spaceflight and private space actors. The intricacies of these new actors and their unique skills and obligations may not be adequately taken into account by the Agreement. The Agreement's acceptance in 1968 led to a considerable evolution of space operations. New ventures such as space travel, asteroid mining, and lunar exploration raise fresh problems that can necessitate alternative legal systems to address difficulties that the Rescue Agreement does not sufficiently address.

4 Future of Global Governance of Rescue and Return Norms

4.1 Commercialisation of Space Activities

The global space economy reached US$447 billion in 2020, up from $428 billion in 2019 and up by 4.4 per cent from a decade earlier, according to the most recent full annual report by the Space Foundation, a non-profit organisation devoted to education, information sharing, and collaboration to support space exploration and space-inspired industries. In 2020, commercial space activity increased by 6.6 per cent to nearly $357 billion, accounting for about 80 per cent of the entire space economy.[40] According to Morgan Stanley, the global space business might earn more than $1 trillion or more in revenue by 2040, up from its current $350 billion.[41]

In this new 'space race', satellite-based operations are a special commercial focus.[42] With the cost of 'access to space' at an all-time low due to fierce competition

[40] 'Space Foundation Releases "The Space Report 2021 Q2"' (*Space Foundation*, 15 July 2022) <www.spacefoundation.org/2021/07/15/global-space-economy-rose-to-447b-in-2020-continuing-five-year-growth/> accessed 10 October 2023.

[41] 'Space: Investing in the Final Frontier' (*Morgan Stanley*, 24 July 2020) <www.morganstanley.com/ideas/investing-in-space> accessed 10 October 2023.

[42] David M Ashford, 'New Commercial Opportunities in Space' (2007) 111 AeronautJ 61; Paul B Larsen, 'Commercial Operator Liability in the New Space Era' (2019) 113 AJILUnbound 109; Hanneke

among commercial space launch service providers and financial assistance from governments throughout the world, an increasing number of private companies are now able to launch satellites into orbit. The ability of low-orbit space satellites to use cutting-edge communications technologies to improve access to broadband services, online healthcare, and consultations in remote areas and communities, as well as more precise weather and extreme event forecasting, is one of this trend's most immediate observable effects.

The current space law regime was established at the beginning of the space age, when only governments had a presence in space and the commercial use of space was a far-off fantasy. As the private space industry expands in these new and exciting ways, it is starting to outgrow this regime.[43] The Rescue Agreement is in a very different situation today, especially in the era of space tourism. Space X, Virgin Galactic, and Blue Origin have all performed passenger-carrying missions safely. It is crucial to look at the terminology used in these provisions in order to confirm the Agreement's applicability to the contemporary environment, where non-governmental spaceflights are taking precedence over conventional ones. Regardless of the type of spaceflight, the Contracting Parties should be held to their rescue responsibility. To add to the general confusion, the provisions are typically worded in a very wide and ambiguous manner, in keeping with the general pattern of space treaties.

One of the most important legal questions with space tourism is whether or not the responsibility to save astronauts and bring back errant spacecraft would apply to such endeavours. The safety of their visitors and the capability of spacecraft recovery will be top priorities for tourism companies as they get ready to launch their flights.[44] The tourist business model will succeed if it can maintain a consistent stream of customers, which will only be achievable if people believe the flights are safe. The owner of a private spacecraft will want to be able to reclaim that craft in case it veers off course and arrives in another country, for purposes other than rescuing the crew and passengers. For a space tourism business, the possibility of losing a spaceship may be disastrous for two reasons. First, building a new vehicle might be too expensive, and if the crashed spacecraft is still operable or repairable, replacing it might not be necessary. Secondly, any confidential information that is acquired by a hostile government runs the risk of being stolen and subsequently made available to rival businesses. A business will want to swiftly recover its lost spacecraft for both of these reasons.

van Traa-Engelman, 'The Commercial Exploitation of Outer Space: Issues of Intellectual Property Rights and Liability' (1991) 4 LJIL 293; George Holt Thomas, 'Commercial Aeronautics' (1917) 21 AeronautJ 369; Lincoln P Bloomfield, 'Outer Space and International Cooperation' (1965) 19 IntlOrg 603.

[43] Michael Byers and Aaron Boley, 'Space Tourism' in *Who Owns Outer Space?: International Law, Astrophysics, and the Sustainable Development of Space* (CUP 2023) 11.

[44] ibid (n 43).

However, a foreign country in possession of the spacecraft might not want to give it up. For instance, if the spacecraft broke the nation's aviation laws, the foreign government might want to seize it. Refusal by a foreign government to return a high-tech spacecraft could potentially have more sinister motives because an accidental landing could be a once-in-a-lifetime chance for some nations to obtain such technology through reverse engineering. To stop such technology theft, the space treaties stipulate that spacecraft must be returned to the state that launched them. If commercial space businesses were satisfied that the advantages of this component of space law extended to their vehicles as well as government spacecraft, it would give them considerable reassurance. The Outer Space Treaty, the Rescue Agreement, and the Moon Agreement all have provisions relating to the obligation to return space assets, just as they do for the responsibility to save. Article VIII of all the treaties, including the Outer Space Treaty, states that 'objects or component parts found beyond the limits of the State Party to the Treaty on whose registry they are carried shall be returned to that State Party'.

This article is written broadly enough to mandate the return of errant space objects whether they are discovered on Earth, in outer space, on a celestial body, or on the high seas. Article 5 of the Rescue Agreement expands and explains this obligation in many ways. Article 12(2) of the Moon Agreement merely extends the provisions of Article 5 of the Rescue Agreement to assets located on the Moon by referencing it: 'Vehicles, installations and equipment, or their component parts, found in places other than their intended location, shall be dealt with in accordance with Article 5 of the Rescue Agreement.'[45] This article has little practical impact. First, only a small number of nations are subject to the Moon Agreement due to the lack of widespread ratification. Secondly, the Moon Agreement's obligation to return space objects does not go beyond the obligations established by the Outer Space Treaty and the Rescue Agreement, which, as was previously claimed, already applied to lunar activities. According to the Moon Agreement, a State must inform the launching State when it learns of an unplanned Moon landing, Fewer issues arise when the responsibility to return errant spacecraft is used for private tourism endeavours than when the obligation to rescue is used.

4.2 Space Tourism and Treaty Obligations on the Duty to Rescue and Return

The guidelines for the formation, operation, and interpretation of treaties are laid out in the Vienna Convention. The Convention's interpretation guidelines, found in Articles 31 to 33, offer a methodical procedure for evaluating the meaning of

[45] Moon Agreement (n 35); Outer Space Treaty (n 6).

treaty provisions.[46] According to Article 33(4) of the Vienna Convention, 'the meaning which best reconciles the texts, having regard to the object and purpose of the treaty, shall be adopted when a comparison of the authentic texts discloses a difference of meaning'. The relationship between the Outer Space Treaty and the Rescue Agreement needs to be understood in order to discuss whether the obligation to rescue and return calls for the rescue of space tourists and commercial spacecraft. The Outer Space Treaty is applicable 'only to the extent that its provisions are compatible' with the Rescue Agreement, in accordance with the *lex posteriori* norm in Article 30 of the Vienna Convention.

That the Rescue Agreement was intended to supersede the Outer Space Treaty with respect to the duty to rescue and return is clear. The rescue and return regulations outlined in Article V of the Outer Space Treaty are expanded upon, supplemented, and, at times, changed by the terms of the Rescue Agreement. There is no question that these modifications were meant to replace the prior regulations. Even though the Outer Space Treaty and the Rescue Agreement are mentioned in the Preamble, and it is stated that the goal of the Treaty is 'to develop and give further concrete expression' to the duty to rescue and return contained in the Outer Space Treaty, this does not go as far as explicitly tying the Rescue Agreement to the Outer Space Treaty. Therefore, when the terms of the Outer Space Treaty and the Rescue Agreement conflict, the Rescue Agreement must take precedence under the application of the *lex posteriori* rule. Even if the legal need to rescue extends to commercial spaceflight, it can be derived that states are required to save only crew members and private passengers.

4.3 Artemis Accords and the Rescue Agreement

The Artemis Accords established in 2020 by the National Aeronautics and Space Administration (NASA) in coordination with the US Department of State, and the other seven founding members, reaffirm the United States and signatory nations' adherence to the Registration Convention, the Rescue and Return Agreement, and the ethical standards that NASA and its partners uphold, including the open dissemination of scientific findings.[47] Every responsible civil space programme must include the provision of emergency aid to people in need. The Artemis Accords

[46] Marina Foltea, 'The VCLT Rules of Interpretation' in *International Organizations in WTO Dispute Settlement: How Much Institutional Sensitivity?* (CUP 2012) 88–124; Christian Djeffal, 'The Rule of Interpretation in the VCLT: Method and Methodology' in *Static and Evolutive Treaty Interpretation: A Functional Reconstruction* (CUP 2015) 81–202; Panos Merkouris, '(Inter)Temporal Considerations in the Interpretative Process of the VCLT: Do Treaties Endure, Perdure or Exdure?' (2014) 45 NYIL 121; Anne van Aaken, 'The Cognitive Psychology of Rules of Interpretation in International Law' (2021) 115 AJILUnbound 258; Ahmad Ali Ghouri, 'Determining Hierarchy Between Conflicting Treaties: Are There Vertical Rules in the Horizontal System?' (2012) 2 AsianJIntlL 235.

[47] Rosanna Deplano, 'The Artemis Accords: Evolution or Revolution in International Space Law?' (2021) 70 ICLQ 799:

place a high priority on protecting astronauts. Participating nations pledge to do everything in their power to keep astronauts safe and to offer them the right medical attention and assistance if they are hurt or become unwell while on a mission. The Artemis Accords uphold the idea that in the event of an emergency endangering an astronaut's life or well-being, partnering nations shall commit to taking all reasonable steps possible to assist astronauts in distress. This entails not only the responsibility of the government that sent the astronauts, but, if necessary, collaboration and cooperation with other States. The Agreement on the Rescue of Astronauts, the Return of Astronauts, and the Return of Objects Launched into Outer Space is therefore reaffirmed by the Artemis Accords, which also include partner countries of NASA. Additionally, as part of the Accords, NASA and its partner countries agree to do all in their power to help astronauts who are in need.[48]

5 Conclusion

The next phase of human spaceflight will be led by space tourism businesses. The biggest problem for these businesses will be ensuring the safety of their clients. These businesses will encounter numerous technological, financial, and regulatory hurdles in the early stages of their operations. The issue of passenger safety has several facets and calls for safe technology, appropriate training for the flight crew, as well as passenger screening and education. The capacity of a corporation to save its passengers in the case of an emergency during a flight would also be crucial to the survival of both the passengers and the company. A compelling case can be made that the parties to the Rescue Agreement must rescue space tourists in order to help businesses ensure the safe recovery of their passengers and to recover and return private spacecraft, even spacecraft used to transport passengers into space. The Space Shuttle *Columbia* accident in 2003 caused a major shift in public perceptions of in-orbit rescues, and NASA responded by preparing the STS-3xx or Launch on Need missions to arrange for rescue in specific situations. There is also a debate on the application of the duty to return and rescue errant spacecraft in the

By requiring the signatory States to determine which of them should register any relevant space object, the Artemis Accords indirectly refer to the provisions of Article V (return of space objects and astronauts in the event of accident, distress or emergency landing) and Article VII (liability of the launching State for damage caused by the space object) of the Outer Space Treaty.

[48] 'Signatories of the U.S.-Led Artemis Accords Meet in Person for the First Time' (2023) 117 AJIL 133; Artemis Accords: Principles for Cooperation in the Civil Exploration and Use of The Moon, Mars, Comets, and Asteroids for Peaceful Purposes (adopted 13 October 2020) <www.nasa.gov/specials/artemis-accords/img/Artemis-Accords-signed-13Oct2020.pdf> accessed 10 October 2013 (Artemis Accords) s 6: 'The Signatories commit to taking all reasonable efforts to render necessary assistance to personnel in outer space who are in distress, and acknowledge their obligations under the Rescue and Return Agreement.'

era of private tourism ventures to protect the security of their customers and prevent spacecraft misappropriation.

In relation to the Outer Space Treaty, the Rescue Agreement addresses the numerous duties that States have to provide rescue, support, and return of astronauts, which fall somewhere between a general humanitarian commitment and political and national security considerations. The return of objects and the cost of rescue and return operations in outer space can be compared to similar provisions under the 1982 United Nations Convention on the Law of the Sea and the 1989 Convention for the Unification of Certain Rules of Law Respecting Assistance and Salvage at Sea.[49] The Rescue Agreement has been used on multiple occasions in relation to the retrieval and return of space objects, but never in relation to mishaps or the distress of astronauts or cosmonauts. Finally, it is important to address current issues such as the privatisation and commercialisation of space activity. This covers the rise in private interest in space travel and the debate over whether suborbital flights are covered by the Rescue Agreement. An evolving interpretation and implementation of the Agreement can be used to address several legal issues raised by technological advancement. As Ambassador Goldberg said in the General Assembly on 19 December 1967:

> This Agreement bears witness to the fact that the United Nations can make a real contribution to extending the rule of law to new areas and to insuring the positive and peaceful ordering of man's efforts in science and the building of a better world.[50]

The Rescue and Return Agreement also represents an example of global collaboration and accountability in the wide universe where human exploration and scientific discoveries are pushing the limits of our understanding. This Agreement, which was developed during a period of intense space race competition, represents humanity's commitment to the safety and welfare of astronauts, regardless of their nationality or mission goals. It established a model for cooperation in the face of difficulty by putting aside political differences and putting the worth

[49] Convention for the Unification of Certain Rules of Law Respecting Assistance and Salvage at Sea (adopted 23 September 1910, entered into force 1 March 1913) USTS 576, Annex IV; Bingbing Jia, 'Unsaid Rules of UNCLOS: Essential Elements for its Proper Interpretation?' in Wenhua Shan, Sheng Zhang, and Jinyuan Su (eds), *China and International Dispute Resolution in the Context of the 'Belt and Road Initiative'* (CUP 2021) 251; Edward A Laing, 'A Perspective on Provisional Measures under UNCLOS' (1998) 29 NYIL 45; Sara McLaughlin Mitchell and Andrew P Owsiak, 'Judicialization of the Sea: Bargaining in the Shadow of UNCLOS' (2021) 115 AJIL 579.

[50] 'United Nations Endorses Text of Agreement on Rescue and Return of Astronauts and Space Vehicles' (1968) 58 DeptStBull 80, 83–84.

of protecting human life even in outer space. The Agreement proves that spacefaring nations can indeed cooperate successfully to safeguard their astronauts. The principles set forth in this Agreement remain as applicable as ever, even as our aspirations for space travel increase and projects like the exploration of Mars loom in the distance.

6

STATE RESPONSIBILITY AND INTERNATIONAL LIABILITY

Sandeepa Bhat B.

1 Introduction

Norms on State responsibility and international liability are the two most significant aspects of international space law. Dealing with international liability independent of State responsibility under the Outer Space Treaty[1] indicated an appreciation of the distinction between the two by the States. However, it was doubtful whether the distinction was clearly understood, which is evident in the negotiating history and in the texts of the Outer Space Treaty available in other languages. The fact that the French, Spanish, Russian, and Chinese texts of the Outer Space Treaty use the term 'responsible' or its equivalent under both provisions[2] is an indicator of this confusion. Even the English text's division is not based on the modern-day distinction between the two concepts as evidenced in the works of the International Law Commission.[3] The Outer Space Treaty goes only by the logic that international liability necessarily involves payment of compensation, which is not an essential element of State responsibility.[4]

Over the years, distinctions between State responsibility and international liability have become more evident.[5] State responsibility essentially involves a breach of an international obligation, which is not a part of international liability.[6] International liability arises out of damage caused by any activity, which is not per se a breach of an international obligation.[7] Hence, in simple terms, State responsibility arises out of a 'wrongful act' and international liability emanates out

[1] Treaty on Principles Governing the Activities of States in the Exploration and Use of Outer Space, Including the Moon and Other Celestial Bodies (adopted 27 January 1967, entered into force 10 October 1967) 610 UNTS 205 (Outer Space Treaty).

[2] Bin Cheng, 'Article VI of the 1967 Space Treaty Revisited: "International Responsibility", "National Activities", and the "Appropriate State"' (1998) 26(1) JSpaceL 7, 10.

[3] See Sompong Sucharitkul, 'State Responsibility and International Liability under International Law' (1996) 18 LoyLAIntl&CompLRev 821.

[4] Motoko Uchitomi, 'State Responsibility/Liability for 'National' Space Activities' (2002) 44 ProceColloqLOuterSpace 51, 51.

[5] See Rebecca MM Wallace, *International Law* (3rd edn, Universal Law Publishing 2003) 203.

[6] See Malcolm N Shaw, *International* Law (7th edn, CUP 2014) 569–70.

[7] Wallace (n 5).

Sandeepa Bhat B., *State Responsibility and International Liability* In: *International Space Law in the New Space Era.* Edited by: Sandeepa Bhat B., Dilip Ukey, and Adithya Variath, Oxford University Press. © Sandeepa Bhat B. 2024.
DOI: 10.1093/9780198909415.003.0006

of 'damage' caused by lawful activities. Accordingly, the consequences of the two are also different on two major counts. First, international liability necessitates the payment of compensation to undo the damage but State responsibility may or may not warrant payment of compensation. The latter may not involve the payment of compensation, especially when no damage is caused to others, even though there is a breach of an international obligation.[8] Secondly, the attraction of State responsibility mandates the cessation of the act concerned since it involves a breach of an international obligation. However, international liability does not stand in the way of continuation of the relevant act by the State.[9] It is pertinent to note here that the Outer Space Treaty was enacted in an era when the distinctions between the two concepts were unclear. Accordingly, the modern-day distinction between State responsibility and international liability cannot be found in the Outer Space Treaty.

The above dilemma in regard to State responsibility and international liability norms under space law is further complicated by the fact that the space treaties were adopted at a time when the States were the sole actors in the space sector. More significantly, very few States had the space technology to carryon space activities, and no private player was involved in the space sector. Hence, the space treaties are tailor-made for dealing with State activities in outer space. After the entry of private players into commercial space activities, international treaties were not successfully negotiated in this field. Accordingly, a web of uncertainty exists in the application of responsibility and liability regimes, which were adopted in connection with State activities, to private space activities. Thus, responsibility and liability norms in international space law need detailed analysis, in particular to comprehend their differences in comparison with their operations under public international law as well as in domestic law.

2 State Responsibility under Space Law

2.1 Interpreting the First Sentence

Article VI of the Outer Space Treaty covers the aspects of international responsibility in three sentences. The first sentence reads as follows:

> States Parties to the Treaty shall bear international responsibility for national activities in outer space, including the Moon and other celestial bodies, whether such activities are carried on by governmental agencies or by non-governmental

[8] UNGA Res 56/83 (28 January 2002) UN Doc A/Res/756/83 (ARSIWA) art 30.

[9] For a detailed discussion, see Sandeepa Bhat B. and P Ishwara Bhat, 'Legal Framework of State Responsibility and Liability for Private Space Activities' in Sandeepa Bhat B. (ed), *Space Law in the Era of Commercialisation* (Eastern Book Company 2010) 131–49.

entities, and for assuring that national activities are carried out in conformity with the provisions set forth in the present Treaty.

2.1.1 Nature of the Responsibility

A careful reading of the first sentence clearly indicates that the States have undertaken direct responsibility for both public and private space activities. This is distinct from international law. International law makes a distinction between primary responsibility and secondary responsibility of States.[10] While the former is the direct responsibility of States for their wrongful acts or omissions, the latter is an indirect responsibility in connection with any wrong committed by their individuals. The primary responsibility of a State arises when an international obligation is breached by someone with the authority of the State concerned, Such a breach might occur either through the commission of an act contrary to the State's obligation[11] or due to the failure of the State to take necessary steps to comply with an obligation.[12]

International law does not impose primary responsibility on States for the wrongful acts of individuals in the absence of State authorisation, since the notion of imputation or attribution of individual acts to States is the essence of primary responsibility.[13] In case of individual acts, the State concerned has the duty to ensure that its individuals are not involved in the commission of internationally wrongful acts. However, this is just a due diligence requirement, and the practical difficulty in preventing the commission of internationally wrongful acts by its individuals would not attribute responsibility to a State. But again, States cannot turn blind eye to the wrongful acts by their individuals`, especially when such acts have transnational effects. The State concerned has the responsibility of prosecuting and punishing the offender to render justice to the victim. Failure to discharge this duty results in the secondary responsibility of States. Hence, the secondary responsibility of States arises only in the case of failure to discharge the due diligence requirement, coupled with the failure to prosecute and punish the offender.[14]

[10] See Luca Ferro, 'Brothers in Arms: Ancillary State Responsibility and Individual Criminal Liability for Arms Transfers to International Criminals' (2015) 54(1) MilL&LWarRev 139.

[11] eg use of force by a State contrary to the obligations under the UN Charter. See Ian Brownlie, 'International Law and the Use of Force by States Revisited' (2000) 21 AustYBIL 21–38.

[12] eg failure of the State to take steps to rescue an astronaut in case of an emergency landing in its territory. Since such an obligation of the States Parties to the Rescue Agreement 1968 is very compelling as a treaty norm, failure to discharge this obligation would be a breach attributing State responsibility. For a detailed analysis of strong rescue obligations, see Paul G Dembling and Daniel M Arons, 'The Treaty on Rescue and Return of Astronauts and Space Objects' (1968) 9(3) William&MaryLRev 630–63; see also Stephen Gorove, 'Legal Problems of the Rescue and Return of Astronauts' (1969) 3(4) IntlLawyer 898, 901.

[13] See Kenneth B Hoffman, 'State Responsibility in International Law and Transboundary Pollution Injuries' (1976) 25(3) ICLQ 509, 541; see also James Crawford, *Brownlie's Principles of Public International Law* (9th edn, OUP 2019) 526–27.

[14] See *Janes Claim, US v Mexico* (1926) 4 RIAA 82; see also *B.E. Chattin v The United Mexican States* (1927) 4 RIAA 282.

In contrast with international law, the Outer Space Treaty imposes primary responsibility on States for both governmental and non-governmental activities in outer space. Though some scholars have argued that this higher level of responsibility is voluntarily undertaken by States,[15] it seems unrealistic in the current era of extensive private space activities. An argument is even made that space activities involve a high element of risk, and therefore that States need to behave with a high sense of responsibility in both their public and private space activities.[16] Whatever is said and done, imposing primary responsibility on States for private space activities reflects an understanding of space activities as something within the exclusive domain of States, with private entities playing only a supplementary role. This is also reflected in the assertion of the Soviet Union during the negotiations of the Outer Space Treaty.[17] However, with the changing scenario resulting in a wide array of private space activities, the primary responsibility of States for private space activities has become a matter of concern.

2.1.2 National Activities in Outer Space

The words 'national activities' and 'in outer space' are significant and in need of proper interpretation to understand the sphere of application of State responsibility under the Outer Space Treaty. Governmental activities, whether directly carried on by the government of a State or with the authorisation granted to private players, would invariably come within the definition of national activities.[18] However, the wording of the provision indicates that there are also other activities of private players that would fall within the meaning of national activities. Whether such activities include those by nationals of a concerned State, or by persons located in the concerned State, or involving the space objects registered in the concerned State, or, alternatively, all of these, is a debatable question. Bin Cheng argues that since responsibility is based on competence, the 'national activities' under Article VI of the Outer Space Treaty should be confined to those activities that fall within the competence of the concerned State.[19] In simple terms, space activities are considered to be national activities of a State when that State has the competence or legal power to regulate those activities.

We should understand that, in certain circumstances, competence to regulate space activities may lie with multiple States. In case of a conflict between the States on the regulation of space activities due to the competence of multiple States, Bin

[15] Christian Joseph Robison, 'Changing Responsibility for a Changing Environment: Revaluating the Traditional Interpretation of Article VI of the Outer Space Treaty in Light of Private Industry' (2020) 5(1) UBolognaLRev 1, 8–9.

[16] John M Kelson, 'State Responsibility and the Abnormally Dangerous Activity' (1972) 13 HarvIntlLJ 197, 216.

[17] See William B Wirin, 'Practical Implications of Launching State—Appropriate State Definition' (1994) 37 ProceColloqOuterSpace 109, 110.

[18] Cheng (n 2) 20.

[19] ibid 23.

Cheng argues for territorial nexus as the major connecting factor, which overrides quasi-territorial and personal connections. In the absence of a territorial nexus, the quasi-territorial nexus overrides the personal connection, is his opinion.[20] Bin Cheng also seeks to buttress this argument with the help of Article 14(1) of the Moon Agreement,[21] which stipulates the regulation of activities of 'non-governmental entities under their [States'] jurisdiction'. However, in practical terms, it is not that easy to create such a hierarchy. While using the Moon Agreement to interpret the provisions of the Outer Space Treaty is on shaky ground, even the aspect of jurisdiction under the Outer Space Treaty does not recognise the hierarchy between the quasi-territorial and personal jurisdictions. Article VIII of the Outer Space Treaty confers jurisdiction on the State of registration of space objects 'over the space object' and 'any personnel thereof'. The quasi-territorial jurisdiction is reflected in connection with the 'space object' and personal jurisdiction is reflected in connection with the 'personnel', and no hierarchy is stipulated between the two under the provision. Hence, without creating any hierarchy, considering that the expression 'national activities' is wide enough to cover any activity connected with States through territorial, quasi-territorial or personal nexus is appropriate. Accordingly, all such States bear international responsibility for the space activities concerned.

The next aspect which calls for interpretation is the geographical or cosmographical location of activities covered under Article VI of the Outer Space Treaty, as it attributes responsibility to States for 'activities in outer space'. A plain reading of these words may provide an impression that the Outer Space Treaty does not impose responsibility on States for activities which are not per se conducted in outer space.[22] Going by this understanding, the incidental activities on Earth, such as manufacturing satellites or launch vehicles, establishing ground stations or launch facilities, processing of signals or data, and data transmission do not fall under the provision. Even activities such as the launch and return of space objects may fall outside the ambit of the provision when the space object is in the airspace during the process of launch or return. Such an approach would result in dealing with space activities piecemeal, which would be detrimental to the proper regulation of space activities.

In the early literature, space activities have been evidently defined to include even activities that are not per se conducted in outer space. All incidental activities that are related to activities in space have been part of the definition of space

[20] ibid 25; see also Sergio Marchisio, 'National Jurisdiction for Regulating Space Activities of Governmental and Non-Governmental Entities' (United Nations/Thailand Workshop on Space Law,16–19 November 2010) 1, 4 <https://www.unoosa.org/pdf/pres/2010/SLW2010/02-02.pdf> accessed 9 August 2023.

[21] Agreement Governing the Activities of States on the Moon and Other Celestial Bodies (adopted 18 December 1979, entered into force 11 July 1984) 1363 UNTS 3 (Moon Agreement).

[22] The debate on this issue can be seen during the 1980s in the context of responsibility for remote sensing activities. For a detailed discussion, see Bin Cheng, 'The Commercial Development of Space: The Need for New Treaties' (1991) 19(1) JSpaceLaw 17, 39–40.

activities.[23] Commercial space activities are also defined widely and classified to include: (1) accessory, (2) principal, (3) industrial and (4) complementary activities.[24] A range of activities, including data broadcasting, remote sensing, telecommunications, celestial resource mining, space manufacturing, space research, space transportation and collection of solar energy in space, fall within the ambit of these categories.[25] In addition, the activities that relate to the conduct of commercial space activities, for example space insurance, would also fall under the category of complementary activities. Even the national space laws of space-faring nations have defined space activities more broadly.[26]

The 2013 Model Law on National Space Legislation,[27] prepared by the International Law Association (ILA), defined 'space activity' as 'the launch, operation, guidance, and re-entry of space objects into, in and from outer space and other activities essential for the launch, operation, guidance and re-entry of space objects into, in and from outer space'.[28] The ILA Model Law is a guideline for enacting national space laws and for regulating space activities in furtherance of Article VI of the Outer Space Treaty. Hence, definition of 'space activities' under the ILA Model Law should have the same meaning as 'activities in outer space' stipulated under Article VI of the Outer Space Treaty. This leads to the obvious conclusion that activities in outer space are not necessarily restricted to those conducted in outer space, but also include certain activities that are planned, directed or executed on the Earth. Thus, the true meaning of national activities in outer space has evolved over time to be reflected adequately in State practices to include any activity related to activities in outer space.[29]

2.1.3 Ensuring Conformity

Since Article VI of the Outer Space Treaty is a compromise between the United States' support for and the Soviet Union's opposition to allowing private space activities, the compromised text obligates States to ensure that all space activities adhere

[23] See Youri M Kolossov, 'Outer Space Activities as an Object of International Space Law' (1985) 28 ProceColloqOuterSpace 232, 233; see also Arthur M Dula, 'Regulation of Private Commercial Space Activities' (1983) 23(2) JurimetricsJ 156, 160.

[24] See generally Philippe Kahn, 'Situations d'un droit commercial spatial', in Philippe Kahn (ed), *L'exploitation commerciale de l'espace: Droit positif, droit prospectif* (LexisNexis 1992) 93.

[25] See Karl-Heinz Bockstiegel, 'Commercial Space Activities: Their Growing Influence on Space Law' (1987) XII AnnAir&SpaceL 175–92.

[26] See National Aeronautics and Space Act 1958 (US); Law of the Russian Federation on Space Activities, Decree No 5663-1, 20 August 1993; Space Industry Act 2018 (UK); and Space Development Promotion Act 2005 (South Korea).

[27] See UNCOPUOS, 'Information on the Activities of International Intergovernmental and Non-governmental Organizations relating to Space Law' Legal Subcommittee, Fifty-second Session (Vienna, 8–19 April 2013) UN Doc A/AC. 105/C.2/2013/CRP.6 (ILA Model Law).

[28] ibid art 2; Stephan Hobe, 'The ILA Model Law for National Space Legislation' (2013) 1 GermJAir&SpaceL 81, 89.

[29] See Frans G von der Dunk, 'Scoping National Space Law: The True Meaning of "National Activities in Outer Space" of Article VI of the Outer Space Treaty' (2020) 62 ProceColloqOuterSpace 227, 231.

to the provisions of the Outer Space Treaty.[30] Accordingly, States are required to devise appropriate mechanisms to ensure treaty compliance by both public and private space entities. Despite the evidence of States losing control over private space activities in the current era of commercialisation,[31] States' obligation to ensure compliance remains intact under the Outer Space Treaty. Since States have undertaken the primary responsibility for private space activities, the obligation to ensure such compliance is absolute and cannot be discharged by mere compliance with due diligence standards.

2.2 Interpreting the Second Sentence

The second sentence of Article VI of the Outer Space Treaty focuses further on the activities of non-governmental entities. It carries forward from the first sentence the requirement of ensuring that space activities are in conformity with the Outer Space Treaty provisions. Accordingly, it stipulates that '[t]he activities of non-governmental entities in outer space, including the moon and other celestial bodies, shall require authorization and continuing supervision by the appropriate State Party to the Treaty'. Thus, the onerous obligation of States to ensure that all space activities are in conformity with the Outer Space Treaty could be discharged by adopting a vibrant system of authorisation and continued supervision of space activities. This again stems from the fact that though the United States could successfully argue for allowing non-State actors to carryon space activities, the USSR could achieve the inclusion of a provision for the authorisation and continued supervision of such activities.[32]

As discussed in the analysis of the first sentence, a reference to 'activities in outer space' here should also include all 'space activities', whether conducted in outer space or elsewhere. However, there are two more interpretational concerns in this sentence, which are elaborated below.

2.2.1 'Shall Require Authorization and Continuing Supervision'

The expression [space activities] 'shall require authorization and continuing supervision' raises the following question: on whom does the primary responsibility of subjecting private space activities to authorisation and continuing supervision

[30] See Vladimír Kopal, 'Treaty on Principles Governing the Activities of States in the Exploration and Use of Outer Space, Including the Moon and Other Celestial Bodies' (*United Nations Audiovisual Library of International Law*, 2008) <https://legal.un.org/avl/pdf/ha/tos/tos_e.pdf> accessed 19 August 2023.

[31] Several authors have expressed concerns regarding loss of State control over private space activities, especially with the technological developments. See eg Richard Berkley, 'Space Law versus Space Utilization: The Inhibition of Private Industry in Outer Space' (1997) 15(2) WisIntlLJ 421–44.

[32] Eng Teong See, 'Commercialization of Space Activities—The Laws and Implications' (2017) 82(1) JAirL&Com 145, 159.

rest? Does it rest primarily on States or on the private space actors who would like to start a space activity? The answer to this question is significant, since if the primary responsibility is on the private entity to seek authorisation and subject its activities to a State's supervision, the States Parties to the Outer Space Treaty may discharge their burden of authorisation and continuing supervision by complying with the due diligence requirement. Enactment of a national space law and asking every private entity to seek authorisation as well as to comply with the requirements of continued supervision would have resulted in discharging the States' obligation under the second sentence. This could have been an ideal scenario for the States to avoid unnecessary hardship, which might arise out of unauthorised as well as illegal activities of their private entities.[33]

Unfortunately, the scheme of the Outer Space Treaty does not indicate any support for the above possibility. This is evident from the fact that the Outer Space Treaty fails to deal with private space activities in any other provision of the Treaty. Even responsibility and liability for damage caused by any space activity are not imposed on the private entities. Hence, assuming that Article VI of the Outer Space Treaty attributes primary responsibility to the private entities to seek authorisation and subject space activities to continued State supervision is not appropriate. The obligation of authorisation and continued supervision flows directly from the obligations under the first sentence, which involves a high burden of responsibility on States. Hence, States are duty-bound to check the private space activities and subject them to a mandatory process of authorisation and continued supervision.[34] Any failure in the practical implementation of authorisation and continued supervision results in attracting State responsibility.[35]

2.2.2 Appropriate State Party

The onerous obligation on States to authorise and continuously supervise private space activities brings forward the next question as to which State should discharge this burden when multiple States are connected to a space activity. This question gains additional significance due to the use of the singular, 'appropriate State Party to the Treaty', under the provision.[36] A parallel provision under the Moon Agreement, Article 14(1), indicates a collective responsibility of States and stipulates that 'States Parties shall ensure that non-governmental entities under their jurisdiction shall engage in activities on the Moon only under the authority and

[33] See Gbenga Oduntan, 'Aspects of the International Legal Regime concerning Privatization and Commercialization of Space Activities' (2016) 17(1) GeoJIntlAff 79, 83.

[34] Neni Ruhaeni, 'The Concept of Allocation of Certain Risk and its Construction in the Formulation of National Legislation on Liability in Outer Space Activities' (2018) 307 AdvSocSciEdu&HumRes 342, 344.

[35] Anne-Sophie Martin and Steven Freeland, 'Artificial Intelligence—A Challenging Realm for Regulating Space Activities' (2020) XLV AnnAir&SpaceL 275, 299.

[36] See Michelle LD Hanlon, 'Adapting the ISS Code of Conduct to Form the Foundation of Astrolaw' (2019) 21(1) SanDiegoIntlLJ 105, 120.

continuing supervision of the appropriate State Party'. However, both provisions seem assume that either each space activity is carried out within the exclusive domain of one State party,[37] or, alternatively, that there are defined and objective parameters for the determination of the appropriate State Party.[38] Some authors have as far as stating that whenever several States are involved in a space activity, they should collectively designate one State to authorise and continuously supervise the private space activity concerned.[39]

An examination of the five space treaties reveals that terms like 'launching State' and 'State of registry' are defined but not 'appropriate State Party'. Even the criteria for the determination of 'appropriate State Party' are not found in the space treaties. Factors such as launching, registration, State of nationality of private space player and other jurisdictional connotations might be relevant, but none of them can be considered decisive especially when they are fragmented between multiple States.[40] Hence, a presumption that an 'appropriate State Party' can be narrowed down to one State in a multistate connected space activity is not workable. Thus, attributing a meaning to 'appropriate State Party' to denote one single State would result in chaos since each State could point towards another on the basis of different connecting factors. This position is also supplemented by the fact that the Liability Convention[41] attributes joint and several liability to launching States for damage caused by multistate connected space activities.[42] It indicates that all those States which come within the range of liability need to have effective mechanisms to regulate such space activities. Hence, resorting to authorisation and supervision mechanisms becomes pertinent to all of them, which again points towards the plurality of 'appropriate State Party' in multistate connected space activities.

The Outer Space Treaty only stipulates the requirement of authorisation and continued supervision. How these need to be implemented is left to the States themselves. States have resorted to national space laws in this connection. Since States are not barred from going beyond the international mandate in regulating their private space activities, the extent of regulation of private space activities is subject to the policy decisions of States.[43] Hence, the schemes of authorisation and

[37] David I Fisher, 'Injury to Rights of Personality Caused by Satellite Programme Contents—Prospects of Relief under the Law of Outer Space' (2000) 39 ScandinavianStudL 419, 422.

[38] See Setsuko Aoki, 'The Concept of Launching State Revisited' (2002) 15 JAerospacePoly&L 123, 128.

[39] Rochus Moenter, 'The International Space Station: Legal Framework and Current Status' (1999) 64(4) JAirL&Com 1033, 1041; see also Luca Erhart and Maria Boutovitskai, 'Transforming Article VI of the Outer Space Treaty into an Effective Mechanism of Space Debris Mitigation' (Proceedings of 8th European Conference on Space Debris, 2021) <https://conference.sdo.esoc.esa.int/proceedings/sdc8/paper/223/SDC8-paper223.pdf> accessed 19 August 2023.

[40] See Kumar Abhijeet, 'State Practices Regarding International Responsibility for National Activities in Outer Space' (2020) 44(2) JSpaceL 352, 363–65.

[41] Convention on International Liability for Damage Caused by Space Objects (adopted 29 March 1972, entered into force 1 September 1972) 961 UNTS 187 (Liability Convention).

[42] ibid art V.

[43] Michel Bourély, 'Quelques réflexions au sujet des législations spatiales nationales' (1991) XVI AnAir&SpaceL 245, 247.

continued supervision vary from State to State. The authorisation process usually includes testing of the applicant's capabilities in terms of space activities, finances (including liability coverage), compliance with international norms and State policies etc. The requirement of supervision is implemented by mandating periodic reports, establishing a supervising authority, inspection of facilities etc. Though the effectiveness of a system of authorisation and continued supervision is not explicitly mentioned under the Outer Space Treaty, attribution of direct responsibility to States for private space activities compels States to have an effective system.

2.3 Interpreting the Third Sentence

The last part of Article VI of the Outer Space Treaty deals with the responsibility for the space activities of international organisations. It reads as follows:

> When activities are carried on in outer space, including the Moon and other celestial bodies, by an international organization, responsibility for compliance with this Treaty shall be borne both by the international organization and by the States Parties to the Treaty participating in such organization.

This provision indicates the collective responsibility of the organisation as well as the individual responsibility of States participating in the organisation to comply with the Outer Space Treaty when space activities are conducted by international organisations. There is no hierarchy set between the organisational responsibility and participating States' responsibility. So the participating States cannot escape from their direct responsibility under the pretence of the organisational responsibility. Cocca considers this as a departure from the traditional interpretation of an international organisation as a separate legal subject of international law.[44] However, such a departure is also found in other international organisations including the United Nations.[45] Moreover, retaining the responsibility of participating States is significant to prevent any State from escaping responsibility for breaching the Outer Space Treaty, which stands as the *magna carta* of international space law.

The extent of responsibility of the international organisation and each of the participating States, as well as the extent of legal consequences to be attributed to each of them in case of failure to comply with the Outer Space Treaty, are not

[44] Aldo Armando Cocca, 'Prospective Space Law' (1998) 26(1) JSpace 51, 55.

[45] The collective responsibility of States under the banner of the United Nations does not mean that the States are exonerated from individual responsibility. See *Legality of the Use by a State of Nuclear Weapons in Armed Conflict*(Advisory Opinion) [1996] ICJ Rep 66; see also Olga Gerlich, 'Responsibility of International Organizations under International Law' (2013) 2(2) Folia Iuridica Wratislaviensis 11, 55.

stipulated under the provision. Hence, the practical implementation of this provision will meet with several interpretational concerns. Despite its limitations, this provision is important in reflecting the responsibility of all those connected with any space activity. It also stands as a supplementary argument to interpret 'appropriate State Party' under the second sentence to include more than one State in a multistate connected space activity.

3 International Liability Norms

3.1 Outer Space Treaty

Space activities are inherently risky and any mishap is known to potentially cause catastrophic damage.[46] Since space activities are not subject to territorial limits, the activity of one State has the potential to cause damage in another State. The latter, not being connected with the space activity of the former, becomes the innocent victim of such disasters. Hence, the need to address liability for damage caused by space activities strictly was felt to be necessary since the early stages of the development of space law. This is reflected in dealing with international liability initially under UN General Assembly Resolution 1962 of 1963[47] and, subsequently, under Article VII of the Outer Space Treaty. Article VII of the Outer Space Treaty reads as follows:

> Each State Party to the Treaty that launches or procures the launching of an object into outer space, including the moon and other celestial bodies, and each State Party from whose territory or facility an object is launched, is internationally liable for damage to another State Party to the Treaty or to its natural or juridical persons by such object or its component parts on the Earth, in air or in outer space, including the moon and other celestial bodies.

The provision includes four categories of States that are liable for damage caused by any space activity, which are later defined as 'launching State' under the Liability Convention.[48] Each category, except the category of territory, may involve multiple States within it. The launch facility, for example, may be jointly owned by several States. In such a scenario, the liability may be attributed to several States with respect to damage caused by a space object under category four[49] of the definition of

[46] Kirsten Schmalenbach, 'Convention on International Liability for Damage Caused by Space Objects' in Peter Gailhofer and others (eds), *Corporate Liability for Transboundary Environmental Harm* (Springer 2023) 523–36, 523–24.

[47] UNGA Res 1962 (XVIII) (13 December 1963).

[48] See Liability Convention (n 41) art I(c).

[49] Category four involves the 'facility' from which the space object is launched.

launching State. As a limitation, Article VII stipulates the liability for damage only when the victim State is a party to the Outer Space Treaty.[50] This suggests a desire to increase the number of ratifications of the Outer Space Treaty rather than to adequately compensate the victims.

Added to the above limitation of liability, the basis of liability and apportionment of liability are not addressed under the Outer Space Treaty. Since the liability for damage caused in outer space and celestial bodies is also covered under Article VII of the Outer Space Treaty, which State would be liable becomes a matter of debate as both States in question would be involved in space activities. If the basis of liability is strict or absolute in such circumstances, the result becomes absurd. Similarly, in case of joint ventures, how the liability was to be fixed and apportioned among multiple States also remained unclear under the Outer Space Treaty. Hence, there needed to be a supplementary convention to clarify the unanswered questions of liability under the Outer Space Treaty.[51] This objective is attained substantially through the Liability Convention.

3.2 Fixing the Basis of Liability

The Liability Convention is argued to be a victim-oriented convention, in particular due to the absence of a precondition that the victim/claimant State be a party to it in order to receive compensation.[52] It divides the liability of the launching State into two parts. For any damage by the space object 'on the surface of the earth or to aircraft in flight', the launching State is absolutely liable to pay compensation.[53] In contrast, fault-based liability is applicable with respect to situations covered under Article III of the Liability Convention, which states as follows:

> In the event of damage being caused elsewhere than on the surface of the earth to a space object of one launching State or to persons or property on board such a space object by a space object of another launching State, the latter shall be

[50] See Bin Cheng, 'Liability for Spacecraft' (1970) 23 CLP 216–39.

[51] See Adrian Taghdiri, 'Flags of Convenience and the Commercial Space Flight Industry: The Inadequacy of Current International Law to Address the Opportune Registration of Space Vehicles in Flag States' (2013) 19(2) BostonUJSci&TechL 405, 412; see also Motoko Uchitomi, 'Sustainable Development in Outer Space—Applicability of the Concept of Sustainable Development to Space Debris Problems' (2001) 43 ProceColloqOuterSpace 71, 73–74.

[52] The victim-oriented approach of the Liability Convention has been highlighted by several scholars. See Michael Tse, 'One Giant Leap [Backwards] for Mankind—Limited Liability in Private Commercial Spaceflight' (2013) 79(1) BrookLRev 291–320; see also Dan St John, 'The Trouble with Westphalia in Space: The State-Centric Liability Regime' (2012) 40(4) DenvJIntlL&Pol 686, 699; Sergio Marchisio, 'The 1983 Italian Law N. 23 on the Compensation for Damage Caused by Space Objects' (2005) 54(2) GermJAir&SpaceL 261, 263.

[53] Liability Convention (n 41) art II.

liable only if the damage is due to its fault or the fault of persons for whom it is responsible.

A careful perusal of both provisions indicates that the liability for damage on the surface of the Earth is absolute, while that in outer space is fault-based, unless the claimant State's object is an aircraft flying in outer space. When the damage is caused in the airspace, the liability may be absolute or fault-based depending on the kind of object of the claimant State. If the claimant State's object is an aircraft, absolute liability under Article II of the Liability Convention is applicable. In contrast, if the claimant State's object is a space object, fault-based liability under Article III of the Liability Convention is applicable. This makes the legal characterisation of the claimant State's object most crucial for the fixation of liability for damage in airspace. Unfortunately, determining whether a flying object is an aircraft or a space object has become extremely difficult, especially in the current era of aerospace objects.[54] Hence, there will be an inevitable tug-of-war between the launching State and the victim/claimant State in attributing a legal character to the claimant State's object. The former will try to establish the claimant State's object as a space object to trigger liability under Article III, which is contingent on establishing fault. The latter will always assert that its object is an aircraft to receive the benefit of absolute liability of the launching State under Article II.

There are several other interpretational concerns under these provisions of the Liability Convention. Both provisions stipulate damage caused 'by the space object'. Whether such damage should be confined to cases of physical impact of the space object or extended to include any damage caused by the space activity concerned is unclear. Though the narrow interpretation is against the victim-oriented approach of the Liability Convention, such a possibility cannot be ruled out, especially in light of past experiences in dealing with damage caused by aircraft under aviation law.[55] Another closely connected question relates to dealing with damage caused to persons or property on board an aircraft in flight under Article II of the Liability Convention. This question arises as Article III of the Liability Convention explicitly stipulates the coverage of damage caused to persons or property on board a space object. However, Article II of the Liability Convention fails to do so in connection with an aircraft in flight. As a logical extension of this issue, even Article III suffers from limitation since it mentions only the person or property 'on board' the space object and thereby fails to cover the situation where the person or property is outside the space object.

[54] Katherine M Gorove, 'Delimitation of Outer Space and the Aerospace Object—Where is the Law' (2000) 28(1) JSpaceL 11–28.

[55] States have adopted a restricted approach to confine the meaning of damage caused 'by the aircraft' to include only the damage by physical impact of the aircraft. See generally Gerd Rinck, 'Damage Caused by Foreign Aircraft to Third Parties' (1962) 28(4) JAirL&Com 405, 408–09.

The wording of Article II also poses serious doubts about its applicability to damage caused below the surface of the Earth. Similarly, under Article III, attributing liability to the launching State for the damage due to the 'fault of persons for whom it is responsible' brings an equal amount of vagueness. As discussed previously, identifying the responsible State(s) for individual activities in the realm of outer space is extremely difficult. Given the infinite possibilities of connecting an individual space actor to different States, which are the States that would be held responsible for the fault of persons under Article III ignites unending debate. Finally, a matter of great concern under Article III of the Liability Convention is the establishment of fault. In the absence of repetitive practices, applying the test of reasonableness to space activities is difficult.[56] The determination of fault without the test of reasonableness is unrealistic. The failure to claim any compensation in the Cosmos-Iridium collision of 2009 is the best example of our inability to determine fault under Article III of the Liability Convention in practical terms.[57]

The absolute liability under Article II is subject to exoneration under Article VI of the Liability Convention. The launching State is exonerated from absolute liability either fully or partially in case the damage is caused 'either wholly or partially from gross negligence or from an act or omission done with intent to cause damage on the part of a claimant State or of natural or juridical persons it represents'.[58] Unfortunately, there is no agreed standard for determining 'gross negligence' or 'act or omission done with intent to cause damage'. This results in varied interpretations in different jurisdictions. A similar concern has also been evidenced in aviation law while interpreting the words 'wilful misconduct or. . . such default'[59] in determining the extent of the carrier's liability by applying Article 25 of the Warsaw Convention 1929.[60] The last part of Article VI(1) of the Liability Convention adds further confusion in terms of natural or juridical persons represented by the claimant State. Whether such representation would be connected with some sort of express or implied authorisation by the claimant State or with some other connecting factors such as nationality, citizenship, residence, registration etc. of natural or juridical persons is unclear.

The exoneration from absolute liability under Article VI(1) of the Liability Convention is contingent upon the activities of the launching State being in conformity with international law including, in particular, the Charter of the United

[56] Paul G Dembling, 'Establishing Liability for Outer Space Activities' (1971) 13 ProceColloqOuterSpace 87, 88.

[57] See Henry R Hertzfeld and Ben Basely-Walker, 'Legal Note on Space Accidents' (2010) 59(2) GermJAir&SpaceL 230; see also Lauren Bressack, 'Addressing the Problem of Orbital Pollution: Defining a Standard of Care to Hold Polluters Accountable' (2011) 43(4) GWashIntlLRev 741.

[58] See Liability Convention (n 41) art VI(1).

[59] See Giuseppe Guerreri, 'Wilful Misconduct in the Warsaw Convention: A Stumbling Block?' (1960) 6(4) McGillLJ 267; see also Juan E. Acosta, 'Wilful Misconduct Under the Warsaw Convention: Recent Trends and Developments' (1965) 19(4) UMiamiLRev575.

[60] International Convention for the Unification of Certain Rules relating to International Carriage by Air (adopted 12 October 1929, entered into force 13 February 1933) 137 LNTS 11.

Nations and the Outer Space Treaty.[61] The wording of this provision again creates doubt in the interpretation. Though the Charter of the United Nations and the Outer Space Treaty are the basic documents of international law and outer space law respectively, a specific mention of them under this provision raises a question on the status of these two documents vis-à-vis other norms of international law. In simple terms, specific reference to the Charter of the United Nations and the Outer Space Treaty may indicate the requirement of absolute adherence to them, with the remaining norms of international law being relegated to a secondary level in terms of compliance by the launching State. Though it might be unintentional, the attempt to emphasise the need to comply with the Charter of the United Nations and the Outer Space Treaty has resulted in an interpretational predicament.

3.3 Multistate Liability

The Liability Convention also addresses the situation of liability in cases where multiple launching States are involved in causing damage. Articles IV and V of the Liability Convention attribute joint and several liability to launching States in two different circumstances. Article IV addresses the situation of collision between two or more space objects elsewhere than on the surface of the Earth, which results in damage to a third State. It attracts joint and several liability of the first two States (launching States of space objects) to compensate the third State. Article V applies the joint and several liability when the damage is caused by a space object which is launched jointly by multiple States. For this purpose, the State from whose territory or facility a space object is launched is considered as a launching State on an equal footing with the other two categories of States.[62] This negates any argument of hierarchy in attributing liability among the four categories of launching States in the wake of two subdivisions made in the definition clause.[63]

Since the concept of joint and several liability operates, the victim States can claim compensation from any of the launching States that are jointly and severally liable for damage caused. Though the launching State paying compensation can ask other liable States to share the burden, no prejudice can be caused to the right of the victim State to receive compensation from any or all of the launching States under the umbrella of joint and several liability. The apportionment of liability under Article IV is to be made on the basis of the extent of the fault of the

[61] Liability Convention (n 41) art VI(2).

[62] See ibid art V(3).

[63] In defining the launching State, both art I(c) of the Liability Convention and art I(a) of the Registration Convention mention the State that 'launches' or 'procures the launching' in the first sub-category and the State whose 'territory' or 'facility' is used for launching in the second sub-category. It should not be misconstrued as creating a hierarchy between the two sub-causes while fixing liability on launching States.

launching States.[64] In case the extent of fault of the States cannot be established, equal apportionment of the burden of compensation among the launching States is stipulated. Under Article V, the State paying the compensation can seek indemnification from other participating States in the joint launching. The participants of the joint launching can also enter into agreements regarding the apportionment of their financial obligation to compensate the victim State.[65]

3.4 Non-Application of the Liability Convention

Article VII of the Liability Convention excludes its application in two situations. First, when the damage is caused to the nationals of the launching State of the space object that has caused damage. Secondly, when the damage is caused to foreign nationals by a space object of a launching State 'during such time as they are participating in the operation of that space object from the time of its launching or at any stage thereafter until its descent, or during such time as they are in the immediate vicinity of a planned launching or recovery area as the result of an invitation by that launching State'. It is presumed that in both the above-mentioned situations, remedies for victims should be available at the domestic level of the launching State.

Though the operative presumption of providing a remedy to some victims at the domestic level was not a significant concern when the space treaties were entered into, it is of concern in the modern globalised world. There are a significant number of nationals of one State who have become permanent residents of other States.[66] Except for the nationality connection, they may have no other connection with their State of nationality. In case of any damage caused to them in their State of residence by the space objects of their State of nationality, seeking justice and obtaining an adequate remedy by approaching the courts of their State of nationality would not be practicable. Taking away the application of the Liability Convention in such cases is a departure from the victim-oriented approach of the Liability Convention.

In addition to the norms of liability, the Liability Convention also provides for a dispute settlement mechanism. While diplomatic negotiation is the primary means of settling claims of compensation for damage,[67] upon failure the settlement has to be reached by establishing a Claims Commission.[68] The Claims Commission must

[64] See Liability Convention (n 41) art IV(2).
[65] See ibid art V(2).
[66] The World Migration Report 2022 counts 281 million international migrants in the world in 2020. See 'World Migration Report 2022' (*IOM UN Migration*, 2022) <https://publications.iom.int/books/world-migration-report-2022> accessed 26 August 2023.
[67] See Liability Convention (n 41) art IX.
[68] See ibid art XIV.

consist of a member appointed by the claimant State, a member appointed by the launching State and a Chairman appointed jointly by both parties.[69] The decision of the Claims Commission would be final and binding only when the parties agree in advance. In the absence of such an agreement between the parties, the award of the Claims Commission is only recommendatory.[70] Hence, the entire process of settlement of the claim or resolution of any dispute under the Liability Convention is based on the cooperation between States.

4 Conclusion

State responsibility and international liability for space activities are of the utmost significance in the conduct of space activities. Though the Outer Space Treaty deals with them separately, the understanding of the modern-day distinction between the two concepts is absent under the Treaty. While the provision on State responsibility is not further supplemented, the provision on international liability is supplemented by the Liability Convention. However, interpretational concerns relating to both the State responsibility and international liability provisions of the space treaties stand as stumbling blocks in their practical implementation. This problem is further aggravated by fundamental changes in space activities over time. Hence, there is an urgent need to reevaluate the norms relating to State responsibility and international liability under international space law.

69 See ibid art XV.
70 See ibid art XIX(2).

7
REGISTRATION OF SPACE OBJECTS

Mahulena Hofmann

1 Introduction

The registration of space objects is an important trust-building measure. Since the beginning of the space age, information about space objects has been required, first, based on UNGA Resolution 1721 of 1961, and later on the basis of Article XI of the 1967 Outer Space Treaty (OST). Registration in the narrow sense was introduced by the 1975 Registration Convention, which has served as its basis until now, requiring its States Parties to establish a national registry of space objects and to deliver the relevant information to the UN Register held by the UN Secretary-General. Recent developments raise several new legal questions—the transformation of the present UN Register to an online one, the question of registration of large constellations, and the parallel registration of radio stations on board space objects within the framework of the ITU.

2 The Objective of the Registration of Space Objects

The primary objective of international registration of space objects is their identification; this supports mutual trust-building and collision avoidance. The idea of registering space objects can be traced back to the first years of space activities. Already with the first launching of satellites by the Soviet Union in 1957 and by the United States (US) in 1958 questions arose in relation to tracking space objects for the purpose of avoiding collisions between satellites and aircraft, as well as for the identification of their components during the de-orbiting phases.[1] An additional purpose of the international registration of space objects—as formulated by one of the founders of international space law, Isabella Diederiks Verschoor—is the attributability of the spacecraft to its home country, having as a consequence

[1] Bernhard Schmidt-Tedd and Stephan Mick, 'Article VII' in Stephan Hobe, Bernhard Schmidt-Tedd and Kai-Uwe Schrogl (eds), *Cologne Commentary on Space Law*, vol I (Carl Heymanns Verlag 2009) 149.

Mahulena Hofmann, *Registration of Space Objects* In: *International Space Law in the New Space Era.*
Edited by: Sandeepa Bhat B., Dilip Ukey, and Adithya Variath, Oxford University Press. © Mahulena Hofmann 2024.
DOI: 10.1093/9780198909415.003.0007

its domestic protection, as well as the guarantee of respecting national and international law on the part of the State of registration.[2]

The idea of registering spacecraft did not come from the blue—it was inspired by the long-standing praxis of registration of ships and aircraft. As mentioned in 1967 by one of the pioneers of airspace law, Nicolas Mateesco Matte, in the 1960s the norms on registration applicable in air law tended to be accepted for satellites,[3] albeit with some important modifications. In international air law, the Chicago Convention of 1944[4] provided that aircraft carry the nationality of the State in which they are registered; double registrations are void.[5] In contrast to the later regime of space objects, 'State carriers' such as airplanes used by the military are exempted from obligatory international registration; for the purpose of registration, at least the main part of the aircraft must be owned by natural or legal persons of the State of registration. In maritime law, according to the 1958 Convention on the High Seas replaced in 1994 by the United Nations Convention on the Law of the Sea,[6] ships can sail under only one flag; in contrast to space objects, a genuine link—a special relationship between a flag State and the ship having the right to fly that flag[7]—is required for the registration.[8]

In international and national space law, the registration of a space object is a significant fact with important legal consequences. As expressed in a nutshell by Bernhard Schmidt-Tedd and Stephan Mick, the present international legal framework of registration of space objects creates a chain of attribution between the launching State, the space object, international responsibility for space activities and jurisdiction and control by the registering State, without the direct purpose of establishing liability.[9] At the national level, the registration of a space object in the national registry is usually a public law act regulated by domestic legislation.

It should be added that the international and national registration of space objects is accompanied by several other regimes of registration: radio stations on board satellites are registered nationally and by the International Telecommunication Union (ITU). There are national catalogues detecting space debris; other registries are planned—such as those listing safety zones on celestial bodies to be used for

[2] Isabella H Ph Diederiks-Verschoor, 'Some Observations on Registration of Spacecraft' (1968) 11 ProceColloqLOuterSpace 133.

[3] Nicolas M Mateesco Matte, *Aerospace Law* (Sweet and Maxwell 1969) 329.

[4] Convention on International Civil Aviation (adopted 7 December 1944, entered into force 4 April 1947) 15 UNTS 295, art 17.

[5] ibid art 18.

[6] Convention on the High Seas (adopted 29 April 1958, entered into force 30 September 1962) 450 UNTS 11, art 6 (Convention on the High Seas). See also Convention on the Law of the Sea (adopted 10 December 1982, entered into force 16 November 1994) 1833 UNTS 397, art 92 (Convention on Law of the Sea).

[7] Convention on Law of the Sea (n 6) art 91.

[8] Convention on Law of the Sea (n 6) art 94.

[9] Schmidt-Tedd and Mick (n 1) 147–48.

future space mining. All these measures serve to collect information on space objects and prevent conflicts in outer space activities.

3 The 1961 UN Resolution and its Impact

The first international legal document adopted with respect to the registration of space objects was Resolution 1721 B (XVI) of the UN General Assembly of 1961 on 'International Co-operation in the Peaceful Uses of Outer Space'.[10] Pursuing the aim of international cooperation and transparency, this document created a voluntary regime of international registration of space objects. Rather than creating a new international space organization, instead the United Nations (UN) became the focal point for assembling information from space-faring States. The resolution charged the UN Secretary-General with the task of maintaining a public registry of launchings into outer space. Through the Secretary-General, the States were invited to 'promptly furnish' information about the 'objects launched into orbit or beyond' to the UN Committee on the Peaceful Uses of Outer Space (COPUOS) (para 1).

Adopted in 1961, this resolution has remained relevant until today. In the first years of space activities, States were following its principles to deliver data about their space objects. Today, it still serves as a legal basis for the registration of space objects for those States who have not ratified the 1975 Registration Convention.[11] To give an example: before ratifying the Registration Convention in 2020,[12] Luxembourg was registering numerous communication satellites according to this resolution. The information furnished on the basis of this resolution is kept by the UN Office for Outer Space Affairs (UNOOSA) in a separate register. The use of the indicator 'promptly' in the text of the operative part of the Resolution is interesting: this is a stronger condition than the criterion applied in the 1975 Registration Convention, which requires submitting information on space objects somewhat vaguely 'as soon as practicable' (Article IV).

4 The 1963 UN Declaration of Legal Principles

In 1963, the Declaration of Legal Principles Governing the Activities of States in the Exploration and Use of Outer Space[13] (Declaration of Principles) was adopted

[10] UNGA Res 1721 B (XVI) (20 December 1961).

[11] Convention on the Registration of Objects Launched into Outer Space (adopted 14 January 1975, entered into force 15 September 1976) 1023 UNTS 15 (Registration Convention).

[12] Law of 15 December 2020 on the Law Approving the Convention on the Registration of Objects Launched into Outer Space, Adopted by the General Assembly of the United Nations in New York on 12 November 1974, Mémorial A N° 1087, 28 December 2020.

[13] UNGA Res 1962 (XVIII) (13 December 1963).

by the UN General Assembly. As Stephan Hobe remarks, this resolution marked the first attempt by the international community to enact legal principles for activities in outer space in a formalized manner.[14] Concerning the registration of space objects, the international registry of space objects established by Resolution 1721 B (XVI) of 1961 on 'International Co-operation in the Peaceful Uses of Outer Space' was enlarged by adding an important national dimension.

The 1963 Declaration itself did not create but rather already presupposed the existence of national registries of space objects and even space vehicles. According to its paragraph 7, the 'State on whose registry an object launched into outer space is carried' shall retain jurisdiction and control over such object, and any personnel thereon, while in outer space. The astronauts—as envoys of mankind—who make an emergency landing on the territory of a foreign State or on the high seas, should be safely and promptly returned to the 'State of registry of their space vehicle' (para 9).

With some minor modifications, this Declaration became the antecedent of the 1967 OST. In relation to the registration of space objects, the principle of the non-changeability of jurisdiction and control remained practically untouched by the OST; concerning the astronauts, the criterion of a 'registry of their space vehicle' was kept with identical wording.

5 The 1967 Outer Space Treaty

The OST entered into force in October 1967. Concerning the registration of space objects, it did not introduce many novelties, leaving the detailed regulation of registration to the later 1975 Registration Convention. In this respect, its provisions can be divided into two categories.

The first category is composed of the provisions reflecting the international practice based on the 1961 UNGA 1721 B (XVI) resolution. Article XI OST amended these recommendatory rules by stipulating a legal obligation of the States Parties to inform the UN Secretary-General as well as the public and the international scientific community of the nature, conduct, location and results of their activities in outer space. To achieve consensus, this broad obligation was watered down by the vague requirement that this information must be provided 'to the greatest extent feasible and practicable'. Despite this lack of precision, the provision proved to be useful in specific cases, that is, as regards informing the UN Secretary-General and the community about facts not covered by the 1975 Registration Convention,

[14] Stephan Hobe, 'Historical Background' in Stephan Hobe and others (eds), *Cologne Commentary on Space Law*, vol I (Carl Heymanns Verlag 2009) 12.

e.g. transfer of ownership of space objects when in outer space (see later in the chapter).[15]

The second category of provisions of the OST dealing with registration consists of several articles which rely on the fact of national registration of a space object as an indication of legal status or of a place of return:

Article VIII of the OST repeats the wording of paragraph 7 of the 1963 Declaration of Principles, adding to the element 'while in outer space' the territorial indication 'and on a celestial body'. The adopted text of the provision of the OST now stipulates that a State Party 'on whose registry an object launched into outer space is carried shall retain jurisdiction and control over such object, and over any personnel thereof, while in outer space or on the celestial body'. Vladlen S. Vereshchetin interpreted this term as the power to exercise legislative, judicial and administrative authority over the object.[16]

The term 'object launched into outer space' is not defined in Article VIII OST. Admittedly, there is no legal definition of outer space in international space law. However, practice considers the reaching of the Earth's orbit as fulfilling this criterion. Also, the form of the national registry is left to the discretion of the States Parties; it can be defined in the domestic space legislation and may take the form of a formalized or informal registry, originally in paper form, but more and more frequently replaced by an electronic registry.

From the formulation of Article VII it flows that, similarly to paragraph 7 of the 1963 Declaration of Principles, the establishment of national registries of space objects was not the purpose of Article VIII; the provision only confirms that the jurisdiction and control are not affected by the stay of the object in outer space. The national registry is considered to be one of the indicators of jurisdiction and control over a space object when in outer space. A State which launches an object into outer space can have jurisdiction and control over a space object without having registered it nationally or internationally. In other words, the reference to a national registry in the OST does not necessarily need be interpreted as an obligation to establish such a registry without which the jurisdiction and control would not exist; in our opinion, it is a declaratory provision which has to be read in combination with Article VI OST dealing with the international responsibility of State Parties for their national activities in outer space. There are also other views, such as that of Bernhard Schmidt-Tedd who considers that the registration of a space object has a constitutive effect: according to him, the appropriate State of registry 'gains' jurisdiction and control over the space object through the registration.[17]

[15] Jean-François Mayence and Thomas Reuter, 'Article XI' in Stephan Hobe and others (eds), *Cologne Commentary on Space Law*, vol I (Carl Heymanns Verlag 2009) 201.

[16] Vladlen S Vereshchetin, 'International Space Law and Domestic Law: Problems of Interrelations' (1981) 31 JSpaceL 32.

[17] Bernhard Schmidt-Tedd, 'Registration of Space Objects' in *Oxford Research Encyclopedia of Planetary Science* (OUP 2017) <https://ui.adsabs.harvard.edu/abs/2017oeps.book...95S/abstract> accessed 1 August 2023.

The consequence of national registration of the space object is the fact that, as long as the object is formally registered in one State Party, that State retains jurisdiction and theoretically also control over the object despite the fact that the object might have been transferred to an operator belonging to another jurisdiction. Today, this mechanism may be decisive in the discussion about the jurisdiction over space debris and especially in cases of active debris removal.

Two further provisions of the OST which also rely on the registration of space objects are Article V dealing with the return of astronauts and Article VIII regulating the return of space objects. Article V regulates the legal status of astronauts and the assistance which they should be given in case of an accident, emergency landing on the territory of another State Party or on the high seas. The astronauts who have made such a landing should be safely and promptly returned to the 'State of registry of their space vehicle'. Concerning space objects and their component parts, those found beyond the borders of the State Party to the Treaty on 'whose registry they are carried' shall be returned to that State Party, which shall, upon request, furnish identifying data prior to their return (Article VIII OST). This provision on space objects was new compared to the 1963 Declaration of Principles; both provisions were formulated in more detail in the 1968 Rescue Agreement.[18]

In sum, the OST presupposes the existence of a national registry of space objects and vehicles but neither requires nor establishes it; this task has been shifted to the later Registration Convention. Concerning the international exchange of information, Article XI prepared the ground for the specific provisions of the Registration Convention, but it also remains effective on its own for cases that the Registration Convention does not cover.

6 The 1968 Rescue Agreement

In contrast to the 1963 Declaration and the OST, the 1968 Rescue Agreement abandoned the term 'registry'; instead, it uses the notion of 'launching authority'.

This is different from the 1963 Declaration of Principles, which required that the objects launched into outer space or component parts found beyond the limits of the State of registry should be returned to that State (Principle 7). Similarly, the astronauts who had made an emergency landing were to be safely and promptly returned 'to the State of registry' of their space vehicle (Principle 9). Also, the OST connects the return of astronauts (Article V) and space objects (Article VIII) with their State of registry.

[18] Agreement on the Rescue of Astronauts, the Return of Astronauts and the Return of Objects Launched into Outer Space (adopted 22 April 1968, entered into force 3 December 1968) 672 UNTS 119 (Rescue Agreement).

7 The 1975 Registration Convention

According to its Preamble, the Convention on Registration of Objects Launched into Outer Space (REG) created a mandatory system for the registration of space objects, contributing to their identification. In relation to the OST, the Registration Convention is not only a *lex specialis* specifying some of its provisions, but also brought in new regulations which were not to be found in the OST. It is composed of two interrelated State obligations—to keep a national registry of space objects and to furnish information on registered space objects to the UN Register maintained by the UN Secretary-General.

The central term of the Convention is the 'State of registry': according to Article I(c) of the Convention, this is a 'launching State on whose registry a space object is carried'. It should be seen that the reference to a 'launching State' narrows down the pool of possible States of registry, excluding those who are or were not launching States but acquired the space object later after its launching. In reverse, this raises the question as to the entitlement of non-launching States to register a space object, that is, after a transfer of jurisdiction over a satellite when in orbit.[19]

According to Article VII OST, Article I of the 1972 Liability Convention[20] and Article I REG, a 'launching State' is a State which launches or procures the launching of a space object, as well as a State from whose territory or facility the object is launched. Any of these categories can theoretically register the space object; in case of two or more launching States, they shall jointly determine which one of them shall register the object (Article II(1) REG). There is a voluminous literature on the flexible term 'procures' launching,[21] which in case of its involvement in the launching phase, could enable a successor State to enter the acquired space object into its national registry.

The first obligation, to register space objects domestically, arises at the moment 'when a space object is launched into Earth orbit or beyond it' (Article II(1) REG); consequently, according to international law, an earlier domestic registration of a space object is a voluntary act. What should be registered is covered by the term 'space object', which is not very precisely defined in the Convention: the constantly recurring circular definition contained in Article I(b) REG and also Article I(d) of the Liability Convention states that 'space object includes component part of a space object as well as its launch vehicle and parts thereof'. In practice, a separate

[19] See Mahulena Hofmann and Andreas Loukakis (eds), *Ownership of Satellites* (Nomos/Hart 2017) 296.

[20] Convention on International Liability for Damage Caused by Space Objects (adopted 29 March 1972, entered into force 1 September 1972) 961 UNTS 187 (Liability Convention).

[21] References in Bernhard Schmidt-Tedd and others, 'Article II (National Registries/ Registration Obligation)' in Stephan Hobe and others (eds), *Cologne Commentary on Space Law*, vol 2 (Carl Heymanns Verlag 2013) 251.

registration of the launcher and the payload, or registration of different components of a larger structure is accepted.

Concerning the legal framework of the domestic registry of space objects, the Parties to the Registration Convention enjoy discretion when determining the contents of the information kept and the conditions under which it is maintained (Article II(3)). However, they are obliged to inform the UN Secretary-General of the fact of the establishment of the registry (Article II(1)). The list of national registries is maintained by the UNOOSA.[22]

The second obligation—to furnish information about space objects to the UN Secretary-General—is defined by Article IV REG. There are two categories of information which must be delivered to the UN Secretary-General by the State of registry of a space object. First, information on a functioning space object carried in the national registry (Article IV(1)). This information consists of the name of the launching State or States, a designator of the space object or its national registration number, the date and territory or location of launch, as well as its basic orbital parameters, such as the nodal period, inclination, apogee and perigee. This information must be accompanied by a description of the general function of the space object. There is no fixed deadline for delivering this information: the provision requires these data 'as soon as practicable'; this vague term weakens this obligation substantially.

Furthermore, the State of registry has the duty to inform the Secretary-General about the space objects which have been but are no longer in Earth orbit (Article IV(3)). This information about space objects already registered serves as information about satellites or their parts de-orbited from the Earth's orbit. It probably does not extend to objects orbiting other celestial bodies. Also, the impact of this provision is weakened by the fact that the information should be rendered 'to the extent feasible and as soon as practicable'.

The information delivered by the State of registry is maintained by the UN Secretary-General in a register which is open to the general public (Article III). Today, the UN Register of Objects Launched into Outer Space has an electronic form and is publicly available on the website of the UNOOSA.[23]

As already mentioned, in case of more than one launching State of a space object, they shall jointly determine which one of them shall register the object in the domestic national registry (Article II(2)). In addition to such joint launches, the Convention is open to application by international intergovernmental organizations (IGOs) conducting space activities (Article VII). Until now, four organizations have declared acceptance of the rights and obligations provided for in the

[22] 'Index of Notifications by Member States and Organizations on the Establishment of National Registries of Objects Launched into Outer Space' (*UN Office of Outer Space Affairs*) <www.unoosa.org/oosa/en/spaceobjectregister/national-registries/index.html> accessed 1 August 2023.

[23] 'Online Index of Objects Launched into Outer Space' (*UN Office of Outer Space Affairs*) <www.unoosa.org/oosa/osoindex/search-ng.jspx?lf_id=> accessed 1 August 2023.

document—Eumetsat, ESA, Eutelsat and Intersputnik are registering space objects on their own behalf; these organizations also keep their own 'domestic' registries of space objects.

In sum, the Registration Convention has created a broad system of exchange of information on launched space objects. Its weak points are the limitation of the obligation to register space objects to launching States, as well as the vague time criteria for providing information to the UN Secretary-General. However, it cannot be excluded that this vagueness is the result of a political compromise and that it is the only way in which this international treaty could have entered into force.[24]

8 The 1979 Moon Agreement

The 1979 Agreement Governing the Activities of States on the Moon and Other Celestial Bodies[25] does not address the question of registration of space objects or vehicles. Concerning the jurisdiction and control over personnel, vehicles, equipment, facilities, stations and installations on the Moon and other celestial bodies, it only confirms that States Parties 'retain jurisdiction and control' over these (Article 12).

9 The 2007 Resolution on Registration

In 2007, taking into account the development of new technologies and the relevant practice, the texts of the 1975 Registration Convention and Article VIII OST were completed by UNGA Resolution 62/101 titled 'Recommendations on Enhancing the Practice of States and International Intergovernmental Organisations in Registering Space Objects' (the Registration Practice Resolution). The recommendations were formulated on the basis of reports submitted by the Member States of the COPUOS on their practice in registering space objects.[26] Interestingly, the Resolution states in its Preamble that nothing in the document constitutes an authoritative interpretation of or a proposed amendment to the Registration Convention.

The Resolution suggested several measures leading to the harmonization of registering practices of Member States. It recommended using the international

[24] Ratified by 74 States Parties, with four declarations of acceptance of obligations by four IGOs, UN Doc A/AC.105/C.2/2022/CRP.10, 28 March 2023, plus ratification of Romania and Paraguay as of 15 June 2023.

[25] Agreement Governing the Activities of States on the Moon and Other Celestial Bodies (adopted 18 December 1979, entered into force 11 July 1984) 1363 UNTS 3 (Moon Treaty).

[26] UNGA Res 62/101 (17 December 2007) (Recommendations on enhancing the practice of States and international intergovernmental organisations in registering space objects).

designator of space objects as elaborated by the Committee on Space Research (COSPAR),[27] the Coordinated Universal Time as the reference for the date of launch, and the use of kilometres, minutes and degrees as the standard units for orbital parameters.[28]

However, the crucial contribution of the Resolution is the part dealing with the change in supervision of a space object in orbit (para 4): As mentioned above, according to the wording of the REG and the 1962 UNGA Resolution, the shift in supervision from the launching State to a non-launching State could never be reflected in the change of registration. The only 'clean' path permitting official information about such change to the UN would be Article XI OST, or a more cumbersome exercise in interpreting the non-launching State as a State which in some way procured the launching.

The truth is that a change in the supervision of a space object is not prohibited by international law. In relation to the registration of the space object, it can relate to two distinct legal situations: if the space object changes its jurisdiction to another launching State, a modification of the information in the domestic and international registries is unproblematic. However, if the space object were to change its control from the launching to a non-launching State, a solution allowing the Registration Convention to be interpreted in a flexible manner would be needed.

The solution presented by the Resolution is derived from the fact that the new, acquiring State becomes an 'appropriate' State in the sense of Article VI OST and as such is responsible for its national space activities. This factual shift of control and supervision of the space object is fully in line with Article VI OST. Based on this fact, the Resolution recommends that the State of registry, in cooperation with the new, appropriate State according to Article VI OST, furnish to the UN Secretary-General information on the date of change in supervision, the identification of the new owner or operator, any change of orbital position and any change in the function of the space object. Admittedly, this additional information does not lead directly to the change of registration; it contributes, however, at least to the identification of a State responsible under Article VI OST.

The next important element of the 2007 Resolution is mandating the UNOOSA to develop a model registration form assisting States and international IGOs in their submission of registration information. The form[29] was created with the aim

[27] According to the Annex to the Online Index form, Definition of terms, p 3, COSPAR international designator is an alphanumeric designator established by the Committee on Space Research (COSPAR) for space objects that successfully reach Earth orbit or beyond (*UN Office for Outer Space Affairs*) <www.unoosa.org> accessed 1 August 2023.

[28] Bernhard Schmidt-Tedd, Niklas Hedman and Anne Hurtz, 'Paragraph 2 (Recommendation: Harmonization of Practice)' in Stephan Hobe and others (eds), *Cologne Commentary on Space Law*, vol 3 (Carl Heymanns Verlag 2015) 432 .

[29] 'United Nations Register of Objects Launched into Outer Space' (*UN Office for Outer Space Affairs*) <http://www.unoosa.org/oosa/SORegister/resources.html> accessed 1 August 2023.

of reaching uniformity in the information provided; it also invited States to provide information on the change of supervision of space object[30] and to submit additional information on it.

10 United Nations Practice in Registering Space Objects

On behalf of the UN Secretary-General, the UNOOSA maintains an international register of space objects. The Register is based on UNGA Resolution 1721 B (XVI) for those States which are not yet Parties to the Registration Convention,[31] on the 1975 Registration Convention for States Parties and on the generally applicable 2007 Registration Resolution.

On the basis of information provided by the States, an online index of space objects has been developed. As pointed out in the literature,[32] the UNOOSA has no mandate to interpret or make decisions concerning the State of registry, and subsequently the Office does not make any modifications to information which is provided to the UN without prior consultation with the Member States.

To date, approximately 86 per cent of all satellites, probes, landers, crewed spacecraft and space station flight elements launched into Earth orbit or beyond have been registered with the UN Secretary-General.[33] For the rest, as the Registration Convention does not define the deadline within which a space object should be registered—they should be registered as soon as practicable—the UNOOSA can only informally approach the States concerning the objects which, therefore, might remain unregistered for a long time.

The Online Index of Objects Launched into Outer Space[34] includes both official documents submitted to the UNOOSA and unofficial information which is marked green and enclosed in square brackets.[35] The Registration submission form is composed of four parts: Part A invites the submission of information in conformity with the Registration Convention or General Assembly Resolution 1721 B (XVI): this part includes information on the launching State or international IGO, the designator of the space object which can be a COSPAR international designator[36] and/or the national designator as used by the State of registry,[37] date and

[30] Natercia Rodriguez and Kamran Memon, 'United Nations Practice in Registering Space Objects' in M Hofmann and A Loukakis (eds), *Ownership of Satellites* (Nomos/Hart 2017) 89.
[31] Online Index of Objects (n 23).
[32] Rodriguez and Memon (n 30) 90.
[33] UN Register of Objects (n 29).
[34] Online Index of Objects (n 23).
[35] The US Starlink satellites not yet being registered with the UN Registry in 2023.
[36] Annex to the Online Index form (n 27).
[37] ibid (national designator is the designator or number assigned to a space object by the State of registry).

territory or location of launch, basic orbital parameters, general function of space object, as well as the information on the change of status—date of decay, entry or de-orbit of the object.

Part B is devoted to additional information according to General Assembly Resolution 62/101 (Registration Practice Resolution). This part collects voluntarily submitted data on the change of status in operations, including the date when a space object is moved to a disposal orbit. Information relating to the change of supervision of a space object as recommended by the Resolution is collected under Part C of the Online Index. This deals with the date of change in supervision, the identity of the new owner, change in orbital position and the change of function of the space object. Part D gathers 'additional voluntary information for use in the UN Register of objects launched into Outer Space': The States may submit, for example, information on a space object's owner or operator, on the launch vehicle or on the celestial body the space object is orbiting.

Several cases of change of supervision are documented in the literature[38] which were dealt with by the UNOOSA. It seems that the Member States are taking different approaches to providing information on this topic. In this context, Natercia Rodiguez and Kamran Memon raise the question of whether the present Online Index should be modified to facilitate the intuitive inclusion of additional information, including on the change of supervision that the UNOOSA started to receive under Sections C and D of its registration template.[39]

Also, the time span for the fulfilment of the registration obligation differs substantially. There are cases where it took from a few months to a few years to register the launch of space objects.[40] In some cases, the UNOOSA has identified the fact that some objects launched into outer space have never been formally registered.[41] As Srinivasan Venkatasubramanian points out, the same is true of the registration of transfer of satellites from one State to another, and also the time taken to provide information on decommissioning of satellites often amounts to many years. One of the reasons for these delays is the fact that the Registration Convention does not specify any time limit for registration. There is no legal consequence for failure to register a space object as the State can argue that the registration was not yet practicable, as required by the Registration Convention.

[38] Rodriguez and Memon (n 30) 91.

[39] ibid 95.

[40] Srinivasan Venkatasubramanian, 'Interaction between Space Objects and the Protection of Frequency Assignment' in M Hofmann and A Loukakis (eds), *Ownership of Satellites* (Nomos/Hart 2017) 182.

[41] 'Practice of States and International Organizations in Registering Space Objects' Forty-fourth Session of the COPUOS Legal Subcommittee (Vienna, 4–15 April 2005), UN Doc A/AC.105/C.2/L.255, 19.

11 Registration by an International Organization: Example of the European Space Agency

The European Space Agency (ESA) is an international IGO which, through its declaration of acceptance of the Registration Convention, is committed to the same obligations as any State of registry of a space object.[42]

The literature[43] reports registration of an *Artemis* satellite which was registered by ESA. When in orbit, this satellite was transferred to an entity incorporated under the laws of England and Wales. The ESA provided the UN Secretary-General with the information on the change of supervision of the satellite; formally, ESA remained the 'State of registry' of the satellite.[44]

ESA also serves as a registration authority for space programmes carried out by the European Union (EU) as the EU has not yet acceded to any of the UN space treaties and, therefore, is unable to register its numerous satellites. Consequently, many of the *Galileo* satellites launched during the initial phase of this navigation programme were officially or unofficially registered under the authority of the ESA.[45] ESA also registered the *Sentinel* satellites used for Earth observation in its ESA Space Object Register and furnished the UN Secretary-General with the information required by the Registration Convention, with the EU designated as the owner of these space objects.[46]

12 National Legislation

Several States have adopted specific rules on the registration of space objects in their national legislation. To give some examples, the legislation of the US, Russian Federation, China and Luxembourg are shortly presented below.

The US has a long tradition of registering space objects, the REG having been signed as early as January 1975. The information to be furnished to the UN Secretary-General is located by the Office of Space and Advanced Technology at the Department of State.[47] According to the *Cologne Commentary on Space Law*,

[42] Ioanna Thoma, 'Transfer of Satellites in Orbit: the ESA Experience' in M Hofmann and A Loukakis (eds), *Ownership of Satellites* (Nomos/Hart 2017) 109.

[43] Rodriguez and Memon (n 30) 94.

[44] The potential liability from damage was solved by the ESA Council Resolution which stipulated that in case of damage the United Kingdom would substitute ESA in a claim or compensate ESA for any loss suffered as a result of such claim.

[45] Andreas Loukakis, 'EU as Owner of Galileo Satellites, Consequences for Registration and Liability' in M Hofmann and A Loukakis (eds), *Ownership of Satellites* (Nomos/Hart 2017) 129.

[46] In case of liability for damage, the EU will indemnify the ESA and hold harmless the ESA and ESA Participating States. See Thoma (n 42) 108.

[47] Bernhard Schmidt-Tedd and others (n 21) 270.

this information is updated with data provided by the Strategic Command, which maintains a catalogue of space objects.[48]

14 Code of Federal Regulation, § 417.19 requires that each launch operator must provide to the Federal Aviation Administration (FAA) the information required by paragraph (b) of this section for all objects placed in space by a licensed launch, including a launch vehicle and any components, not later than thirty days following the conduct of a licensed launch, namely the international designator of the space object(s), as well as date and location of launch, general function of the space object and final orbital parameters. Only objects owned and registered by the US government and objects owned by a foreign entity are exempted from this obligation.

The Russian Federation succeeded to the international obligation of the Soviet Union which was one of the original signatories of the REG in 1975. The national registry of space objects is maintained by the Russian Space Agency (ROSCOSMOS). The legal basis for the registration is the 1993 Law on Space Activities,[49] and the Administrative Regulation of the Russian Federal Space Agency on the implementation of the State Function of Maintaining the Registry of Space Objects Launched into Outer Space by the Russian Federation.[50] The Annex to the Administrative Regulation enumerates the categories of information to be submitted, both for the purpose of a national registry and the transfer of this information to the Ministry of Foreign Affairs for submission to the UN Secretary-General.

China, a Party to the REG since 1988, bases the registration of space objects under its jurisdiction on the 2001 Regulation for the Administration of Objects launched into Outer Space.[51] The main authority is the State Administration of Science, Technology and Industry for National Defense (SASTIND), which registers space objects through the Chinese Space Administration (CNSA).[52] The information, including the owner of the space object, the date and territory of the launch, basic orbital parameters and the function of the space object, has to be delivered within sixty days of the launch of the space object. International registration is carried out through the Ministry of Foreign Affairs within sixty days after the national registration took place.

In parallel with the ratification of the REG in 2020,[53] Luxembourg has included a provision for registration of space objects under its general 2020 Law on Space Activities.[54] Article 15 of the Law created a National Register of Space Objects

[48] ibid.

[49] No 5663-1 of 20 August 1993, as amended in November 2021.

[50] Order of the Federal Space Agency of 22 March 2010, No 44.

[51] Order No 6 of the Commission for Science, Technology, Industry and National Defense and the Ministry of Foreign Affairs of the People's Republic of China, 8 February 2001 (see Schmidt-Tedd and others (n 21) 292).

[52] 'Australian Navigational Guide Explaining Laws for Space' (*SpaceLaws*) <www.spacelaws.com> accessed 7 August 2023.

[53] Journal officiel, Mémorial A, N° 1087, 28 December 2020.

[54] Journal officiel, Mémorial A, N° 1086, 15 December 2020.

for which Luxembourg assumes a registration obligation under Article VIII OST and Article II REG. The operator of a space object taking the initiative to launch or to procure the launch must provide the Minister in charge of Space Policy and Legislation with all information allowing identification of the object, such as its launch, its position in outer space, launching date and territory or place, the main parameters of the orbit and the general function of the object. The operator must also inform the Minister without delay of any change or risk of change in the parameters of the space object, particularly the danger of its de-orbiting.

These few examples show the main elements of the national legislation dealing with the registration obligation. They all mention the entity responsible for maintaining the national registry, explain the procedure and define the category of information to be submitted. In contrast to the international registration framework, these legal acts can be enforced, and the violation of these provisions is sanctioned.

13 Registration of Space Objects Forming Part of a Satellite Constellation

In the early years of space activities, satellites were launched one by one, and also the international legal framework was shaped to reflect this fact. The evolution of technology led to multiple launches, and later to the utilization of networks of satellites. Today, satellites can operate in groups known as satellite constellations.[55] The term 'very large constellations' (VLCs) refers to a network of hundreds or even thousands of small satellites orbiting the Earth and communicating with one another as part of a more extensive system of telecommunication satellites.[56]

With this growing phenomenon, the question of the registration of such a large constellation arose. The Legal Subcommittee of the COPUOS Working Group on the Status and Application of the five UN Treaties on Outer Space presented in March 2023 a Non-paper by its Chair called 'Revised draft recommendations concerning the submission of registration information on space objects forming part of a satellite constellation'.[57] In this document, the following measures are recommended.

The State of registry could inform the UNOOSA, as part of the registration process, whether the space object being registered formed part of a

[55] Suwijak Chandaphan, 'Very Large Constellations (VLCs)' in M Hofmann and PJ Blount (eds), *Concise Encyclopedia of Space Law* (forthcoming, Edward Elgar 2024).

[56] Abdul Sami, 'A Shift from Small Constellation System of Large Satellites Towards Large Constellation System of Small Satellites in Space Architecture' (*Traductor de Ciencia*, 25 May 2020) <https://traductordeciencia.es/a-shift-from-small-constellation-system-of-large-satellites-towards-large-constellation-system-of-small-satellites-in-space-architecture/> accessed 28 February 2023.

[57] 'Annotated Provisional Agenda' (Vienna, 20–31 March 2023) Sixty-second Session of the COPOUS Legal Subcommittee (1 February 2023) UN Doc A/AC.105/C.2/L.323.

satellite constellation, for example by including such information in Part D of the Registration submission form. To avoid difficulties in the multitude of space object registration information submissions and to allow for their efficient processing, the State of the registry could use the first submission on a space object forming part of a satellite constellation to provide additional appropriate information and refer to this first submission when registering additional space objects forming part of the same satellite constellation. Without prejudice to the formal submission of registration information on such space objects, the State of registry could consider appropriate means to make available additional updated information on the status of the satellite constellation.

14 Other Registries

14.1 Master International Frequency Register

Not space objects as such, but onboard radio stations providing various satellite communication services and the frequencies used by them are the focus of the regulation of the ITU, one of the UN specialized agencies and one of the oldest IGO. Its main purpose is to prevent and avoid harmful interference between radio stations, including radio stations on space objects. The main methods are the allocation of bands of the radio-frequency spectrum, allotment of radio frequencies and registration of radio-frequency assignments, and for space services of any associated orbital positions in the geostationary orbit (GSO) or of any associated characteristics of satellites in other orbits.[58] The central instrument for the registration of frequency assignments is the Master Frequency International Register (MFIR) maintained by the ITU Radiocommunication Bureau.[59]

The recording is the result of complex procedures based on the interplay between the operators, the national administrations and the ITU. These procedures are defined by the Radio Regulations (RR), one of the binding 'instruments' of the ITU (Article 4 of the ITU Constitution), especially Articles 9, 11 and Appendices 30, 30A and 30B of the RR. The space services are divided into planned and non-planned: The non-planned services are based on a first-come-first-served principle; the planned services consist of an a priori planning of a fixed amount of spectrum for use by all countries at any time without the need to make filings and coordinate a satellite network; however, similar approaches as for non-planned services are applied to the modification of these uses.[60]

[58] Constitution of the International Telecommunication Union (adopted 22 December 1992, entered into force 1 July 1994) 1825 UNTS 330, art 1(2).

[59] Convention of the International Telecommunication Union (adopted 22 December 1992, entered into force 1 July 1994) 1825 UNTS 330, art 12.

[60] Venkatasubramanian (n 40) 179.

At the national level, the envisaged frequency for satellite networks or services is filed by the operator and submitted to the national administration. This application has to take into account the ITU Table of Allocations (Article 5 RR) which explains which services can use which frequency bands.

All procedures have fixed time schedules and schemata: the national administration has to provide the necessary characteristics of the new assignment[61] to the ITU Radiocommunication Bureau for advance publication in the International Frequency Information Circular in a fixed time framework (Article 9 RR). Numerous categories of space services[62] need to be coordinated with other administrations. The notices should not reach the Bureau earlier than three years before the assignments are brought into use (Article 11.25 RR). The Bureau examines the notice on its conformity with respect to the probability of harmful interference that may be caused to assignments already recorded. When the examination leads to a favourable finding, the respective frequency assignment can be recorded in the MFIR. An unfavourable finding of the notice would be returned to the notifying administration with an indication of the appropriate action to be taken.

The recognition of the frequency assignment takes place on a provisional basis and must be followed by their utilization,[63] the 'bringing into use'. If it is not completed in a timely manner (seven years),[64] the frequency assignment may be cancelled.

The recording of the assignment in the MFIR is connected with international rights and obligations (Article 7 RR): any frequency assignment recorded in the MFIR shall have the right to international recognition. This means that—in order to avoid harmful interference—administrations of other countries have to take them into account when making their own assignments and to avoid and immediately eliminate any harmful interference to stations using this assignment.

A challenge for the system—as in the case of registration of space objects—is the growing popularity of large non-GSO satellite constellations. In the past, it was sufficient that one satellite of the constellation fulfilled the criteria necessary for ITU registration. This does not work in the case of large constellations: To prevent warehousing of a spectrum assigned to the constellation of numerous satellites, the ITU adopted a 'milestone approach',[65] according to which the 'bringing into use' of all satellites belonging to the network can take place in several phases: the non-GSO

[61] See Radio Regulations of the International Telecommunications Union (adopted 22 December 1992, entered into force 1 July 1994) 1825 UNTS 330, art 11 (Radio Regulations) (Enumeration of the frequency assignments to be notified to the Bureau).

[62] See ibid art 9.

[63] Elina Morozova, 'Non-Geostationary Satellite Systems: New Rules of Bringing them into Use and Phasing Their Deployment', in PJ Blount and M Hofmann (eds), *Space in a Networked World* (Brill/Nijhoff 2023) 144.

[64] Radio Regulations (n 61) art 11.44.

[65] RES35-1, RESOLUTION 35 (WRC-19) A milestone-based approach for the implementation of frequency assignments to space stations in a non-geostationary-satellite system in specific frequency bands and services.

satellite systems will have to deploy 10 per cent of their constellation within two years of the end of the current regulatory period for bringing into use, 50 per cent within five years and they must complete the deployment within seven years.

The milestone-based approach provides a regulatory mechanism to ensure that the MIFR reasonably reflects the actual deployment of such non-GSO satellite systems in specific radio-frequency bands and services. It also seeks to strike a balance between the prevention of radio-frequency spectrum warehousing, proper functioning of coordination mechanisms and the operational requirements related to the deployment of non-GSO systems.[66]

14.2 Registration of Space Resources

The recent plans to use resources found in outer space led to a discussion about a new international regime which would guarantee that the minerals are used 'for the benefit of all peoples' and are the 'province of all mankind'.[67] The UNOOSA has established a Working Group on this subject and is discussing future arrangements at the universal level.[68]

Some deliberations take part in parallel at the academic and operational levels: the Hague International Space Resources Governance Working Group was established in 2015 and assembled industry leaders, scientists, diplomats, as well as political and legal experts from across the globe. The outcome of its work is 'Building Blocks for the Development of an International Framework for the Governance of Space Resources Activities',[69] which was presented in the UNCOPUOS.

The ITU system of registering priority rights to frequencies was an inspiration for Article 7 of the Building Blocks (BB). This stipulates that the future international framework should enable the attribution of priority rights to an operator to search for and/or recover space resources upon registration in an international registry. Moreover, the States and international organizations should provide information on their space resources activities through an international database, including information on the termination of their activity. Both these information sources—the registry and the database—should be maintained by an international

[66] 'ITU Members Agree to New Milestones for Non-Geostationary Satellite Deployment' (*ITU*) <www.itu.int/hub/2020/01/itu-members-agree-to-new-milestones-for-non-geostationary-satellite-deployment/> accessed 3 August 2023.

[67] Mark Sundhal and Antonino Salmeri, 'The Registration of Lunar Activities: Recommendations from the Registration Project' in *2021 Proceedings of the International Institute of Space Law* (Eleven 2022) 91.

[68] 'Working Group on Legal Aspects of Space Resource Activities under the Legal Subcommittee of the Committee on the Peaceful Uses of Outer Space' (*UN Office of Outer Space Affairs*) <www.unoosa.org/oosa/en/ourwork/copuos/lsc/space-resources/index.html> accessed 1 August 2023.

[69] Olavo de O Bittencourt Neto and others, *Building Blocks for the Development of an International Framework for the Governance of Space Resources Activities: A Commentary* (Eleven 2020).

body or bodies responsible for the consideration and promotion of best practices and the monitoring and review of the international framework (BB 18). This approach clearly favours the ITU model of registering priority rights to frequency assignments before the UNCLOS deep sea resources licensing and distribution system.[70]

14.3 The UNIDROIT Space Assets Protocol

The 2021 UNIDROIT Space Assets Protocol[71] to the Cape Town Convention[72] aims to establish an international register of securities in space assets and an internationally harmonized regime for dealing with default remedies and procedures in order to ease the financing of certain commercial space operations.[73] Until now, the Protocol has not entered into force.[74]

The core of the Cape Town Convention is an international register of security interests over various categories of mobile assets, including spacecraft and satellites.[75] The Protocol on Space Assets pursued the Convention model, by creating 'registrable interests'. It is designed to deliver an international registry system that enables the parties which are situated in different jurisdictions and involved in providing and securing international interests, to take it into account when looking to ensure effective enforcement over property in the satellite. According to Lesley Jane Smith,[76] the registered creditors would rank in order of their registration, followed by unsecured creditors, insofar as there has been a registration of an international interest, or prospective international interest, in a space asset.[77]

15 Conclusion

The registration of space objects is an important trust-building measure and one of the tools for identifying space objects. It has many advantages, such as a solid legal

[70] ibid 48.

[71] Protocol to the Convention on International Interests in Mobile Equipment on Matters Specific to Space Assets, 9 March 2012, not in force, UNIDROIT Doc DCME-SP-Doc 43.

[72] Convention on International Interests in Mobile Equipment, UNTS vol 2307, 285.

[73] Frans von der Dunk, 'Transfer of Ownership in Orbit: from Fiction to Problem' in M Hofmann and A Loukakis (eds), *Ownership of Satellites* (Nomos/Hart 2017) 33.

[74] <www.unidroit.org> accessed 12 June 2023. For signatures, see <www.unidroit.org/instruments/security-interests/space-protocol/status/>.

[75] Lesley J Smith, 'The Cape Town Convention and the Space Assets Protocol: Considerations after the Event' in M Hofmann and A Loukakis (eds), *Ownership of Satellites* (Nomos/Hart 2017) 272.

[76] ibid 276.

[77] For details, see 'The Interaction between International Private Law and Space Law and Its Impact on Commercial Space Activities' in *2021 Proceedings of the International Institute of Space Law* (Eleven 2022) 153.

basis composed of the rules of the OST, the Registration Convention and the 2007 Resolution on Registration. The disadvantages of this mechanism are the lack of clear time limits for delivering the information by the launching States which can be interpreted extensively and the lack of any international supervising framework equipped with executive powers. This is not surprising as the registration of a space object in the UN does not have any direct legal consequences.

The situation is different in the case of registration of the frequencies and orbital positions effectuated by the ITU. The entry of an assignment in the MIFR is in the interest of the operators as it is followed by its international recognition and protection in cases of harmful interference. To avoid spectrum hoarding, the ITU is entitled to cancel an entry if it has not been used for a specific period.

It is unlikely that the more or less voluntary regime of registration of space objects in the UN Register will be modified in a similar direction in a near future. Not only has no space treaty been adopted since 1979, but also the character of space activities as an unconditional area of international cooperation has changed recently. Let us see whether the growing interest in extracting space resources from celestial bodies and the potential conflicts which may arise over the choice of the appropriate site, or its protection, will lead to a pragmatical need to coordinate the activities, not only bilaterally but also multilaterally. Inspired by the ITU MIFR, the creation of a new international register of the use of areas on celestial bodies for space resources activities may be the next step in this direction.

8
THE MOON AGREEMENT—A SWAN SONG

Michelle LD Hanlon

1 Introduction

The Agreement Governing the Activities of States on the Moon and Other Celestial Bodies[1] (Moon Agreement) is something of a proverbial 'ugly duckling'—the largest and last to be hatched and the first to be scorned.[2] The fifth and final of the quintet of space-related agreements negotiated at the United Nations Committee on the Peaceful Uses of Outer Space (UNCOPUOS), opened for signature on 18 December 1979. Like its four predecessors, it required five ratifications before it entered into force. Unlike its predecessors, which entered into force mere months after opening for signature, it took the Moon Agreement nearly five years to gain them. It finally entered into force on 11 July 1984, the thirtieth day following the deposit by Austria of the fifth instrument of ratification. It was not an auspicious start, but it proved to be prophetic. In the ensuing decades, the Moon Agreement has gained only 18 adherents, significantly fewer signatories and ratifications than its counterparts. It also has the distinction of being the only international space instrument to have a formal withdrawal. On 5 January 2023, Saudi Arabia, which had acceded to the Moon Agreement in 2012, gave notice to the United Nations Secretary General of its intent to withdraw in January 2024.

[1] Agreement Governing the Activities of States on the Moon and Other Celestial Bodies (adopted 18 December 1979, entered into force 11 July 1984) 1363 UNTS 3, art 11(1) (Moon Agreement).

[2] See Hans Christian Andersen, 'The Ugly Duckling' <http://pinkmonkey.com/dl/library1/tale120.pdf> accessed 28 July 2023. In the fairy tale, a mother duck patiently awaits the birth of her largest and final egg only to find the hatchling to be 'very large and ugly'. The other ducks and farmyard animals immediately ostracized the misshapen creature, and it was forced to fend for itself in wilderness. Close to death, the ugly bird was saved by a kind stranger who protected the fowl through the harsh winter. Come spring, the bird was released back into the wild whereupon it found itself 'no longer a dark, gray bird, ugly and disagreeable to look at, but a graceful and beautiful swan'. The ugly duckling was not a duckling at all, but a swan so beautiful that all 'the old swans bowed their heads before him'. ibid. With acknowledgment to PJ Blount and Anonymous who, to this author's knowledge, were the first to refer to the Moon Agreement as an 'ugly duckling' in writing in their opinion piece: 'Another Pyrrhic Victory: The White House's Latest Executive Order on Space Mining' (*SpaceWatch.Global*).<https://spacewatch.global/2020/04/another-pyrrhic-victory-the-white-houses-latest-executive-order-on-space-mining/> accessed 23 September 2023.

Michelle LD Hanlon, *The Moon Agreement—A Swan Song* In: *International Space Law in the New Space Era.*
Edited by: Sandeepa Bhat B., Dilip Ukey, and Adithya Variath, Oxford University Press. © Michelle LD Hanlon 2024.
DOI: 10.1093/9780198909415.003.0008

In the tale of the ugly duckling, after months of scorn and ridicule, it turns out that the duckling had emerged from a misplaced egg. It looked and sounded different from the other ducks because it was actually a swan, in the tale, a far more elegant and powerful bird. Could there be a similar fairytale ending for the Moon Agreement? There is certainly an argument to be made that it is a treaty misplaced in time. With the Soviet Union and the United States (US) racing to the Moon, the need for regulations governing lunar resource utilization certainly had a sense of imminence attached. When humans placed the first footsteps on the lunar surface in 1969, our potential was seemingly boundless. More than 70 States contributed to Messages of Peace which were part of the memorial items left on the lunar surface. Nearly every message spoke of using the technology 'which has made it possible and the resources which may be discovered . . . for the benefit of all [hu]mankind[3] irrespective of terrestrial divisions of race or creed or levels of development'.[4] Given that humanity went from its first satellite in 1957 to the first human celestial travellers in barely more than a decade, it was not at all fanciful to imagine we would be mining the Moon well before the turn of the century.

This chapter provides an overview of the Moon Agreement. Section 1 puts the development of the Agreement in a historical context. Section 2 provides an article-by-article review of the Agreement. Section 3 assesses the future of the Moon Agreement and considers its legacy. While the Moon Agreement may not take elegant flight as a swan into the future in its current form, it certainly offers a critical lesson for the future development of a regulatory framework for space resource utilization. So long as we are unable to truly understand and know the extent and feasibility of access to space resources, it is unwise to attempt a characterization. The swan that emerges from the Moon Agreement is not a celebrated framework, but a fundamental lesson on how concepts born in Earthly bonds cannot be imposed in a vacuum.

If nothing else, it is important to consider and understand the Moon Agreement as a tool for interpreting the Outer Space Treaty, since the subsequent language adopted by the negotiators of the Moon Agreement provides insight as to what additions or revisions had to be made where it was perceived that the Outer Space Treaty needed clarification.[5]

[3] Necessary changes are made throughout the chapter to respect gender neutrality.

[4] Linden Forbes Sampson Burnham (Prime Minister of Guyana), 'Messages of Peace: Guyana' (*For All Moonkind*) <www.forallmoonkind.org/moonkind-mission/messages-of-peace/> accessed 28 July 2023.

[5] Vienna Convention on the Law of Treaties (adopted 23 May 1969, entered into force 1980) art 31(3)(a). (In the interpretation of treaties 'there shall be taken into account, together with the context: . . . subsequent agreement between the parties regarding the interpretation of the treaty of the application of its provisions.')

2 Historical Context—Where did the Moon Agreement Come From?

2.1 Establishing the Nest and the Ducklings

The 1960s and 70s bore witness to a relatively rapid development of binding multilateral agreements designed to regulate human activities in space. The post-World War II race to space was perceived at the time to be balanced enough to warrant negotiation, and even compromise. It was perhaps because of the dangerous and potentially violent dichotomous Cold War environment that good sense prevailed and serious discussion was had to prevent terrestrial conflict from entering space. More importantly, both Cold War parties likely had a strong interest in ensuring that the first mover was not able to establish an enduring strategic advantage. The rest of the world went along because:

> The enthusiasm and hope that the beginning of the space age kindled in human hearts were very quickly followed by the fear that these amazing products of human will and genius might be turned into instruments which could destroy, or seriously endanger, the peace of the entire human race.[6]

The 1967 Treaty on Principles Governing the Activities of States in the Exploration and Use of Outer Space, including the Moon and other Celestial Bodies[7] (Outer Space Treaty), the backbone of the outer space governance regime, established outer space as the 'province of all [hu]mankind' 'free for exploration and use by all States'.[8] It is noteworthy here that this concept of 'providence' is 'but an embryonic form of the later "common heritage of mankind" concept; it does not yet carry with it the latter's distinct connotation of direct inclusive administration of the resource so labelled'.[9] Even so, it was unequivocally established that '[o]uter space, including the [M]oon and other celestial bodies, is not subject to national appropriation by claim of sovereignty, by means of use or occupation, or by any other means'.[10] The negotiators agreed that in this domain that falls outside any sovereign jurisdiction, international law shall nevertheless apply.[11] The Treaty falls short of a binding operative provision mandating that all of space shall

[6] Enrico Scifoni, 'Principle Res Communis Omnium and the Peaceful Use of Space and of Celestial Bodies' (1964) 7 ProceColloqLOuterSpace 50.

[7] Treaty on Principles Governing the Activities of States in the Exploration and Use of Outer Space, Including the Moon and Other Celestial Bodies (adopted 27 January 1967, entered into force 10 October 1967) 610 UNTS 205 (Outer Space Treaty).

[8] ibid art I.

[9] Siegfried Wiessner, 'Access to a Res Publica Internationals: The Case of the Geostationary Orbit' (1986) 29 ProceColloqLOuterSpace 147, 148.

[10] Outer Space Treaty (n 7) art II.

[11] ibid art III.

be used exclusively for peaceful purposes. Instead, the Treaty prohibits the placement of weapons of mass destruction and nuclear weapons anywhere in space. It also clearly establishes that the Moon and other celestial bodies shall be used exclusively for peaceful purposes.[12]

The next three treaties to hatch from the accommodating nest of the 1970s UNCOPUOS essentially fleshed out and detailed key provisions of the Outer Space Treaty. The Agreement on the Rescue of Astronauts, the Return of Astronauts and the Return of Objects Launched into Outer Space[13] (ARRA) expands on the provisions of Article V and Article VIII of the Outer Space Treaty.[14] The Convention on International Liability for Damage Caused by Space Objects[15] (Liability Convention), which opened for signature in March 1972 and entered into force in September of that year, expands on Article VII of the Outer Space Treaty.[16] The fourth and final proper little duckling treaty is the Convention on Registration of Objects Launched into Outer Space[17] (Registration Convention). Affirming the concept, established in Article VIII of the Outer Space Treaty regarding ownership and jurisdiction over objects launched into space, the Registration Convention establishes an international registry of space objects for the purpose of, among other things, the identification and tracking of space objects. Notably, this was the first treaty to need more than one year to gain the five adherents it required to enter into force.

It can be observed that these four treaties followed logical paths to development. Once fundamental principles were established under the Outer Space Treaty, the delegates of the Legal Subcommittee (LSC) of the UNCOPUOS seemed intent on working out at least some of the details related to the agreed concepts and provisions.

2.2 An Ugly Duckling Emerges—Common Heritage of Humankind?

The Moon Agreement was from the beginning a different beast. The Outer Space Treaty makes no attempt to regulate or even categorize the actual use of space

[12] ibid art IV.

[13] Agreement on the Rescue of Astronauts, the Return of Astronauts and the Return of Objects Launched into Outer Space (adopted 22 April 1968, entered into force 3 December 1968) 672 UNTS 119 (Rescue Agreement).

[14] Art V states in pertinent part that astronauts 'shall be safely and promptly returned to the State of registry of their space vehicle'. Art VIII states that States retain jurisdiction and control over objects they launch into space and that all such objects shall be returned to their owner if found. See Outer Space Treaty (n 7) arts V and VIII.

[15] Convention on International Liability for Damage Caused by Space Objects (adopted 29 March 1972, entered into force 1 September 1972) 961 UNTS 187 (Liability Convention).

[16] Art VII of the Outer Space Treaty establishes international liability for damage caused to another State Party by a space object. Outer Space Treaty (n 7) art VII.

[17] Convention on Registration of Objects Launched into Outer Space (adopted 14 January 1975, entered into force 15 September 1976) 1023 UNTS 15 (Registration Convention).

resources. The Outer Space Treaty is filled with general aspirations stressing the themes of cooperation and freedom of access and exploration. Cast above all else is an overarching imploration that all activities in outer space 'shall be carried out for the benefit and in the interests of all countries'.[18] However, the word 'resource' is not even mentioned. Thus, in developing and negotiating a treaty to govern resource utilization—unlike the return and registration of space objects or the process concerning liability—the Outer Space Treaty offers no direct starting point.

Nevertheless, deliberation on the use of outer space resources began, as one scholar wrote, before 'the ink was hardly dry, on the Outer Space Treaty'.[19] Consideration of the utilization of space resources was listed as an agenda item for the 1967 meeting of the LSC and discussion was permitted to continue at the 1969 meeting, which took place from 9 June to 4 July. The LSC report acknowledges proposals submitted by Poland, Argentina and France concerning the '[c]onsideration of the question of the rules which should govern man's activities on the [M]oon and other celestial bodies, including the legal regime governing substances coming from the [M]oon and other celestial bodies'.[20]

In retrospect, it feels that a decision in 1969 to determine how to regulate space resource utilization was premature. After all, here we are more than 50 years later, and the technology to mine any celestial body still has not been realized. But it is really not that surprising:

> By 1969, outer space activity was beginning to capture the attention of thoughtful people. Foremost was the successful landing on the [M]oon. An awareness existed that tangible [M]oon rocks were being returned to [E]arth. The possibility was being explored whether mineral and other tangible and intangible resources could be exploited. The role of communications in space was becoming very real. There was talk of establishment of human habitations on the [M]oon. It was being accepted that even the temporary presence of humans would require the exploitation of such resources.[21]

When, on 20 July 1969, the first human set foot on the Moon, it was 'a major step toward making the future occupation and exploitation of the [M]oon a reality'[22] and an alarm bell for the international community, or more specifically,

[18] Outer Space Treaty (n 7) art I.

[19] Carl Q Christol, 'The 1979 Moon Agreement: Where is it Today' (1999) 27 JSpaceL 1, 6. For example, in June 1969, France proposed that the LSC consider, among other things, 'questions relating to the legal regime governing man's activities on the surface of the [M]oon and of other celestial bodies'. UNCOPUOS, 'France Proposal' UN Doc A/AC.105/C.2/L.64 (27 June 1969).

[20] UNCOPUOS, 'Report of the Legal Subcommittee on the work of its Eighth Session' (9 June–4 July 1969) UN Doc A/AC/105/58 Annex I.

[21] Christol (n 19) 5.

[22] Nancy L Griffin, 'Americans and the Moon Treaty' (1981) 46 JAirL&Com 729, 729 (citations omitted).

the non-space-faring States who wanted to assure themselves a share of the 'fruits of space exploration'.[23]

Perhaps as a result, in 1970, Argentina submitted to the LSC a 'draft agreement on the principles governing activities in the use of natural resources of the [M]oon and other celestial bodies'.[24] It is here for the first time that it is suggested that the 'natural resources of the Moon and other celestial bodies shall be the common heritage of all [hu]mankind'.[25] The short proposal goes on to clarify that: the term 'natural resources' shall include '[a]ll substances originating in the Moon or other celestial bodies';[26] natural resources used *in situ* will be treated differently from those brought back to Earth for use;[27] and that benefits accrued from the use of the natural resources shall be made available to all, taking into account the needs of developing countries as well as the 'rights of those undertaking these activities'.[28] The 1970 meeting of the LSC was focused on drafting of the Liability Convention and consideration of Argentina's draft was deferred until 1971.[29]

Then, on 27 May 1971, the Soviet Union transmitted a draft agreement directly to the United Nations General Assembly (UNGA) for consideration at the upcoming session.[30] The 'Draft Treaty Concerning the Moon' contains 15 articles.[31] The following are noteworthy points:

- The draft purports to only cover the Moon, recognizing that it is 'the only natural satellite of the Earth [with] an important role to play in the conquest of outer space'.[32]
- It reiterates many foundational points from the Outer Space Treaty including, among other things, that international law shall apply to activities conducted on the Moon or in cislunar space;[33] the Moon shall be used exclusively for peaceful purposes;[34] activities shall be conducted with principles of due regard, cooperation, and freedom of exploration;[35] retention of ownership of

[23] Bin Cheng, 'The Moon Treaty: Agreement Governing the Activities of States on the Moon and Other Celestial Bodies within the Solar System Other than the Earth, December 18, 1979' (1981) AnnChinSoc'yIntlL 1, 3.

[24] UNCOPUOS, 'Argentina: Draft Agreement on the Principles Governing Activities in the Use of Natural Resources of the Moon and Other Celestial Bodies' UN Doc A/AC.105/C.2/L.71 in UNCOPUOS, 'Report of the Legal-Subcommittee on the Work of its Ninth Session' (8 June–3 July 1970) UN Doc A/AC.105/85, Annex II.

[25] ibid art 1.

[26] ibid art 2.

[27] ibid art 3.

[28] ibid arts 4 and 5.

[29] Christol (n 19) 8.

[30] 'Union of Soviet Socialist Republics: Request for the inclusion of an item in the provisional agenda of the twenty-sixth session' UN Doc A/8391* (4 June 1971).

[31] UNGA, 'USSR: Draft Treaty Concerning the Moon' UN Doc A/C.1/L.568 (5 November 1971).

[32] ibid Preamble.

[33] ibid art I.

[34] ibid art II.

[35] ibid arts III and V.

space objects and liability for damage caused thereby;[36] and the prevention of adverse changes in the lunar environment.[37]

- With respect to the use of resources, the draft does not label the resources as the common heritage of humankind, but indicates that '[n]either States, international intergovernmental or nongovernmental organization and national organizations having the status of juridical persons or not, nor natural persons, may claim the surface or subsoil of the Moon as their property'.[38] Also '[p]ortions of the surface or subsoil of the Moon may not be the object of concession, exchange, transfer, sale or purchase, lease, hire, gift or any other arrangements or transactions with or without compensation between States, international intergovernmental and non-governmental organizations or national organizations having the status of juridical persons or not, or of arrangements or transactions between natural persons'.[39]

There were machinations occurring behind the scenes as well. A 'Memorandum for the Record' issued by the US National Reconnaissance Office indicates that a bilateral meeting was held between the US and the Soviet Union in which the 'Soviets were soliciting US support on their proposed Moon Treaty'.[40] The Memo notes that the US considered this proposed treaty to be 'redundant in many regards to the [Outer Space Treaty]' and thus that the Soviet Union had proposed it 'as a time consumer'. With this understanding, the US supported the proposal.[41]

Thus, with the US and Soviet Union in agreement, the UNGA, '[t]aking into account the advances made in recent years in the exploration of outer space, including those resulting from extensive lunar research programmes on the basis of modern science and technology', formally requested the COPUOS 'to consider, as a matter of priority, the question of the elaboration of a draft international treaty concerning the Moon'.[42]

At the 1972 meeting of the LSC, Agenda Item 3 was labelled simply 'Questions relating to the [M]oon'.[43] A Working Group was established which had, as its starting

[36] ibid arts IX and XI.

[37] ibid art IV.

[38] ibid art VIII.

[39] ibid.

[40] Harold S Coyle, Jr, 'Memorandum For the Record, 1 Sept. 1971' <www.nro.gov/Portals/65/documents/foia/declass/Archive/NARP/1971%20NARPs/SC-2021-00001_C05134201.pdf> accessed 15 July 2023.

[41] ibid. See Stefan-Michael Wedenig and Jack Wright Nelson, 'The Moon Agreement: Hanging by a Thread?' (*Institute of Air & Space Law*, 26 January 2023) <www.mcgill.ca/iasl/article/moon-agreement-hanging-thread> accessed 15 July 2023 (suggesting that this memorandum indicates a desire to delay COPUOS action that may have impeded US or Soviet space activities and thus demonstrates a violation of the good faith principle universally recognized as indicated by the Vienna Convention on the Law of Treaties).

[42] UNGA Res 2779 (XXVI) (29 November 1971).

[43] UNCOPUOS, 'Report of the Legal Subcommittee on the Work of its Eleventh Session' (10 April–5 May 1972) UN Doc A/AC.105/101 (11 May 1972).

point, the proposals submitted by Argentina and the Soviet Union, as well as 18 working papers submitted by the United States. The United Kingdom, Sweden, Australia, Belgium, France, Italy, Egypt, India, and Bulgaria also submitted papers, individually and in some cases in concert with one another.[44] A draft treaty including 21 provisions was quickly formed, though much of the language remained bracketed, an indication that consensus had not been reached on the exact verbiage. Chief among the issues of debate were: the characterization of natural resources as the common heritage of humankind; the scope of the agreement, that is, whether it should just apply to the Moon or to other celestial bodies as well; and the timeliness of sharing information regarding missions to the Moon.[45]

The Soviet Union in particular was stubborn in its opposition to extending the Treaty to celestial bodies other than the Moon and also to the concept of 'common heritage of humankind', which the Soviet Union found troubling in its lack of definition.[46] Negotiations over the new treaty, including the meaning of the principle of common heritage were the backdrop of seven years of annual LSC meetings.

Bin Cheng aptly describes the next phase of this ugly duckling's evolution as '[t]he mystery and miracle of birth'.[47] With no movement towards agreement after seven years of formal negotiation, an informal working group was established to consider the Moon Agreement draft. This Working Group held four meetings between 26 June and 3 July 1979.[48] As Bin Cheng reports, these meetings were not published, and 'the public is in fact debarred from knowing what happened then'.[49] However, it is reported that language was agreed 'after informal consultations among members on the main outstanding issue', namely the characterization of natural resources on the Moon as the common heritage of humankind.[50]

The compromise reached is found in Article 11(1) of the Moon Agreement, which states that 'the [M]oon and its natural resources are the common heritage of [hu]mankind, which finds its expression in the provisions of this agreement and, in particular, in paragraph 5 of this article'.[51] With this, and several clarifications, the COPUOS agreed that its work on the Agreement was completed and it submitted the draft to the UNGA for consideration.[52]

44 ibid. For an excellent summary of their positions, see Christol (n 19) 10.

45 See Christol (n 19) 10–12; David Everett Marko, 'A Kinder Gentler Moon Treaty: A Critical Review of the Current Moon Treaty and a Proposed Alternative' (1993) 8 JNatResources&EnvtL 30; Cheng (n 23).

46 See Christol (n 19) 10–12; Marko (n 45) 300.

47 Cheng (n 23) 4.

48 UNGA, 'Report of the Committee on the Peaceful Uses of Outer Space' UNGAOR 34th Session Supp No 20, UN Doc A/34/20 (1979) ¶ 57. See ibid.

49 Cheng (n 23) 5.

50 UNGA (n 48) ¶ 60.

51 ibid.

52 ibid ¶ 66.

The Moon Agreement received unanimous approval without a vote by the UNGA on 14 December 1979.[53] Yet it took nearly five years to enter into force and, as of July 2023, has only 18 ratifications, with one withdrawal scheduled for January 2024. The following article-by-article analysis will suggest that the Moon Agreement was well ahead of its time and created unspecified obligations related to the untested concept of common heritage of humankind. This ultimately made it difficult for States to clearly explain to its constituents what obligations ratification would trigger.

3 A Twenty-First-Century Analysis of the Ugly Duckling

3.1 Purpose and Scope

The purpose of the Moon Agreement, as set forth in its Preamble, focuses, as initially proposed by the Soviet Union, on the important role the Moon has to play in the exploration of outer space. This is an undeniable fact. The Moon is the Earth's closest neighbour and essentially a proving ground for human activities in space. We will use the Moon to learn how to live in low gravity conditions, test techniques for building shelters out of *in situ* materials, and ultimately mine and reduce water ice to hydrogen to fuel further exploration to Mars and beyond. That the Preamble also notes 'the achievements of States in the exploration and use of the Moon and other celestial bodies'[54] is a testament to the feeling, prevailing at the time, that the ability to properly harness space resources for exploration and even for commerce was not far away.

The Preamble also indicates a determination 'to promote on the basis of equality the further development of cooperation among States in the exploration and use of the Moon and other celestial bodies',[55] a reaffirmation of the Outer Space Treaty's statement, found in its Article I, that the Moon and other celestial bodies 'shall be free for exploration and use by all States . . . on a basis of equality'.[56]

Not surprisingly, the Preamble also echoes the desire 'to prevent the Moon from becoming an area of international conflict', a parallel to which can be found in the Outer Space Treaty's mandate that the Moon and other celestial bodies shall be used 'exclusively for peaceful purposes'.[57]

[53] UNGA Res 34/68 (5 December 1979).
[54] Moon Agreement (n 1) Preamble.
[55] ibid.
[56] Outer Space Treaty (n 7) art I.
[57] Outer Space Treaty (n 7) art IV.

The provisions of the Outer Space Treaty reference 'outer space' multiple times without offering a definition of the term, thereby suggesting that the Outer Space Treaty applies at all times when activities are not occurring on Earth or in its environment, wherever they may be otherwise. In the development of the Outer Space Treaty, there was a natural tendency to create an all-encompassing agreement which was magnified by the fact that the international community was, at the same time, codifying the law of the sea. While it has become popular to try and create parallels between maritime law and space law, this is a dangerous and indolent error. Certainly, both are domains that the international community has deemed beyond national sovereignty, but there the similarities end. The high seas are a finite domain teeming with unique marine life and intertwined in ways we are just beginning to understand with the human ecosystem and the health of our Earth. The Convention on the Law of the Sea itself was built—not hastily in the wake of the first sailboat to circumnavigate the world—but after centuries of human activity had created tradition and custom to be detailed and codified.

The scope of the Moon Agreement is not narrowed to the Moon itself, but it does, wisely, abandon the hubris of the Outer Space Treaty which purports to cover all activities of humans wherever they may be in outer space. Again, this is not surprising. The Outer Space Treaty is an aspirational document offering guidelines and principles. The Moon Agreement is an attempt to build a more durable framework, probably with the Convention on the Law of the Sea as a model. With this in mind, the drafters agreed that its terms should apply to the Moon and all 'other celestial bodies within the solar system, other than the Earth, except insofar as specific legal norms enter into force with respect to any of these celestial bodies'.[58] This narrowing of scope both allows for greater flexibility for future endeavours and acknowledges that we cannot begin to know yet what lies beyond or how we might use what we find there. It is an important notion that we must keep in mind as we continue to evolve space law.

In recognition of humanity's growing understanding of space and the myriad ways we can harness its resources, the Agreement also clarifies that the term 'Moon shall include orbits around or other trajectories to or around it'.[59] Interestingly, it is clear that it 'does not apply to extraterrestrial materials which reach the surface of the Earth by natural means'. While this is certainly a common-sensical position, it is fundamentally ironic. The Moon Agreement seems to suggest that if you invest funding and labour to bring a natural resource back to Earth, your enjoyment of that resource will be regulated. However, if you come upon a natural resource by happenstance on Earth, you may, presumably, sell it to the highest bidder with no question of how benefits might be shared. Of course, the Agreement counts on the

[58] Moon Agreement (n 1) art 1.
[59] ibid.

fact that any natural resource that reaches the Earth's surface will be sufficiently small as to be negligible, but the dichotomy still exists.

3.2 Mirroring the Outer Space Treaty

Article 2 of the Moon Agreement takes much of Article III of the Outer Space Treaty verbatim,[60] adding emphasis on the concept of cooperation by highlighting 'the Declaration on Principles of International Law concerning Friendly Relations and Cooperation among States in accordance with the Charter of the United Nations, adopted by the General Assembly on 24 October 1970', and reinforcing the concept of due regard encapsulated in Article IX of the Outer Space Treaty. Operationally these changes do not change the general thrust of the provision but are helpful for interpretive purposes.

Article 3 confirms and expands on the idea that the Moon and other celestial bodies shall be used 'exclusively for peaceful purposes'. A reminder here that, pursuant to the first Article of the Moon Agreement, any reference to the Moon includes all other celestial bodies in the solar system. Article 3 somewhat superfluously bans the placement of nuclear weapons or weapons of mass destruction around the Moon, as these are banned in outer space as a whole by the Outer Space Treaty. Nevertheless, this provision adds more specifics, banning the threat of force or any hostile act, the establishment of military bases and the use of the Moon as a platform from which to use or threaten the use of force.

Article 4 also reiterates or attempts to strengthen many of the concepts in the Outer Space Treaty. It affirms Article I of the Outer Space Treaty, in proclaiming that 'exploration and use of the Moon shall be the province of all [hu]mankind and shall be carried out for the benefit and in the interests of all countries, irrespective of their degree of economic or scientific development'.[61] The concepts of cooperation and mutual assistance are also reinforced, with an additional nod to sustainability that does not exist in the Outer Space Treaty. The Moon Agreement encourages parties to pay due regard 'to the interests of present and future generations as well as to the need to promote higher standards of living and conditions of economic and social progress and development in accordance with the Charter of the United Nations'. This clause gives more purpose to the concept of 'shared benefits'; however, it is the introduction of the concept of intergenerational equity which heralds an important evolution. The Outer Space Treaty focuses on present activities and suggests guidelines, essentially, for the exploitation of space. This

[60] Art III of the Outer Space Treaty (n 7) states in full: 'States Parties to the Treaty shall carry on activities in the exploration and use of outer space, including the Moon and other celestial bodies, in accordance with international law, including the Charter of the United Nations, in the interest of maintaining international peace and security and promoting international cooperation and understanding.'

[61] Moon Agreement (n 1) art 4.

provision of the Moon Agreement suggests that with utilization comes responsibility and stewardship.

Article 5 of the Moon Agreement adds depth and breadth to the information provisions of Article XI of the Outer Space Treaty, which avers broadly that States Parties have agreed to inform the UN of their space activities. The text of the Moon Agreement provision is far more detailed, useful—and forceful.

First, States Parties 'shall inform . . . to the greatest extent feasible and practicable, of their activities concerned with the exploration and use of the Moon'. It is not just information. The Article requires data on 'the time, purposes, locations, orbital parameters and duration', of each mission to the Moon, to be shared 'as soon as possible'[62] with ongoing update requirements. While the Moon Agreement does not go so far as to create or suggest the implementation of safety zones, it heightens the requirement to consult should multiple parties find themselves in the same location. This was very forward-thinking, and again demonstrates the belief, at the time, that humanity would be active on the Moon within a short period of time.

Article 10 supports the return and rescue concept contained in Article V of the Outer Space Treaty and requires the parties to take all 'practicable' measures to save lives, including offering shelter. Interestingly, the drafters felt the need, in Article 10, to confirm that 'any person on the Moon [is] an astronaut within the meaning of Article V' of the Outer Space Treaty. Again, perhaps presciently, the Moon Agreement was looking forward to lunar tourism and the development of thriving communities on the lunar surface. Regardless, it is somewhat uplifting to see the concept of an astronaut, undefined in the Outer Space Treaty, broadened to include anyone on a celestial body. Of course, this leaves in a funny legal limbo the status of non-State personnel on human-made objects in space, leaving the rights of humans in space in general to be tackled by twenty-first-century jurists and policymakers.

Article 12 of the Moon Agreement carries forward the concept captured in Article VIII of the Outer Space Treaty that States retain jurisdiction and control over 'their personnel, vehicles, equipment, facilities, stations and installations on the Moon' and references again the return and rescue obligations as set forth in the ARRA.[63]

Finally, Article 13, taken together with Article 14, mirrors the liability provisions captured in Article VII of the Outer Space Treaty as buttressed by the Liability Convention. Article 14 itself parallels the authorization and supervision concepts in Article VI of the Outer Space Treaty almost verbatim:

> States Parties to this Agreement shall bear international responsibility for national activities on the Moon, whether such activities are carried on by governmental

[62] ibid art 5.

[63] ibid art 12.

> agencies or by non-governmental entities, and for assuring that national activities are carried out in conformity with the provisions set forth in this Agreement. States Parties shall ensure that non-governmental entities under their jurisdiction shall engage in activities on the Moon only under the authority and continuing supervision of the appropriate State Party.[64]

3.3 Adding Details not Covered in the Outer Space Treaty

Building on the principles and guidelines contained in the Outer Space Treaty, the Moon Agreement makes a valiant, if untimely, attempt to fill in details left open by the broad strokes of its precursor.

Article 6 not only asserts the freedom of scientific investigation; it also provides some details on what may be permitted. Among other things, Parties

> have the right to collect on and remove from the Moon samples of its mineral and other substances. Such samples shall remain at the disposal of those States Parties which caused them to be collected and may be used by them for scientific purposes . . . States Parties may in the course of scientific investigations also use mineral and other substances of the Moon in quantities appropriate for the support of their missions.[65]

Note that this provision allows for removal and use without question for scientific investigation and to support the missions of the States Parties. This provision can be read extremely permissively as, arguably, any manner of gaining knowledge about the resources on the Moon can be considered science. What is not so clear is how broadly the term 'scientific purpose' can be scoped. It is likely that the intent was to prohibit the commercial use—sale—of space resources but, certainly, arguments can be made that sales can also be for scientific purposes. For example, if a commercial company retrieves samples on behalf of a State Party so that the State may conduct scientific experiments, is that considered a scientific use? What if the State reimburses the company for the cost of the retrieval? How far would this logic extend? May the State in turn sell the resource to a university or academic institution so long as the material continues to be used for scientific purposes?

Article 7 is truly a forward-looking provision and one which this author fervently hopes is captured in future agreements. Here, it is recognized that the Moon and other celestial bodies have unique environments and that adverse changes to those environments ought to be avoided. This is far stronger and broader than the Outer Space Treaty admonition to avoid the 'harmful contamination' of celestial

[64] ibid art 14.
[65] ibid art 6.

bodies.[66] The third clause of Article 7 introduces the concept of designating areas as 'international scientific preserves' with 'special protective arrangements'. The key here is not that development on celestial bodies must be forestalled for environmental concerns but that any such development must be responsible and sustainable with proper respect for the resources. Here we see, as with Article 4, a maturing from the idea of boundless exploitation to the idea of responsible management. The notion of creating scientific preserves is especially laudatory as this would require international agreement and build a sense of common stewardship over certain areas of celestial bodies, allowing for more freedom and use elsewhere.

That freedom of exploration and use is amplified in Article 8, which specifically permits States to '[l]and their space objects on the Moon and launch them from the Moon'; and '[p]lace their personnel, space vehicles, equipment, facilities, stations and installations anywhere on or below the surface of the Moon' and move them freely.[67] The only restriction on this freedom is that such activities 'shall not interfere' with the activities of others. Article 9 supports the establishment of crewed and uncrewed installations on the Moon with the rather milquetoast admonitions that each station 'shall use only that area which is required for the needs of the station' and shall not impede free access to all areas of the Moon.[68] This Jekyll and Hyde approach is not helpful. States may place installations anywhere on the Moon but cannot bar others from accessing the area upon which an installation is situated. Thus, essentially, you can put whatever you want on the Moon, so long as you understand that everyone else will have access to that location. One can easily imagine an instance where multiple parties are attempting to mine a certain area of the Moon. This language will not prevent conflict, only exacerbate it. This is especially worrisome as we consider the effect of lunar dust on human activity. As this author has noted, research 'indicates that upon approach and landing, lunar lander engine exhaust will blow rocks, soil and dust at high velocities. This lander ejecta can severely damage hardware even tens of kilometres away from the landing site.'[69] Simply traversing an area near an installation can cause damage to that installation and any operational activities. The only certainty that the language of Articles 8 and 9 provides is that parties will no doubt resort to the Liability Convention to achieve recompense for any damage done to space objects at the contested location.

[66] Outer Space Treaty (n 7) art IX. This provision is largely read to protect biological discoveries on other celestial bodies and not the environment on those bodies.

[67] Moon Agreement (n 1) art 8.

[68] ibid art 9.

[69] Michelle LD Hanlon and Bailey Cunningham, 'The Legal Imperative to Mitigate the Plume Effect: An "Aggravation and Frustration" That Imperils Our History and Our Future' (2019) 43 JSpaceL 309.

3.4 Article 11—Common Heritage of [Hu]mankind

Some would say the biggest challenge to the implementation of the Moon Agreement is found in four little words found in Article 11(1): common heritage of [hu]mankind. A concept that divided the negotiators of the Moon Agreement from the start, the main objection to this term was its lack of definition. As L. Goldie puts it:

> One caution lawyers, diplomats and statesmen should observe is to avoid trying to treat layman's language as if it were formulated in terms of technical legal concepts . . . On the other hand, the phrase, common heritage of humankind, a layman's formula if ever there was one should be given the greatest respect. While it should not, indeed cannot be viewed as a prescription, it can be accepted as an important hortatory message, a kind of policy directive.[70]

The concept, as S Gorove put it succinctly, 'carries no clear judicial connotation but belongs to the real of politics, philosophy or morality and not law'.[71] In short, the implementation of the concept of common heritage of humankind is not a matter of course, it must be negotiated and enumerated. Unfortunately, the words have taken on a political context that cast them in a particularly disfavourable light for many.

Thus, the biggest challenges for the implementation of the Moon Agreement do not lie in a fundamental disagreement over a concept—they rest in 1) the unfortunate evolution of the political meaning attached to the words 'common heritage of humankind', and 2) the comprehensive uncertainty that the Agreement propagates by not immediately enumerating a definition for purposes of the Agreement.

As noted, the history of the Agreement shows that it was rushed to completion in part to ensure and reaffirm the continued peaceful use and non-militarization of space. It was also drafted at a time when humans were still taking tentative steps into space—there was not yet a clear understanding of the technologies needed (or what they would cost). Clearly, the drafters were quite happy to take a wait-and-see approach with respect to what a common human heritage of outer space regime might look like, and this in and of itself might have been fatal.

[70] L Goldie, 'A General International Law Doctrine for Seabed Regimes' (1973) 7 Int'l Lawyer 796, 819.

[71] Stephen Gorove, 'The Concept of Common Heritage of Mankind a Political, Moral or Legal Innovation' (1972) 9 SanDiegoLR 390, 402. See Rudolph Preston Arnold, 'The Common Heritage of Mankind as a Legal Concept' (1975) 9 IntlLaw 153.

It is generally agreed that there are five elements in the common heritage of humankind:

1. The common heritage of humankind is not subject to appropriation
2. All countries share in the management of the common heritage of humankind
3. The benefits derived from the exploitation of common heritage of humankind resources must be shared with all regardless of participation
4. The common heritage of humankind must be used for peaceful purposes; and
5. The common heritage of humankind must be reserved for future generations.[72]

The Outer Space Treaty already covers the first and fourth points—outer space is not subject to national appropriation and must be used for peaceful purposes. One can hardly think any nation would disagree with the need to use the common heritage of humankind sustainably so as to reserve it for future generations. Thus, it is points two and three that generate the greatest concern.

At first glance, it appears that to implement the concept of the common heritage of humankind, an international body must be created to redistribute wealth and technology among nations. Indeed, in response to implementing provisions in the Law of the Sea Treaty, President Ronald Reagan criticized the concept of international management, stating that 'no national interest of ours could justify hanging sovereign control over two-thirds of the earth's surface over to the Third World'.[73] As to the sharing of benefits, Reagan was definitely set against what he called another 'free ride' at the expense of the US.

Reagan cast the Law of the Sea Treaty as being intentionally designed to promote a new world order—a form of global collectivism—that seeks ultimately the redistribution of the world's wealth through a complex system of manipulative central economic planning and bureaucratic coercion. He blamed this on what he called the distorted interpretation of the noble concept of the Earth's vast oceans as the common heritage of humankind. But note, President Reagan did not suggest that the oceans are *not* the common heritage of humankind, he said the Law of the Sea Treaty had distorted the interpretation of that concept.[74]

The Moon Agreement was collateral damage in this distortion. It could have saved itself—and it tried to do so with Article 11. Paragraph 1 of Article 11 is possibly the most important—not because it calls the Moon and its natural resources the common heritage of humankind, but because it specifically states that the definition, the expression, the implementation of the concept of common heritage

[72] Aline Jaeckel, 'Benefitting from the Common Heritage of Humankind: Expectation to Reality' (2020) 35(4) IJMCL 660.

[73] William P Clark, 'Reagan and the Law of the Sea' (The Heritage Foundation) <www.heritage.org/global-politics/commentary/reagan-and-the-law-the-sea> accessed 15 July 2023.

[74] ibid.

will be determined pursuant to paragraph 5.[75] The Agreement itself recognizes the lack of legal precepts regarding common heritage.

Paragraph 5 states that the States Parties will undertake to establish an international regime to govern the exploitation of natural resources as such exploitation is about to become feasible.[76]

Based on the wording of paragraph 5, some argue that it prevents the exploitation of resources until the regime is in place. Others concede that they may exploit the natural resources of space before a regime is put in place, but fear that they will suffer financial harm after the fact, particularly if the regime is able to see their success before determining the rules. But really, the worst flaw is the uncertainty gap. A gap into which leapt all of the angst and anger about the Law of the Sea regime.

Paragraph 7 outlines the main purposes of this proposed regime to include:

a) the orderly and safe development of the natural resources of the Moon;
b) the rational management of those resources;
c) the expansion of opportunities in the use of those resources; and
d) an equitable sharing by all States Parties in the benefits derived from those resources, whereby the interests and the needs of the developing countries, as well as the efforts of those countries which have contributed either directly or indirectly to the exploration of the Moon shall be given consideration.[77]

Clearly, as we move towards gaining the capability to mine resources in space, we need to have (a) through (c). While the fourth point calls for 'equitable sharing', it does not elaborate further on that point. There remains no defined meaning of equitable sharing.

That said, there is an interesting ambiguity here as it seems to suggest that the resources of space are not the common heritage of humankind, but rather the common heritage of States Parties. Surely, not being a State Party does not disqualify you from being part of humankind.

Finally, a source of controversy in Article 11 is paragraph 3, which states that neither the Moon nor any part thereof or natural resources in place shall become the property of any State, international intergovernmental or non-governmental organization, national organization or non-governmental entity, or of any natural person.[78]

Importantly, the agreement references 'resources in place' thus suggesting that an entity may have a property interest in the resources they extract. So, arguably, the

[75] Moon Agreement (n 1) art 11(1).
[76] ibid art 11(5).
[77] ibid art 11(7).
[78] ibid art 11(3).

Moon Agreement allows for private ownership of natural resources that have been extracted and simply requires the establishment of a governing regime to manage the extraction of space resources which may or may not impose prohibitive fines after a company has already undertaken to mine; and will certainly attempt to implement some sort of sharing, though how this will be shared remains indeterminate. In short, the Treaty will not implement necessary laws until mining is feasible—yet the very structure of the Treaty, and the uncertainty surrounding the establishment of this regime, discourages the research and investment necessary to make it feasible.

It should be noted that it is in the best interests of all parties—whether a developing or a developed nation—to implement a fair and supportive system. Because, indeed, the whole world will benefit from space resource utilization whether directly or indirectly. Consider the use of satellites to image the Earth. The first such satellites deployed were for strategic military purposes. Over time, it became understood that data from satellites could provide significantly more information and services. Today nearly every human on Earth directly benefits from satellites that provide global positioning, communications and environmental and natural resource imaging information. The resources provided by human objects in orbital positions have truly been shared by all.

It is not difficult to predict how space resource utilization will benefit all people on Earth. In the short term, extracting water from the Moon to support human activities will finally allow humans to conduct valuable research on the Moon on a long-term basis. This research will provide more information about our universe, and our role in it. Ultimately, learning to mine space resources could allow us to move the heavy industry of our Earth and return our terrestrial environment to pre-industrial conditions. However, the investment required to develop the tools and understanding necessary to utilize space resources is prohibitive in and of itself. Adding additional costs could unduly delay, if not fully impede, the realization of any benefits.

That said, we also clearly need some sort of regime in place. Space exploration and utilization activities need assurances of safety and stability. They need to afford a certain level of legal security. Those who are involved in space activities need to know the cost of that legal security—the complete level of their financial investment, including any licences and fees.

3.5 Enforcement and Administrative Provisions—Article 15

The final provisions of the Moon Agreement are largely administrative. Article 15 does add an extra transparency provision, allowing for each State to inspect the installations and equipment of other States.[79] There also exists a review provision which requires that review of the Treaty be included in the UNGA Agenda ten

[79] ibid art 15.

years after entry into force of the Agreement[80] and allows for one-third of the States Parties to convene a conference any time after the Agreement has been in force for five years. No review or conference has been conducted or convened.

4 So, Where's the Swan?

4.1 Current Status of the Moon Agreement

As of July 2023, the Moon Agreement had 18 ratifications: Armenia, Australia, Austria, Belgium, Chile, Kazakhstan, Kuwait, Lebanon, Mexico, Morocco, the Netherlands, Pakistan, Peru, Philippines, Saudi Arabia, Türkiye, Uruguay and Venezuela, and four signatories: France, Guatemala, India and Romania.[81] It is interesting to note that Argentina, which first formally introduced the idea of developing the Moon Agreement and consequently suggested that celestial resources be characterized as the common heritage of humankind, is not a signatory. Neither is the United States, Russia or China.

In January 2023, Saudi Arabia notified the UN of its intent to withdraw from the Agreement as of January 2024. They did not give a reason for their withdrawal.

Even so, the Moon Agreement is certainly not 'dead'. The LSC of the UNCOPUOS continues to maintain an agenda item on the 'status and application of the five United Nations treaties on outer space'. Of course, some delegates, including the delegate of the US, refer not to five but to 'four core treaties governing the use of outer space'.[82] Nevertheless, the Moon Agreement continues to be discussed. In March 2023, Austria and Belgium included in their statements that they 'would like to encourage States' to become parties to the Agreement.[83] Also in March 2023, Kenya informed the LSC that it was considering ratifying the Agreement.[84]

[80] ibid art 18.

[81] UNCOPUOS, 'Status of the International Agreements relating to Activities in Outer Space as at 1 January 2022' Sixty-first Session of the Legal Subcommittee (28 March–8 April 2022) UN Doc A/AC.105/C.2/2022/CRP.10* (28 March 2022).

[82] 'Statement by Emily Pierce, U.S. Representative to the Legal Subcommittee of the UN Committee on the Peaceful Uses of Outer Space Agenda Item 5—Status of the International Treaties Governing the Use of Outer Space March 20, 2023' Sixty-second Session of the Legal Subcommittee of the Committee on the Peaceful Uses of Outer Space <www.unoosa.org/documents/pdf/copuos/lsc/2023/Statements/20_PM/05_USA_20_March_PM.pdf> accessed 15 July 2023.

[83] 'Statement by Mr. Michael Friedl, Agenda item 10, General Exchange of Views on potential legal Models for Activities in the Exploration, Exploitation and Utilization of Space Resources' Sixty-second Session of the Legal Subcommittee of the Committee on the Peaceful Uses of Outer Space (20–31 March 2023) <www.unoosa.org/documents/pdf/copuos/lsc/2023/Statements/20_PM/10_Austria.pdf> accessed 15 July 2023; Intervention de la Belgique au titre du point 10 de l'agenda du sous-comité juridique de l'UNCOPUOS, Sixty-second Session of the Legal Subcommittee of the Committee on the Peaceful Uses of Outer Space (20–31 March 2023) <www.unoosa.org/documents/pdf/copuos/lsc/2023/Statements/21_AM/10_Belgium_20_March_PM_1.pdf> accessed 15 July 2023.

[84] 'Statement By Her Excellency Ambassador Marymugwanja, Permanent Representative Of Kenya' Sixty-second Session of the Legal Subcommittee of the Committee on the Peaceful Uses of Outer Space

Indeed, as humanity truly does seem to finally possess the technical capability to mine the Moon and other celestial bodies, the need for a regulatory framework seems as essential and inevitable as it must have felt for the Moon Agreement negotiators in the 1970s. Only this time around, the international community is not putting all of its proverbial eggs solely in the nest of the COPUOS. On 13 October 2020, Australia, Canada, Italy, Japan, Luxembourg, the United Arab Emirates, the United Kingdom and the US announced the Artemis Accords, not a binding treaty, but a 'political commitment' to guide the exploration and use of the Moon and other celestial bodies.[85] As of September 2023, the Artemis Accords boast 29 signatories including Argentina, which drafted the first iteration of the Moon Agreement. Saudi Arabia made the decision to withdraw from the Moon Agreement when it joined the Artemis Accords. However, Mexico and Australia, who are also parties to the Moon Agreement, and France and India, who are signatories to the Moon Agreement, have also signed the Accords. While many have hailed this as the death of the Moon Agreement, quite the contrary is true. So long as Mexico and Australia remain parties to the Moon Agreement, and as of this writing there is no indication of any intent to withdraw, these two States at least are bound by the obligations of the Moon Agreement. Similarly, Venezuela, also a party to the Moon Agreement, has signed on to China's International Lunar Research Station project.[86] While that project has yet to publish any principles or agreements, again Venezuela too remains bound by the obligations of the Moon Agreement. Right there, we have a commonality between two competing missions with the mining of lunar and other space resources at their sites.

Layered on top of this is the establishment, in 2021, of the UNCOPUOS Working Group on Legal Aspects of Space Resource Activities.[87] The five-year remit of the Group does not include the negotiation of a new treaty, but the '[f]inalization of a set of initial recommended principles for [space resource mining] activities for the consideration and consensus agreement by the COPUOS'.[88] States, including Mexico, Austria and Belgium, have affirmatively alluded to the Moon Agreement in their submissions to this Group. Again, so long as they are bound by the Moon

(20–31 March 2023) <www.unoosa.org/documents/pdf/copuos/lsc/2023/Statements/20_AM/3_Kenya_20_March_AM.pdf> accessed 15 July 2023.

[85] Artemis Accords: Principles for Cooperation in the Civil Exploration and Use of The Moon, Mars, Comets, and Asteroids for Peaceful Purposes (adopted 13 October 2020) <www.nasa.gov/specials/artemis-accords/img/Artemis-Accords-signed-13Oct2020.pdf> accessed 15 July 2023 (Artemis Accords).

[86] Andrew Jones, 'Venezuela Signs up to China's Moon Base Initiative' (*Space News*, 18 July 2023) <https://spacenews.com/venezuela-signs-up-to-chinas-moon-base-initiative/> accessed 15 July 2023.

[87] UNGA, 'Report of the Committee of the Peaceful Uses of Outer Space Sixtieth Session (7–16 June 2017)' UNGAOR 72nd Session Supp No 20, UN Doc A/72/20E (2017) ¶ 208.

[88] UNCOPUOS, 'Report of the Legal Subcommittee on its Sixty-first Session' Sixty-fifth Session of the Committee on the Peaceful Uses of Outer Space (Vienna, 1–10 June 2022) UN Doc A/AC.105/1260, Annex II Appendix.

Agreement, they cannot act in opposition to its basic tenets. Since the concept of common heritage of humankind is a political and moral one, and because it has no legal definition, States bound by the Moon Agreement do have the luxury of some interpretive breadth. That said, it is important to note that the current widely accepted characterization of the exploration and use of space is as the 'province of all humankind', something less than the common heritage of humankind, though neither term possesses a legal definition. As far as the natural resources of space themselves are concerned, there remains no accepted characterization.[89]

4.2 A Swan Song

While it is not a correct characterization to say that the ugly duckling Moon Agreement was close to death, it is fair to say that it was not popularly nurtured. Nevertheless, there remains stalwart support for the Agreement—just as there are many admirable concepts captured, though hidden by the controversy over the common heritage of humankind term. The Moon Agreement has suffered a long hard winter—more than 44 years in the cold! As the international community turns its attention again to the development of a framework to regulate space resource activities, whether through the UNCOPUOS, the Artemis Accords or the Chinese-led International Lunar Research Station, will the lake reflect a swan?

Yes. The Moon Agreement is an idealistic and hopeful document that seeks to elevate the human experience and motivate a change in the perception of the natural world around us, here on Earth and in the cosmos. It was the first step in our turning point to understanding the need to conserve and act sustainably, rather than simply exploit with abandon. We need to heed the lessons of our history, and the Moon Agreement tries valiantly to do so.

While arguments can be had over how natural resource benefits may be equitably shared, the Agreement tries to find a balance between freedom and utility. It does not ban space resource mining—quite the opposite in fact. The negotiators were simply prescient enough to realize that one cannot implement an international regime without knowledge of what space resource utilization would actually entail and thus created a mechanism to put an international regime in place once more was understood. Who would argue that the use of space should not benefit all of humankind? Utilizing resources in space may ultimately be the only way we have to return our Earth to its pristine pre-industrial state—would that not be a benefit to all humankind? The fact is that space already offers myriad benefits to all humankind from global positioning satellites to global communications to natural disaster monitoring. It is the close-minded and petty who think of 'benefit' only

[89] Art II of the Outer Space Treaty (n 7) bars a State Party from claiming territory in outer space but is silent as to the use of resources.

from a monetary standpoint. Who would argue that a child, no matter where born, should not have the opportunity to explore and utilize space?

The flaw in the beauty of this swan lies in its failure to address the vast difference between the space domain and any domain we have here on Earth. Borrowing Earth terminology at a time when we little understood what we were labelling was bound to be unsuccessful. Using a term without agreed definitions was a failure, albeit an unsurprising one.

For many, the very concept of an international regime governing space resource utilization is anathema. It is to this author as well. However, this is also a narrow-minded view. Granted, our current experiences with international regimes are filled with stories of corruption, bureaucratic morass, and recalcitrant administrators. But surely, an international regime may be developed with a light touch, over time rather than in speedy haste. Ultimately, we have to accept that the regime that works best for activities that occur beyond our Earth will be something that we have never developed before. It might not even be State-based.

So, in this manner, the Moon Agreement is an incredible swan. One we must allow to take a last elegant flight. The legacy of the Agreement is too entwined with the concept of the common heritage of humankind, and trying to unravel those threads will only send us back in time. We need to look forward. Now we have a very different understanding of the cosmos and our place in it. We have the benefit of nearly five decades of scientific study so that we better comprehend the challenges of accessing, mining and utilizing space resources. We have multiple stakeholders bringing different ideas and innovations to the negotiating table. More importantly, we have 17 nations who remain in the Agreement who can advocate compellingly to have many of the provisions adopted in other agreements and accords. May we use the feathers to line the nest of humanity's future?

9

MOON AGREEMENT: KEY PROVISIONS

Fabio Tronchetti

1 Introduction

The 1979 Agreement on Governing the Activities of States on the Moon and Other Celestial Bodies,[1] also known as the Moon Agreement, is the fifth and final United Nations[2] international treaty regulating human activities in outer space.[3]

The Agreement was specifically designed to lay down rules governing operations on the Moon, and other celestial bodies, including the building of manned stations and the extraction and utilization of resources both for scientific and non-scientific (commercial) purposes.

Unlike the four treaties that preceded it, which have received a notable level of ratification, the Moon Agreement has been ratified by only 17 countries, not including any of the space-faring nations. While the reasons behind the lack of support for the Agreement will be explored later in this chapter, at this stage it suffices to say that among these reasons there is the fact that the Agreement embraces an approach to managing the commercial utilization of lunar resources, that is the doctrine of the 'common heritage of mankind', which is viewed by developed countries, especially the United States, as an outdated approach that goes against profitable lunar activities and undermines the involvement of private operators in lunar undertakings.

Due, in part, to this 'bad reputation', for several decades not much attention was dedicated to the Moon Agreement, also because space actors had little interest in returning to the Moon and carrying out operations therein. However, over the past 15 years or so, the situation has changed drastically; not only has the Moon returned to the forefront of space players' agenda but also subjects of public and

[1] Agreement Governing the Activities of States on the Moon and Other Celestial Bodies (adopted 18 December 1979, entered into force 11 July 1984) 1363 UNTS 3 (Moon Agreement).

[2] On the space law-making process in the United Nations, see Casandra Steer, 'Sources and Law-making Process relating to Space Activities' in Ram S Jakhu and Stephen Dempsey (eds), *Routledge Handbook of Space Law* (Routledge 2017) 3–25; Sergio Marchisio, 'The Evolutionary Stages of the Legal Subcommittee of the United Nations Committee on the Peaceful Uses of Outer Space (COPUOS)' 31(1) JSpaceL 219–42.

[3] For a general analysis of the UN space treaties, see Frans G von der Dunk, 'International Space Law' in Frans G von der Dunk and Fabio Tronchetti (eds), *Handbook of Space Law* (Edward Elgar 2015) 29–126.

Fabio Tronchetti, *Moon Agreement* In: *International Space Law in the New Space Era*. Edited by: Sandeepa Bhat B., Dilip Ukey, and Adithya Variath, Oxford University Press. © Fabio Tronchetti 2024. DOI: 10.1093/9780198909415.003.0009

private nature have expressed their intention to commercially utilize the Moon's resources and to build a permanent presence on the Moon's surface. Considering these plans, States have begun debating about the legal regime applicable to lunar activities, not only at the national and regional levels but also, most notably, at the international level through the establishment of a Working Group on the Legal Aspects of Space Resources Utilization within the United Nations Committee on the Peaceful Uses of Outer Space (UNCOPUOS). While some countries firmly oppose the Moon Agreement, it still seems worth investigating what role, if any, the Agreement can play in shaping the legal framework applicable to the exploration and use of celestial bodies and their resources. On one hand, this analysis is justified by the fact that the Moon Agreement is the only international treaty specifically regulating human activities on a celestial body. On the other hand, as States Parties to the Moon Agreement are taking part in the activities of the Working Group one should expect them to engage in promoting acceptance of the principles that the Agreement enshrines.

2 The Moon Agreement: Key Provisions and Controversial Issues

2.1 Preliminary Remarks

The Moon Agreement was intended to fill the gaps left open by the Outer Space Treaty with respect to the status and utilization of lunar resources as well as the conduct of human activities on a celestial body. Indeed, while the Outer Space Treaty clarified the status of celestial bodies, it left unaddressed that of the resources contained therein and the extent to which the right to explore and use outer space extended to celestial bodies. The Moon Agreement addressed these lacunae by setting forth provisions governing the conduct of activities on the Moon, the construction of facilities on the Moon's surface, and the use of resources for scientific and commercial purposes, the latter to be undertaken under the auspices of an international regime.[4]

2.2 The Outer Space Treaty

The 1967 Outer Space Treaty is the main international instrument regulating human activities in outer space; the Treaty, which has been ratified by over 100

[4] For a general description of the international legal framework applicable to lunar activities and the utilization of the resources contained therein, see Fabio Tronchetti, 'Legal Aspects of Space Resources Utilization' in Frans G von der Dunk and Fabio Tronchetti (eds), *Handbook of Space Law* (Edward Elgar 2015) 769–813; Stephan Hobe and Kuan-Wei Chen, 'Legal Status of Outer Space and Celestial Bodies' in Ram S Jakhu and Stephen Dempsey (eds), *Routledge Handbook of Space Law* (Routledge 2017) 25–41; Stephan Hobe, 'Adequacy of the Current Legal and Regulatory Framework relating to the Extraction and Appropriation of Natural Resources in Outer Space' (2006) 32(1) JSpaceL 115.

countries, represents the foundation of the international space law system.[5] The Outer Space Treaty clarifies the legal status of celestial bodies by providing that they cannot be appropriated by means of use, occupation, or any other means;[6] additionally, the Treaty confers upon States the right to freely access, explore, and use outer space, upon the condition that such exploration and use are for the benefit and in the interests of all countries and the province of all mankind.[7] Furthermore, the Treaty grants States the right to conduct scientific investigation of outer space. The fundamental questions that the Treaty does not address are: (a) whether the non-appropriative nature of celestial bodies also extends to the resources contained therein; (b) whether the right to explore and use outer space also covers the exploration and use of space resources; and (c) how the province of all mankind clause affects space resources' utilization activities.[8] Scholars have offered diverging views on how to answer these questions, ranging from those claiming that celestial bodies' resources cannot be appropriated to those arguing that the right to use outer space extends to space resources.[9] Regardless of this debate, the fact remains that the Outer Space Treaty does not give a clear answer to the legal status of celestial bodies' resources nor to the circumstances and rules according to which these resources might be used.

2.3 The Moon Agreement: Main Provisions

The Moon Agreement was drafted through the 1970s, finalized in 1979, and entered into force in 1984 upon deposit of the fifth instrument of ratification. As indicated in the introduction, no space-faring nation is a Party to it, including those States currently involved in lunar activities.

The Moon Agreement was developed with the purpose of creating the conditions for the peaceful, orderly, and fair development of lunar activities, with particular attention to the interests of less-developed States. To a large extent, the Agreement follows the provisions of the Outer Space Treaty; for example, it

[5] Treaty on Principles Governing the Activities of States in the Exploration and Use of Outer Space, Including the Moon and Other Celestial Bodies (adopted 27 January 1967, entered into force 10 October 1967) 610 UNTS 205 (Outer Space Treaty).

[6] ibid art II. For an analysis of art II, see Virgiliu Pop, 'Appropriation in Outer Space: The Relationship Between Land Ownership And Sovereignty on the Celestial Bodies' (2000) 16 SpacePoly 275; Ram Jakhu and Steven Freeland, 'Article II' in Stephan Hobe, Bernhard Schmidt-Tedd, and Kai-Uwe Schrogl (eds), *Cologne Commentary on Space Law*, vol I (Carl Heymanns 2009) 463.

[7] Outer Space Treaty (n 5) art I.

[8] On these issues, see Fabio Tronchetti, 'The Non-appropriation Principle as a Structural Norm of International Law' (2008) 33(3) Air&SpaceL 277; Philip de Man, *Exclusive Use in an Inclusive Environment* (Springer 2016) 73–136.

[9] For an analysis of these diverging views, see Fabio Tronchetti, 'Title IV—Space Resource Exploration and Utilization of the US Commercial Space Launch Competitiveness Act: A Legal and Political Assessment' (2016) 41(2) Air&SpaceL 143, 145–46.

reaffirms the non-appropriative nature of the Moon and other celestial bodies (as per Article II OST)[10] and reinforces it by making clear that the placement of personnel, space vehicles, facilities, stations, and installations on or below the surface or subsurface of the Moon does not create a right of ownership over the surface or subsurface of the Moon or any areas thereof.

The Moon Agreement also adds some important elements to concepts that were previously provided for in the Outer Space Treaty; for instance, the right of States to explore and use outer space, including the Moon and other celestial bodies, is complemented by the requirement that such exploration and use be carried out with due regard to the interests of present and future generations,[11] and to the necessity of preventing the disruption and contamination of the lunar environment.[12] The latter requirement is particularly relevant, as the Outer Space Treaty lacked dedicated clauses addressing the environmental impact of human activities in outer space.

The truly innovative part of the Moon Agreement is to be found in the provisions dealing with the utilization of resources. Accordingly, the Agreement distinguishes between two forms of use, respectively scientific and commercial (for the purpose of profit). In the context of scientific operations, States can: (a) land objects and personnel;[13] (b) establish manned and unmanned stations on the Moon's surface;[14] (c) collect and use samples of the Moon's minerals and other substances to sustain their scientific missions;[15] and (d) use the portion of the lunar surface necessary to carry out their operations and meet the needs of their stations without, however, altering the freedom to access of all Moon areas.[16] These provisions, expanding upon the freedom of scientific investigation laid down by Article I(3) of the Outer Space Treaty, are generally uncontested, even by those States which are not Parties to the Agreement.

Much more problematic appears to be the section of the Agreement dealing with the exploitation (commercial utilization) of celestial bodies' resources, namely Article 11. Indeed, Article 11(5) indicates that the exploitation of lunar resources is to be undertaken in accordance with an international regime that States Parties undertake to establish once such exploitation is about to become feasible; Article 11(7) lists the main purposes of the international regime, including: (a) the orderly and safe development of the natural resources of the Moon; (b) the rational management of those resources; (c) the expansion of opportunities in the use of those resources; (d) an equitable sharing by all States Parties in the benefits derived from

[10] Moon Agreement (n 1) art 11(2).
[11] See Moon Agreement (n 1) art 4(1).
[12] ibid art 7(1).
[13] ibid art 8(2).
[14] ibid art 9.
[15] ibid art 6(2).
[16] ibid art 9(1)–(2).

those resources, whereby the interests and needs of the developing countries, as well as the efforts of those countries which have contributed either directly or indirectly to the exploration of the Moon, shall be given special consideration. The need to establish an international regime derives from the wording of Article 11(1), which declares the Moon and its natural resources to be 'the common heritage of mankind'; generally speaking, under the concept of the 'common heritage of mankind', an international area, such as outer space, shall be managed collectively by all States, and activities therein shall be undertaken only according to pre-determined rules. In the next section we will analyse the common heritage of mankind of its implications for the lack of success of the Moon Agreement.

2.4 Analysis of the Moon Agreement and Assessing its Lack of Success

Despite its lack of success, the Moon Agreement possesses various positive features.[17] First, it is the only existing legal instrument which specifically tries to regulate the exploration, use, and exploitation of the Moon, other celestial bodies, and their natural resources. This motive alone renders the Agreement worthy of legal analysis. Secondly, since 1979, 17 States have become Parties to it;[18] therefore, it cannot be disregarded. Thirdly, the Agreement sets out practically useful provisions in relation to activities undertaken on a celestial body, such as those requiring the environmental preservation of the lunar environment and enabling the construction and operation of manned and unmanned stations.

However, when one compares the level of ratification of the Moon Agreement with that of the other four space treaties, its lack of success becomes evident. Among its reasons, one can certainly list timing; indeed, when the Agreement was open for ratification in 1979, States had lost interest in the Moon and there was little interest in joining a legal instrument imposing binding obligations on lunar activities. Besides this, the lack of fortune of the Agreement can be largely attributed to the provisions of Article 11, which declare the Moon and its natural resources to be the common heritage of mankind. Broadly speaking, the concept of the common heritage of mankind represents a modification of the traditional regime applicable to international areas, namely areas located beyond national jurisdiction.[19] Traditionally, international areas, like the high seas, while being open

[17] For an extensive analysis of the Moon Agreement, see Fabio Tronchetti, *The Exploitation of Natural Resources of the Moon and Other Celestial Bodies* (Brill 2009) 38 60; Frans G von der Dunk, 'The Moon Agreement and the Prospects of Commercial Exploitation of Lunar Resources' (2007) 31 AnnAir&SpaceL 91.

[18] There were previously 18 Parties to the Moon Agreement; Saudi Arabia, however, on 5 January 2023 announced its intention to withdraw from the Agreement on, even though the withdrawal will only become effective on 5 January 2024.

[19] On the concept of the 'common heritage of mankind', see Rudiger Wolfrum, 'The Principle of the Common Heritage of Mankind' (1983) 43 HJIL 312; Christopher C Joyner, 'Legal Implications

to free exploration and use by all States cannot be appropriated by any country; on the contrary, when an area is labelled as the 'common heritage of mankind', States can no longer explore and use it freely, but their activities must be undertaken in accordance with the rules of an international regime. This approach is based on the belief that when an area is considered the 'common heritage of mankind' it belongs to all countries and, consequently, all countries should share in the management and utilization of its resources to ensure equal opportunities and benefits to all, especially less-developed countries. One should always keep in mind that the doctrine of the common heritage of mankind is a product of the early 1970s, an era characterized by the decolonization process and the demand for a new economic order that would create better opportunities for less-developed countries.

From an international law perspective, the concept of the common heritage of mankind has only been incorporated in two treaties, the 1982 Law of the Sea Convention[20] and the 1979 Moon Agreement. Part XI of the Law of the Sea Convention declares the seabed and ocean floor beyond national jurisdiction as well as the resources contained therein to be the 'common heritage of mankind' (the seabed and its resources are labelled as 'the Area');[21] accordingly, activities in the Area must be undertaken in accordance with the principles of an international regime intended to protect the interests of all countries, particularly developing ones.[22] In this respect, these countries are entitled to receive benefits deriving from the utilization of the Area and other advantages, such as the transfer of technology from developed countries.[23] Activities within the Area are managed by an International Seabed Authority that has the power to authorize the mining of resources (through a system of licences) and the acquisition of ownership over them.[24] When Part XI of the Convention was opened for signature, developed countries, led by the United States, opposed it by claiming that it had a detrimental impact on mining operations and eliminated rewards for investing in deep-sea mining operations. After years of negotiations, in 1994 an Amendment to Part XI was agreed upon to soften its stringent requirements and render the common heritage of mankind idea more acceptable to developed States and private entrepreneurs. For example, the Amendment eliminated the requirement for a mandatory transfer of technology and modified the voting mechanism of the Authority (previously based on a one country, one vote system) to give more weight to the

of the Concept of the Common Heritage of Mankind' (1986) 35 ICLQ 376; Bradley Larschan and Bonnie Brennan, 'The Common Heritage of Mankind Principle in International Law' (1983) 21 ColumJTrannatlL 305.

[20] Convention on the Law of the Sea (adopted 10 December 1982, entered into force 16 November 1994) 1833 UNTS 397, vol 1833, 3; vol 1834, 3; vol 1835, 3, art 92 (Convention on Law of the Seas).

[21] ibid arts 136–37.

[22] ibid art 150.

[23] ibid arts 140 and 144.

[24] ibid Part XI ss 3–4.

States directly involved in seabed mining projects. Once the Amendment to Part XI was agreed upon, the majority of developed countries accepted it and, consequently, became Parties to the Law of the Sea Convention.[25] To this day, however, the United States has still refrained from doing so and has simply signed the Convention without ratifying it.

The Moon Agreement develops the common heritage of mankind idea to a much lesser extent than the Law of the Sea Convention;[26] indeed, unlike the Convention, not only does the Moon Agreement not establish an international authority, it does not even set out a legal regime to govern the exploitation of lunar resources (as indicated above, Article 11(5) requires Parties to establish it once the exploitation is about to become feasible). The Moon Agreement only gives an indication of the principles to be included in such a regime but, essentially, leaves Parties free to determine the content of those principles and the process of formation of the regime. Undoubtedly, the Agreement contains some controversial clauses; for example, Article 11(7), while indicating the 'equitable sharing' by all States Parties in the benefits derived from the natural resources of the Moon as one of the purposes of the regime, fails to clarify what these 'benefits' are and how they should be shared. Furthermore, the Moon Agreement leaves uncertainty as to whether resources might be appropriated once they are removed from their original location, as Article 11(3) states that 'nor natural resources *in place*, shall become the property of any State, international intergovernmental or non-governmental organization, national organization or non-governmental entity or of any natural person'. Importantly, however, the Agreement leaves Parties plenty of leeway to determine how, and in which forms, the common heritage of mankind should apply to the exploitation of celestial bodies' resources; this is an approach that is confirmed by the wording of Article 11(1), which specifies that the common heritage of mankind 'finds its expression in the provisions of this Agreement, in particular in paragraph 5 of this Article'.[27] This is a clear indication that precedents like the one provided by the Law of the Sea Convention should not be used to understand the meaning of the concept of the common heritage of mankind in the Moon Agreement.

[25] Agreement relating to the Implementation of Part XI of the United Nations Convention on the Law of the Sea of 10 December 1982 <www.un.org/depts/los/convention_agreements/texts/unclos/closindxAgree.htm> accessed 19 August 2023; Harminderpal Rana, 'The Common Heritage of Mankind and the Final Frontier: A Revaluation of Values Constituting the International Legal Regime for Outer Space Activities' (1994) 26 RutgersLJ 225; ED Brown, *The International Law of the Sea*, vol 1 (Dartmouth 1994) 277–325.

[26] Carl Q Christol, 'The Common Heritage of Mankind Provisions in the 1979 Agreement Governing the Activities of States on the Moon and Other Celestial Bodies' (1980) 14 IntlL 429; Vladimir Kopal, 'Outer Space as a Global Common' (1997) 40 ProceColloqOuterSpace 108.

[27] For the analysis of the legal meaning of art 11 of the Moon Agreement, see Stanley B Rosenfield, 'Article XI of the Draft Moon Agreement' (1980) 22 ProceColloqOuterSpace 209; Stephan Hobe, 'Common Heritage of Mankind—An Outdated Concept in International Space Law?' (1998) 41 ProceColloqOuterSpace 271; Ricky J Lee, 'Creating an International Regime for Property Rights Under the Moon Agreement' (1999) 42 ProceColloqOuterSpace 409; Kevin V Cook, 'The Discovery of Lunar Water: An Opportunity to Develop a Workable Moon Treaty' (1994) 11 GeoIntlEnvtlLRev 647.

Despite this, developed countries, led by the United States, have repeatedly opposed the Agreement by claiming that its provisions are anti-economic, discourage investment, and put unreasonable burdens on miners.[28] These countries claim that the clauses demanding the setting up of an international regime and the equitable sharing of benefits are counterproductive to the development of the private space mining industry and detrimental to technological innovation and entrepreneurial spirit. Overall, these countries are opposed to the fact that the common heritage of mankind alters the traditional freedom of international domains and substitutes it with international management that supersedes individual States. Such an approach is viewed as unrealistic in the present day. For these reasons, representatives of some of these countries argue that the Agreement should be thrown into the garbage of history. Ultimately, this is why these countries have refused to join the Moon Agreement and have often disregarded its role in governing celestial bodies' undertakings.

3 Recent Regulatory Developments

The past fifteen years or so have witnessed remarkable developments in relation to a possible return to the Moon and the utilization of the resources contained therein. Everything started around 2007–08, when several countries launched unmanned exploratory missions to the Moon to gain better knowledge of its composition and characteristics. This process culminated in the successful collection of samples of Moon rocks by China during the Chang'e-5 mission in 2020.[29]

Besides State activities, one of the most notable aspects of this new Moon era has been the emergence of private companies having as their primary target the recovery and commercial utilization of Moon (and other celestial bodies') resources. Considering the uncertainty surrounding the legality of these projects under the Outer Space Treaty, these companies pushed their respective governments to pass domestic laws that supported and protected their plans. Their lobbying was successful and, indeed, led to major regulatory developments in the form of enactment of domestic space resources utilization laws. This practice was initiated by the United States in 2015[30] and has been followed by Luxembourg in 2017[31] and most recently by the United Arab Emirates (UAE)[32] and Japan.[33] The essence of these laws is pretty similar, in the sense that they all recognize under domestic law

[28] See Lynn Fountain, 'Creating Momentum in Space: Ending the Paralysis Produced by the Common Heritage of Mankind Doctrine' (2003) 35 ConnLRev 1753.

[29] 'Chang'e-5: China's Moon Sample Return Mission' (*Planetary.org*) <www.planetary.org/space-missions/change-5> accessed 19 August 2023.

[30] US Commercial Space Launch Competitiveness Act, HR 2262, 114th Cong (2015) Title IV—Space resources exploration and utilization.

[31] Luxembourg Law 674 on the Exploration and Use of Space Resources of 20 July 2017.

[32] Federal Law No (12) of 2019 on the Regulation of the Space Sector, arts 14 and 18.

[33] Law No 83, Space Resources Act of 15 June 2021.

the right of private entities to collect, own, use, and sell celestial bodies' resources, and they establish (or at least set the foundation for the establishment of) the procedure for the authorization of private space mining activities.[34] Such a procedure is meant to meet the requirements of Article VI of the Outer Space Treaty which requires States to authorize and continuously supervise national space activities, including private ones. Importantly, the enactment of such laws has faced opposition from several countries, such as Germany, Russia, and Brazil; these countries have argued that the private use of space resources for the purpose of profit is inconsistent with the non-appropriative nature of outer space and the idea that the exploration and use of outer space be the province of all mankind. Furthermore, they have stated that, due to the international status of outer space, rules to govern the recovery and utilization of celestial bodies' resources can only be developed by the international community as a whole, not by States acting unilaterally.

Another interesting initiative relevant to the organization of space resources activities was undertaken under the auspices of the so-called Hague International Space Resources Governance Working Group. The Group was established in 2016 with the goal of assessing the need for a governance framework on space resources, and to elaborate a set of 'building blocks' that could potentially form part of a future governance framework. On 12 November 2019, the work on the building blocks was completed and their text was opened for consideration by States and other international organizations.[35] The building blocks are largely inspired by the Moon Agreement as they envision the setting up of an international framework having the task, inter alia, of promoting the safe, sustainable, and efficient utilization of space resources and of ensuring that States and international organizations responsible for space resources activities enable benefit-sharing through participation in space resources activities by all countries, particularly developing ones. Nevertheless, the building blocks seek to balance the original spirit of the Moon Agreement with recent developments and private interests, by recognizing the possibility for operators to acquire priority rights for the recovery and extraction of resources and that resource rights might be acquired through domestic legislation and bilateral/multilateral agreements.

The latter option leads us to a very recent development that has gained a lot of attention both in the press and legal literature, namely the Artemis Accords,[36] a

[34] Tronchetti (n 9); Philip De Man, 'Luxembourg Law on Space Resources Rests on Contentious Relationship with International Framework' (2017) KU Leuven Working Paper No 189; PJ Blount and Christian J Robinson, 'One Small Step: The Impact of the US Commercial Space Launch Competitiveness Act of 2015 on the Exploitation of Resources in Outer Space' (2016) 18(1) NorthCarolinaJL&Tech 160.

[35] UNCOPUOS, 'Building Blocks for the Development of an International Framework on Space Resource, submitted by the Netherlands and Luxembourg' Fifty-ninth Session of the Legal Subcommittee (23 March–3 April 2020) UN Doc A/AC.105/C.2/L.315.

[36] The text of the Artemis Accords, officially signed on 13 October 2020, is available at <www.nasa.gov/specials/artemis-accords/img/Artemis-Accords-signed-13Oct2020.pdf> accessed 19 August 2023.

US-led regulatory initiative aimed at supporting the realization of NASA's Artemis Program.[37] Despite being an American initiative, the Artemis Program has been opened to other willing countries; so far, 24 countries have signed the Artemis Accords, including, inter alia, Japan, the UK, France, the UAE, and Australia.

Both positive and controversial features of the Accords can be highlighted.[38] As to the former, first, the Accords promote the rule of law in the context of space mining operations; this is, per se, an aspect to be commended. Secondly, several of the Accords' principles reflect those of the Outer Space Treaty or provide examples of their concrete implementation.[39] Finally, the Accords include the written commitment by Signatories to participate in international discussions.

Despite these positive elements, there are also controversial sections of the Accords, particularly those dealing with matters not specifically regulated by the Outer Space Treaty. First, this applies to the Accords' approach towards space resources utilization; section 10 of the Accords states: 'The Signatories affirm that the extraction of space resources does not inherently constitute national appropriation under Article II of the Outer Space Treaty, and that contracts and other legal instruments relating to space resources should be consistent with that Treaty.' Evidently, the Artemis Accords thus fully embrace the American view on the utilization of space resources as elaborated in the 2015 Commercial Space Launch Competitiveness Act. As previously explained, the Outer Space Treaty never prohibits but also never explicitly recognizes the right of States (and indirectly of private entities) to engage in space resource utilization, particularly for the purpose of profit. Furthermore, several countries have opposed such a possibility, and some have openly voiced their discontent towards the enactment of national space resource utilization laws and the legality of such instruments.

A second controversial point of the Accords concerns the possibility for Signatories to establish 'safety zones' on a celestial body's surface as a means to avoid interferences with nominal operations.[40] Notably, these 'zones' should be limited in size, temporary in nature, and guarantee 'free access' to all Moon areas.[41]

[37] The Artemis Program aims at returning men to the Moon by 2024. For more information on the Program, see <www.nasa.gov/artemisprogram/> accessed 19 August 2023.

[38] For an analysis of the legal and political implications of the Accords, see Frans von der Dunk, 'The Artemis Accords and the Law: Is the Moon Back in Business?' (*TheBigq*, 2 June 2020) <www.thebigq.org/2020/06/02/the-artemis-accords-and-the-law-is-the-moon-back-in-business> accessed 19 August 2023; Christopher Newman, 'The Artemis Accords and Lunar Exploration—Revolution and Evolution' (*SpaceWatch.Global*, October 2020) <https://spacewatch.global/2020/10/spacewatchgl-opinion-the-artemis-accords-and-lunar-exploration-revolution-and-evolution/> accessed 19 August 2023; Kai-Uwe Schrogl, 'We Must not Overrate the Artemis Accords' (*SpaceWatch.Global*) <https://spacewatch.global/2020/11/spacewatchgl-interviews-kai-uwe-schrogl-we-must-not-overrate-the-artemis-accords/> accessed 19 August 2023.

[39] See eg Artemis Accords: Principles for Cooperation in the Civil Exploration and Use of The Moon, Mars, Comets, and Asteroids for Peaceful Purposes (adopted 13 October 2023) s 5 (Interoperability of Systems) and s 12 (Mitigation of Orbital Debris) (Artemis Accords).

[40] ibid s 11.

[41] ibid s 11(7).

Furthermore, Signatories intend to contribute to multilateral efforts to further develop international practices, criteria, and rules applicable to the definition and determination of safety zones and harmful interference.[42] Despite understanding the practical usefulness of such 'zones', the establishment of 'safety zones' on a celestial body's surface undoubtedly represents the most legally debatable point of the Accords. Indeed, not only did the Outer Space Treaty never foresee such a possibility but also, depending on the Signatories' behaviours, such a practice could result in the appropriation of a portion of a celestial body's surface and subsurface by means of use or occupation, a possibility specifically prohibited by Article II of the Outer Space Treaty. Additionally, the unilateral setting up of such 'zones' could lead to problems of coordination, and possibly to disputes, between the activities undertaken by Signatories to the Accords and third parties, by always taking into account that the latter parties would not be under any obligation to recognize the validity of safety zones established under the Accords.

An additional area of concern is related to the environmental impact of activities undertaken under the umbrella of the Accords. Indeed, the Accords pay exclusive attention to the mitigation of orbital debris, through the passivation and disposal of spacecraft at the end of their mission. This is certainly an important element to be addressed; however, the same level of attention is not paid to the prevention of harmful contamination of the lunar soil and subsoil caused by undertakings.

Besides the above-mentioned unilateral and quasi-multilateral initiatives, there is an ongoing multilateral effort within COPUOS potentially aiming at drafting internationally agreed principles to govern space resource utilization. Notably, since 2017 the UNCOPUOS's Legal Subcommittee has been discussing as an agenda item the legal issues associated with the exploration, utilization, and exploitation of space resources.[43] Until 2021, however, discussions were limited to an exchange of views without any mandate to achieve specific results. This situation changed in 2021 when the Legal Subcommittee's members agreed to establish a Working Group on Legal Aspects of Space Resource Activities.[44] After some initial tasks undertaken in 2022, the Working Group started its official activities in 2023 pursuant to a five-year mandate (2023–27) which includes: (a) the finalization of a set of initial recommended principles for space resources activities for the consideration of and consensus agreement by the Committee, followed by possible adoption by the General Assembly as a dedicated resolution or other action; and (b) the adoption by the Committee of a final report of the Working Group, incorporating the results of each of the activities undertaken by the Working Group in accordance

[42] ibid s 11(6).

[43] UNCOPUOS, 'Report of the Legal Subcommittee on its Fifty-sixth Session' (Vienna, 27 March–7 April 2017) UN Doc A/AC.105/1122.

[44] For information about the Working Group and its activities, see the UN dedicated website at <www.unoosa.org/oosa/en/ourwork/copuos/lsc/space-resources/index.html> accessed 19 August 2023.

with its mandate, as appropriate.[45] Even though it remains to be seen what concrete results the Working Group will be able to achieve (this, of course, will depend on the attitude and willingness of its members, particularly those countries that have passed space resources utilization laws), the fact that countries are addressing this matter at the UN level is important, not only because it reaffirms the central role that COPUOS plays as the law-making forum for space law matters but also because it reflects the awareness that States have of the importance of international agreement on basic principles to govern activities in an international domain (in this case mining activities on the Moon) rather than leaving the matter in the hands of individual States or a limited group of States, as the latter approach is likely eventually to lead to tensions and disputes among countries.

4 What is the Role of the Moon Agreement in the Twenty-First Century?

After discussing the provisions of the Moon Agreement and reviewing recent regulatory developments, it is important to assess what role, if any, the Agreement may play in regulating the upcoming era of lunar activities; in particular, one has to wonder what contribution the Agreement can make to shaping the (international) legal framework intended to govern human settlements on the Moon and the utilization of resources contained therein.

This analysis can be undertaken from different perspectives. It is arguable that from the perspective of those countries that have enacted space resources utilization laws, the Moon Agreement is not meant to play any meaningful role; indeed, not only are these countries not Parties to the Agreement (nor do they plan to join), but rather they have embraced a legal approach (the recognition of private property rights over lunar resources through national legislation) that is the opposite of what the Moon Agreement suggests (an international regime to govern the exploitation of resources). Additionally, some of these countries have also openly manifested their objection towards using the Moon Agreement in any discussion on the legal regime applicable to lunar activities; this is exemplified by a 2020 Executive Order issued by the White House stating:

> The United States does not consider the Moon Agreement to be an effective or necessary instrument to guide nation states regarding the promotion of commercial participation in the long-term exploration, scientific discovery, and use of the Moon, Mars, or other celestial bodies. Accordingly, the Secretary of State shall object to any attempt by any other state or international organization to

[45] UNCOPUOS, 'Report of the Legal Subcommittee on its Sixty-first Session' (Vienna, 28 March–8 April 2022) UN Doc A/AC.105/1260.

treat the Moon Agreement as reflecting or otherwise expressing customary international law.[46]

Much more interesting is the analysis of the Moon Agreement in the context of initiatives like the Artemis Accords, in particular from the perspective of those countries who are simultaneously Party to the Moon Agreement and Signatories of the Artemis Accords. This is the situation of Australia, which acceded to the Moon Agreement in 1986 and joined the Accords in October 2020. From the point of view of the present author, Australia's position is highly controversial and hardly justifiable from a legal perspective.[47] One should always keep in mind that the Artemis Accords were drafted by the United States (in consultation with other countries) and reflect the US approach on the conduct of activities on the Moon and on the right to commercially utilize the resources contained therein. If one considers that the US position is based on the right of States to freely explore and use celestial bodies (and their resources) and on the possibility of authorizing private entities to engage in the recovery of lunar resources, to gain property rights over them, and to use them for the purpose of profit, it should not be hard to see how such an approach contrasts with the one envisioned by the Moon Agreement, where resources are viewed as the 'common heritage of mankind' and their exploitation can only be undertaken in accordance with the rules of an international regime that has, as one of its goals, the equitable sharing of benefits among countries. This motive alone would be sufficient to raise questions on how a State might be part of these two initiatives at the same time. Furthermore, as described above, the United States does not consider the Moon Agreement to be a necessary instrument to guide the exploration and use of the Moon; this renders even more challenging the possibility of justifying the decision of a State Party to the Moon Agreement to join the initiative of a country that has so strongly criticized and rejected the Agreement itself.

The most interesting aspect to be analysed is the contribution to be made by the Moon Agreement to the ongoing activities of the UNCOPUOS Working Group on the Legal Aspects of Space Resources Activities; in that regard, reflection should focus on the behaviour of the States Parties to the Agreement as well as a reference to the Agreement or its provisions in the context of the Working Group's discussions. One should start by pointing out that the initiative to establish a working group as well as that of including the topic of the 'General exchange of views on potential legal models for activities in the exploration, exploitation and utilization

[46] 'Executive Order on Encouraging International Support for the Recovery and Use of Space Resources' (*The White House*, 6 April 2020) <https://trumpwhitehouse.archives.gov/presidential-actions/executive-order-encouraging-international-support-recovery-use-space-resources/> accessed 19 August 2023.

[47] Fabio Tronchetti and Hao Liu, 'Australia's Signing of the Artemis Accords: A Positive Development or a Controversial Choice?' (2021) 75(3) AustJIntlAff 243.

of space resources' in the Legal Subcommittee's agenda back in 2017 came from Belgium (together with Greece), a State Party to the Moon Agreement. Over the years, Belgium has been extremely active in inviting other States to work on the development of an international framework to govern lunar activities, including the utilization of resources, and to consider the benefits deriving from the ratification of the Moon Agreement.[48] Besides Belgium's initiative, one may not fail to notice the lack of a unitary strategy of the 17 States Parties to the Moon Agreement; indeed, while making general statements that praise the setting up of the Working Group on Legal Aspects of Space Resources Activities and that emphasize the importance of internationally agreed principles, these States have not so far acted as a coherent group having the goal of promoting the ratification of the Moon Agreement or the acceptance of its provisions. For instance, in the 2023 session of the UNCOPUOS Legal Subcommittee, Austria, Australia, and the Netherlands, three States Parties to the Moon Agreement, submitted a conference paper intended to provide inputs for the development of a set of initial recommended principles for the exploration, exploitation, and utilization of space resources.[49] This paper, while providing a useful comparison among the provisions of the Moon Agreement, the Outer Space Treaty, and the Artemis Accords, does not go as far as indicating the Moon Agreement as the preferred basis for negotiations or calling upon other countries to join it. Furthermore, if one recalls the provisions of Article 11(5) of the Agreement, requiring States to set up an international regime when the exploitation of lunar resources is about to become feasible, it is arguable that this time has finally come and that the Working Group would offer States Parties the perfect framework to work on such a regime. However, to date, the Parties to the Moon Agreement have not taken any step in this direction. Additionally, these States have not made use of the possibility that is offered to them by Article 18 of the Agreement, according to which States Parties may convene a conference to review the Agreement and to consider the implementation of the provisions of Article 11(5) considering new technological developments. One could also add that Parties to the Moon Agreement have not taken advantage of a statement that some of them prepared in 2008 highlighting the benefits of joining the Moon Agreement.[50] This statement, which emphasizes the flexible nature of the regime that the Agreement proposes and points out the emergence of an updated version

[48] See UNCOPUOS, 'Contribution by Belgium to the General Exchange of Views on Potential Legal Models for Activities in the Exploration, Exploitation and Utilization of Space Resources' Sixty-second Session of the Legal Subcommittee (Vienna, 20–31 March 2023) UN Doc A/AC.105/C.2/L.325.

[49] UNCOPUOS, 'Relevant Considerations for Developing a Set of Initial Recommended Principles for the exploration, Exploitation and Utilisation of Space Resources' Sixty-second Session of the Legal Subcommittee (Vienna, 20–31 March 2023) UN Doc A/AC.105/C.2/2023/CRP.

[50] UNCOPUOS, 'Joint Statement on the Benefits of Adherence to the Agreement Governing the Activities of States on the Moon and Other Celestial Bodies of 1979 by States Parties to that Agreement' Forty-seventh Session of the Legal Subcommittee (31 March–11 April 2008) UN Doc A/AC.105/C/2/2008/CRP.11.

of the common heritage of mankind that does not exclude forms of public and private exploitation nor forbid forms of commercial treatment, could have been used as a valuable instrument to convince UNCOPUOS members of the positive sides of the Moon Agreement and of the contribution that it might make to the formation of a framework to govern space resources utilization.

Another interesting element to be analysed will be the position of those countries which are Parties to both the Moon Agreement and the Artemis Accords, such as the case of Australia discussed above. Considering the striking differences which exist between the two instruments, one has to wonder how Australia would behave in the event of two diverging approaches, one embracing the spirit of the Moon Agreement and the other promoting the Accords' view, and which approach it would support. In order to avoid such a problematic situation, Australia might eventually be 'forced' to pick one of the two sides, in a fashion similar to that of Saudi Arabia which, after joining the Artemis Accords in 2022, decided to withdraw from the Moon Agreement on 5 January 2023 due, among other reasons, to the irreconcilable nature of the two instruments.[51]

Besides the behaviour of the States Parties to the Moon Agreement, it is also important to assess the actions of other COPUOS members. Starting from those countries that have enacted national space resource utilization laws, it is likely that they will oppose any attempt to steer the discussion towards acceptance of the Moon Agreement or its provisions. One has to wonder if these countries, while participating in the Working Group activities, have a real interest in agreeing upon any international principle governing space resource utilization, especially principles that do not reflect their national approach and the interests of the companies they represent. This position is exemplified by the statement given by the United States at the 2023 Legal Subcommittee session which reads as follows: 'At this stage, the United States believes there is neither a need nor a practical basis to create a further elaborated international regime for space resource utilization activities.'

It is also relevant to evaluate the statements given by States that have joined the Artemis Accords but are not Parties to the Moon Agreement, as well as of those countries that are not Parties to either instrument. Beginning with States belonging to the first category, such as New Zealand and Japan, it is interesting to notice that, even though they are signatories to the Accords, they still view the setting up of an international framework agreed upon as the most appropriate means to ensure consistency with international obligations as well as predictability and transparency of behaviours.[52] This approach is even more relevant if one considers that Japan is one of the four countries that have enacted national space resource

[51] Stephan-Michaell Wedenig and Jack Wright Nelson, 'Moon Agreement: Hanging by a Thread?' (*McGill Institute of Air and Space Law*, January 2023) <www.mcgill.ca/iasl/article/moon-agreement-hanging-thread> accessed 19 August 2023.

[52] UNCOPUOS, 'New Zealand—Input to the Working Group on Legal Aspects of Space Resource Activities' Sixty-second Session of the Legal Subcommittee (Vienna, 20–31 March 2023) UN Doc. A/

utilization laws. An interesting statement has been also given by France, which, despite having signed the Accords, has declared that certain principles of the Moon Agreement could inspire the drafting process of the Working Group, in particular those on the transparency of activities conducted, the freedom of scientific research, the protection of the Moon environment and the creation of 'Moon scientific reserves', and the rational management of resources, as well as the principle of the due regard paid to the interests of current and future generations.[53]

Switching attention now to the statements made by those States that have not joined either instrument, it is noteworthy to highlight a growing number of references to the usefulness of the provisions of the Moon Agreement and the need to ensure that the exploitation of space resources offers opportunities, and possibly benefits, to all countries. For example, Russia has emphasized the need for the Working Group on Legal Aspects of Space Resources Activities to ensure mechanisms for the equal and equitable sharing of the benefits arising from the exploitation of space resources and to guarantee that mining operations do not compromise the principle of the free access to all areas of celestial bodies.[54] Germany, has stated:

> When it comes to exploitation of space resources for commercial purposes, we believe that this requires an international legal regime. Such a regime should facilitate the orderly and safe development, rational management and the expansion of opportunities in the use of outer space resources. It should appropriately take into account the efforts of countries contributing to the exploration, exploitation and utilization of celestial bodies and ensure that all countries irrespective of their degree of economic or scientific development can benefit from these activities, without taking away the investment incentives for public and private space flight.[55]

Some of the above statements may offer the key to envisioning the role to be played by the Moon Agreement within the activities of the Working Group on Legal Aspects of Space Resources Activities moving forward, that is to provide

AC.105/C.2/2023/CRP.18; UNCOPUOS, 'Japan—Input to the Working Group on Legal Aspects of Space Resource Activities' Sixty-second Session of the Legal Subcommittee (Vienna, 20–31 March 2023) UN Doc A/AC.105/C.2/2023/CRP.33.

[53] UNCOPUOS, 'France—Input to the Working Group on Legal Aspects of Space Resource Activities' Sixty-second Session of the Legal Subcommittee (Vienna, 20–31 March 2023) UN Doc A/AC.105/C.2/2023/CRP.12.

[54] UNCOPUOS, 'Russia Federation—Input to the Working Group on Legal Aspects of Space Resource Activities' Sixty-second Session of the Legal Subcommittee (Vienna, 20–31 March 2023) UN Doc A/AC.105/C.2/2023/CRP.20.

[55] UNCOPUOS, 'Germany—Input to the Working Group on Legal Aspects of Space Resource Activities' Sixty-second Session of the Legal Subcommittee (Vienna, 20–31 March 2023) UN Doc A/AC.105/C.2/2023/CRP.13.

inputs to the drafting of the principles governing such activities through direct or indirect reference to its provisions. Indeed, due to the opposition of key COPUOS members, it seems unlikely that discussions will lead to the promotion of wider ratification of the Moon Agreement or the acceptance of its provisions in their verbatim formulation. Instead, it seems more reasonable to imagine these provisions providing inspiration for the formulation of the Working Group's principles, possibly with a modified wording, but retaining their original spirit. This seems a realistic goal because some of the most recent regulatory initiatives, namely the Hague Building Blocks for the Development of an International Framework on Space Resources Activities and the Artemis Accords, already incorporate ideas enshrined in the Moon Agreement without, however, specifically recognizing this fact. This is true, for example, of Article 13.1 of the Building Blocks (calling for the international framework to provide for benefit-sharing through the promotion of participation in activities by all countries, a clause resembling Article 11(7) of the Moon Agreement) and section 11 of the Accords (on the prevention of interference among Signatories' activities and the possibility to enter into consultations when this happens, a provision inspired by the wording of Article 8 of the Moon Agreement). Furthermore, the Moon Agreement contains several other provisions, such as those on environmental protection and on criteria for the establishment of lunar stations, that could provide useful ideas within the Working Group's discussions.

Arguably, however, the Moon Agreement will have a better chance of having a meaningful role moving forward only if the States Parties to it will actively promote its provisions and the beneficial contribution that they may make to future lunar endeavours and the framework to govern them; indeed, without a coordinated effort by these States, other COPUOS members could likely stop any attempt to give relevance to the Moon Agreement and any proposal incorporating its principles.

5 Conclusion

Without a doubt, the Moon Agreement is the international space treaty that has raised the greatest level of controversy and debate; intended to promote the orderly management of lunar activities, the Agreement has instead attracted a great deal of opposition because it embraces the common heritage of mankind as the doctrine to govern the commercial utilization of lunar resources. Over the years, such opposition has led to the Agreement becoming the least successful space treaty in terms of ratifications and there are few signs that its fate could significantly change for the better. However, with the renewed interest in the Moon, the emergence of regulatory initiatives, and the attempt to develop an internationally agreed framework within the United Nations, it seems important to analyse what contribution, if any, the Moon Agreement can make to this process. While it is hard to imagine that a significant number of countries would ratify the Agreement in its original form, it appears more realistic for the Agreement to provide a positive contribution, even if indirectly, to the activities of the UNCOPUOS Working Group on Legal Aspects of Space Resources Activities, especially in the elaboration of principles to manage such activities. Notably, in the most recent session of the Working Group, the Moon Agreement has been referred to several times. However, as discussed in this chapter, in order to give the Moon Agreement a fair chance to play a meaningful role in this process it might be necessary for the States Parties to it to act in a coherent and unitary manner so as to convince other COPUOS members of the valuable contribution that the Moon Agreement may make to the regulation of lunar activities in the twenty-first century.

As a concluding remark, one can say that, at the end of the day, reflecting upon the Moon Agreement is worthwhile for several reasons; first, it is the only international treaty specifically attempting to regulate the exploration and use of celestial bodies and their resources; secondly, its provisions represent a compromise that was accepted by all COPUOS members in 1979 (even though the majority of them eventually decided not to ratify it); and thirdly, the Agreement provides several practically useful provisions to guide human activities on the Moon. Ultimately, the Agreement was designed to enable the peaceful, orderly, and safe exploitation of lunar activities; this is a goal that all countries share.

10

SOFT LAWS RELATING TO SPACE ACTIVITIES

Yun Zhao

1 Introduction

After the launch of the first artificial Earth satellite in 1957, the United Nations (UN) played an important role in concluding five space treaties to regulate space activities at the international level. This progressive codification of international space law played a vital role in ensuring orderly development at the start of the space age. These treaties remain major regulatory documents for space activities today, though there is increasingly a need for new legal rules to regulate new types of space activities in the era of space commercialization and privatization. However, with more space-faring nations entering the scene, each with its own national agenda and national interests, the conclusion of binding legal documents proves extremely difficult. This is substantiated by the fact that no more international space treaties have been adopted since the Moon Agreement in 1979.

Consequently, soft law instruments, while not ideal, have become the second-best choice for international society to tackle the dilemma; to a certain extent, these non-binding documents have become the preferred form of defining and refining norms for contemporary issues arising from the commercialization and privatization of outer space.

This chapter examines the development and roles of soft laws in the regulation of space activities. It picks up major instruments to elaborate on how they function to deal with certain aspects of space activities where no binding legal instruments are in place. The chapter concludes that soft laws, while being disputable with regard to their efficiency and effectiveness, can fill in the legal vacuum by providing guidance for new legal issues and contribute to the development of binding rules for space activities in the long run.

2 Development of Soft Laws: Background and Functions

After the launch of the Soviet Union's Sputnik 1, the UN reacted quickly by issuing UN General Assembly (UNGA) resolutions to provide preliminary guidance on

Yun Zhao, *Soft Laws Relating to Space Activities* In: *International Space Law in the New Space Era*.
Edited by: Sandeepa Bhat B., Dilip Ukey, and Adithya Variath, Oxford University Press.
DOI: 10.1093/9780198909415.003.0010

space activities monopolized by the two superpowers, which laid a solid basis for the enactment of the first space treaty in 1967. As such, soft laws, represented by the three UNGA resolutions, were the sole instruments regulating space activities during the first decade of the space era (1957–1966).

Resolutions 1721 A and B (XVI) of 1961 were the first set of soft law instruments calling for 'international cooperation in the peaceful uses of outer space'. This is most meaningful in view of the competition and geopolitical rivalry between the two superpowers (the United States and the Soviet Union) during the Cold War period. The two principles specified in Resolution 1721 A, that is, the application of international law to and non-appropriation of outer space, signify the start of the UN initiatives in establishing the space order. Resolution 1721 B moves further to specific aspects of the future space regime, including the registration mechanism and the central role of the UN Committee on the Peaceful Uses of Outer Space (UNCOPUOS) in promoting space transparency and cooperation.

Resolution 1962 (XVIII) of 1963 is the most important UNGA instrument adopted in the early years of the space era, laying a solid basis for the 1967 Outer Space Treaty. Resolution 1884 of 1963, while not specifically targeting space activities, touches on the principle of peaceful uses of outer space, calling on States 'to refrain from placing in orbit around the earth any objects carrying nuclear weapons or any other kinds of weapons of mass destruction, installing such weapons on celestial bodies, or stationing such weapons in outer space in any other manner'.[1]

No major UNGA resolutions were adopted for space activities from 1964 to 1981, with the adoption of five major space treaties during this period. Soft laws have become more active and important since 1982, partly due to the increasing number of space-faring nations and the legal vacuum, to deal with the issues arising from space commercialization and privatization. More importantly, while UNGA resolutions remain one major form of space law, other entities have also started to take an active role in the formulation of soft laws for space activities.

From the development history, we can discern important functions of soft law instruments in the regulation of space activities. First, soft law instruments lay a solid basis and serve as a testing bed for the development of space treaties. Resolution 1962 (XVIII) is an excellent example in this regard, with all its nine principles reproduced in the 1967 Outer Space Treaty. Resolution 1884 was similarly adopted in the Outer Space Treaty regarding the principle of peaceful uses of outer space.[2] Secondly, soft law rules could develop into customary international law through years of consistent state practice. It has been argued that the unique environment of outer space provides the possibility of the gradual emergence of

[1] UNGA Res 1884 (XVIII) (17 October 1963) para 2(a).

[2] Treaty on Principles Governing the Activities of States in the Exploration and Use of Outer Space, Including the Moon and Other Celestial Bodies (adopted 27 January 1967, entered into force 10 October 1967) 610 UNTS 205, art IV (Outer Space Treaty).

customary international law in certain circumstances, which is referred to as instant customary law.[3] Resolution 1962 (XVIII), preceding the Outer Space Treaty, is again used to exemplify the instant formation of customary international law in the field of space law. Thirdly, soft law instruments assist in the interpretation and enforcement of relevant provisions of the space treaties. For example, the 1996 UNGA Resolution on international cooperation further elaborates the principle of international space cooperation in the 1967 Outer Space Treaty; the 2004 UNGA Resolution on the application of the concept of the 'launching State' provides useful support for the continued application of the Liability Convention in the era of space commercialization.

3 Soft Law Instruments: Selected UNGA Resolutions

3.1 The 1963 Declaration of Legal Principles

The UNGA adopted Resolution 1962(XVIII) entitled 'Declaration of Legal Principles Governing the Activities of States in the Exploration and Use of Outer Space' in December 1963. This resolution declares major legal principles guiding space activities for the first time, laying a solid basis for the adoption of the 1967 Outer Space Treaty. With more than 100 countries becoming Parties to the Outer Space Treaty, most of the principles in this resolution have been considered part of customary international law.

Major principles include the following:

> 1) the exploration and use of outer space shall be carried on for the benefit and in the interests of all mankind; 2) free exploration and use of outer space and celestial bodies on a basis of equality; 3) non-appropriation; 4) the application of international law in the interest of maintaining international peace and security and promoting international cooperation and understanding; 5) international responsibility for national space activities, whether carried on by governmental agencies or non-governmental entities; 6) the principle of international cooperation and mutual assistance with due regard for the corresponding interests of other States; 7) jurisdiction and control of the State of the registry; 8) the international liability of a launching State, and 9) assistance to and prompt return of astronauts in the event of an accident, distress, or emergency landing.[4]

[3] Bin Cheng, 'United Nations Resolutions on Outer Space: "Instant" International Customary Law?' (1965) 5 IJIL 23.

[4] UNGA Res 1962 (XVIII) (13 December 1963).

3.2 The 1982 Broadcasting Principles

Satellite communications were widely used in the 1980s, leading to the trend of space commercialization and privatization. Failing to adopt a binding treaty due to the difficulty in reconciling the principle of State sovereignty with the freedom of information, the UNGA adopted by a majority vote the Principles Governing the Use by States of Artificial Earth Satellites for International Direct Television Broadcasting in 1982.[5] Direct broadcasting by satellites was made possible with technological development, giving rise to challenges in technical, legal, political, cultural and social aspects.

This resolution aims to strike a balance between sovereignty and freedom of information, exploring the compatibility of freedom of information with State sovereignty, in particular the principle of non-intervention.[6] Relevant international law documents shall be applicable to direct broadcasting by satellites, including the UN Charter, the Outer Space Treaty, the International Telecommunication Convention and its Radio Regulations.[7] All States have the right to conduct, or authorize persons and entities under their jurisdictions to conduct, international television broadcasting by satellites and to enjoy the benefits from such activities.[8] This is in line with the principle of freedom to use encapsulated in the Outer Space Treaty. The document further emphasizes that all States shall have indiscriminate access to the technologies on terms to be mutually agreed upon by all concerned.[9] Accordingly, the document dictates substantive equality, instead of formal equality, in the use of relevant technologies among all States.

The principle of international cooperation is reiterated through appropriate arrangements with the aim of accelerating national development.[10] This principle applies directly to the protection of copyright and neighbouring rights involved in the broadcasting services through the conclusion of bilateral or multilateral agreements.[11] International cooperation also requires the peaceful settlement of disputes in accordance with the provisions of the UN Charter,[12] which proposes to settle the disputes through established procedures such as negotiation, mediation, arbitration and judicial settlement.[13]

States shall be held responsible for their broadcasting activities; an international intergovernmental organization, together with its member States, shall

[5] UNGA Res 37/92 (10 December 1982).

[6] Principles Governing the Use by States of Artificial Earth Satellites for International Direct Television Broadcasting, UNGA Res 37/92 (10 December 1982) Principle A (Broadcasting Principles).

[7] ibid Principle B.

[8] ibid Principle C.

[9] ibid.

[10] ibid Principle D.

[11] ibid Principle H.

[12] ibid Principle E.

[13] The Charter of the United Nations (adopted 24 June 1945, entered into force 24 October 1945) XV UNCIO 335, art 22 (UN Charter).

be responsible for such activities.[14] Consultation is important prior to and during direct television broadcasting activities. Prior to the activities, a State shall inform the UN Secretary-General of the nature of such activities, who should disseminate such information immediately and effectively to all relevant entities.[15] In the meantime, the State shall also notify any States that may be affected by such activities and promptly enter into consultation with any States upon request. Relevant agreements or arrangements shall be made before the establishment of broadcasting services; in a sense, prior consent from the receiving State is required for such services. Relevant instruments of the International Telecommunication Union (ITU) shall be followed in cases of unavoidable overspill of radiation to other States.[16] After the establishment of the broadcasting services, the broadcasting State shall enter into consultations promptly with other affected States upon request.[17]

3.3 The 1986 Remote Sensing Principles

Remote sensing activities were first mentioned in the agenda for the eleventh session of the UNCOPUOS Legal Subcommittee.[18] The '1986 Principles Relating to Remote Sensing of the Earth from Outer Space' were adopted by the UNGA in its Resolution 41/65 on 3 December 1986. 'Remote sensing' is defined as 'the sensing of the Earth's surface from space by making use of the properties of electromagnetic waves emitted, reflected or diffracted by the sensed objects, for the purpose of improving natural resources management, land use and the protection of the environment'.[19] The main function of remote sensing includes resource exploration, environment detection and weather broadcast.

This resolution is intended to provide guidance on three major issues: the necessity of prior consent for remote sensing, any necessary limitations on the processing of primary data and the legal status of information related to a State's natural resources. It reiterates the principle of benefits and interests of all countries, with particular consideration for the needs of developing countries.[20] Relevant international law instruments are applicable to remote sensing activities. While emphasizing the principle of free use of outer space, the resolution recalls the sovereignty of all States (specifically those of the sensed State), with the aim of reconciling freedom and state sovereignty.[21]

[14] Broadcasting Principles (n 6) Principle F.
[15] ibid Principle I.
[16] ibid Principle J.
[17] ibid Principle G.
[18] UN Doc A/AC.105/101.
[19] Principles Relating to Remote Sensing of the Earth from Outer Space, UNGA Res 41/65 (3 December 1986) Principle I(a) (Remote Sensing Principles).
[20] ibid Principle II.
[21] ibid Principle IV.

The resolution once again attaches great importance to the principle of international cooperation. It further develops this principle by specifying that 'international cooperation shall be carried out on equitable and mutually acceptable terms, which safeguards the interests of developing countries'.[22] This principle is further substantiated in a specific scenario in the next Principle, which provides a link between achieving the benefits of remote sensing and access to data through agreements or other arrangements. This includes 'the establishment and operation of data collecting and storage stations and processing and interpretation facilities'.[23] Technical assistance shall also be made available to interested States on mutually agreed terms.[24] In this regard, the UN and its agencies shall promote international cooperation in a wide range of remote sensing activities. The sensing State shall provide relevant information to the UN and affected States,[25] which serves as a precondition for international cooperation. The timeline for the provision of information is flexible, requiring only the provision to the greatest extent feasible and practicable to any other State. While no clear guidance on whether prior consent of the sensed State is required, the document does mention the need for bilateral or multilateral consultations with the sensed State, upon request, with the aim of promoting and intensifying international cooperation and enhancing the mutual benefits.[26] This is drafted in such a way that it favours the sensed State.

Protection of the Earth's natural environment is essential for any kind of space activity, including remote sensing. In the meantime, environmental protection relies heavily on remote sensing data which gathers general information showing the changes or any other condition that threatens the natural environment. Remote sensing data can be used not only for environmental protection, but more importantly for predicting potential natural disasters such as earthquakes, tidal waves, landslides and hurricanes, and facilitating necessary disaster management manoeuvres. Accordingly, the sensing State has an obligation to promote the protection of mankind from natural disasters and transmit necessary data and information to affected States as promptly as possible.[27]

Remote sensing data is divided into three categories: primary data, processed data and analysed data. Primary data refers to 'raw data that are acquired by remote sensors borne by a space object and that are transmitted or delivered to the ground from space by telemetry in the form of electromagnetic signals, by photographic film, magnetic tape or any other means'; 'processed data' means the data resulting from the processing of raw data for practical use; 'analyzed data' refers to 'the information resulting from the interpretation of processed data, inputs of

[22] ibid Principle V.
[23] ibid Principle VI.
[24] ibid Principle VII.
[25] ibid Principle IX.
[26] ibid Principle XIII.
[27] ibid Principle XI.

data and knowledge from other sources'.[28] The sensed State shall have access to 'primary data and processed data on a non-discriminatory basis and on reasonable cost terms; the same basis and terms also apply to analyzed data, but particular account should be given to the needs and interests of developing countries'.[29]

The issue of international responsibility, instead of an international liability, is included to hold the sensing State responsible for its remote sensing activities. These activities can be conducted by governmental or non-governmental entities or through international organizations.[30] While the issue of liability is not specifically mentioned, the document does indeed encourage peaceful settlement of disputes through negotiation, mediation, arbitration, litigation or any other peaceful means.[31]

3.4 The Nuclear Power Sources (NPS) Principles

Nuclear power sources, while ideal for space operations, bring with them high risks. In light of the Cosmos 954 incident in 1978 causing serious damage to the Earth through NPS contamination, the international community acknowledged the urgency in developing an appropriate regulatory framework for the use and operation of NPS in outer space. The UNGA adopted the 'Principles Relevant to the Use of Nuclear Power Sources in Outer Space' in December 1992.[32]

International law rules apply to the use and operation of NPS in outer space. In contrast to the definition in the Liability Convention and the Registration Convention, the term 'launching State' is defined as 'State[s] which exercises jurisdiction and control over a space object with nuclear power sources on board at a given point in time relevant to the principle concerned'.[33]

The resolution provides for the strictly applicable scope for NPS in the 'space missions which cannot be operated by non-nuclear energy sources in a reasonable way'. Principle 3, as the most important principle in this resolution, sets out the guidelines and substantive criteria for safe use of NPS regarding radiation protection and nuclear safety, nuclear reactors and radioisotope generators.[34] Safety assessment shall be carried out in compliance with Principle 3 and the Outer Space Treaty prior to launch. In case of malfunctions and the re-entry of malfunctioning space objects, the launching State shall inform the States concerned and the UN

[28] ibid Principle I(b)–(d).
[29] ibid Principle XII.
[30] ibid Principle XIV.
[31] ibid Principle XV.
[32] Principles Relevant to the Use of Nuclear Power Sources in Outer Space, UNGA Res 47/68 (14 December 1992) (Nuclear Power Sources Principles).
[33] ibid Principle 2.1.
[34] ibid Principle 3.

Secretary-General in a timely fashion,[35] and, if requested, conduct consultations with the States concerned promptly.

Prior to the expected re-entry into the Earth's atmosphere of a malfunctioning space object, all States possessing space monitoring and tracking facilities shall provide available information to the UN Secretary-General and States concerned for assessment. After re-entry, 'the launching State shall offer and, if requested by the affected State, provide the necessary assistance to eliminate actual and possible harmful effects'.[36]

Along similar lines to the provision on State Responsibility in the Outer Space Law, this resolution provides for the international responsibility of the launching State for space activities involving the use and operation of NPS. Similarly, the resolution follows relevant provisions on liability in the Outer Space Treaty and the Liability Convention on the issues relating to the liability arising from damages caused by space objects carrying an NPC on board, compensation and reimbursement of expenses.[37] Relevant disputes shall be resolved through negotiations or any other peaceful means.

It is notable that the resolution provides a review and revision mechanism. This arrangement is appropriate in view of the technical guidance provided in this document and the rapid development of space technologies. Moreover, it only addresses the use of NPS for non-propulsive purposes. It is now on the UNCOPUOS agenda for possible revision of the document in the future.

It is to be noted that the UNCOPUOS Science and Technology Subcommittee (STS) and the International Atomic Energy Agency (IAEA) jointly adopted a non-binding document entitled the 'Safety Framework for Nuclear Power Sources Applications in Outer Space' in 2009,[38] in order to provide high-level guidance in the form of a model safety framework. This document has as its safety objective to 'protect people and the environment in Earth's biosphere from potential hazards associated with the relevant launch, operation and end-of-service phases of space nuclear power source applications'.[39] It applies to '1) governments, and relevant international intergovernmental organizations that authorize, approve or conduct space NPS missions; 2) the management of the organization that conducts space NPS missions; 3) the design, development and mission phases of space NPS missions'.[40]

[35] ibid Principle 5.
[36] ibid Principle 7.
[37] ibid Principle 9.
[38] Scientific and Technical Subcommittee of UNCOPUOS and International Atomic Energy Agency, 'Safety Framework for Nuclear Power Source Applications in Outer Space' (19 May 2009) UN Doc A/AC.105/934.
[39] ibid 2.
[40] ibid 3–8.

3.5 The 1996 Declaration on International Cooperation

The principle of international cooperation has been included in various UN documents since the start of the space age. However, earlier documents failed to provide detailed guidance on the implementation of this principle until the UNGA's adoption of the 'Declaration on International Cooperation in the Exploration and Use of Outer Space for the Benefit and in the Interest of All States, Taking into Particular Account the Needs of Developing Countries' in December 1996.[41] This resolution, while still continuing the use of broad language, elaborates in detail the understanding and application of the principle of international cooperation for the first time.

It looks into the forms and substance of international cooperation, with the aim of promoting cooperation in a wide range of space activities for the realization of the common interests of all mankind. It advocates real (and substantive), rather than simply formal, equality in the cooperation among States. International cooperation in space activities shall aim to promote the development of space science and technology, foster the development of appropriate space capabilities and facilitate the exchange of expertise and technology.[42]

Such cooperation shall be carried out for peaceful purposes only. The principle of party autonomy is closely respected in carrying out space cooperation, with no countries being forced into cooperation against their will. The States are free to determine all aspects of their participation in international cooperation as long as the cooperation is carried out on an equitable and mutually acceptable basis.[43] Particular attention shall be given to the special needs and interests of developing countries. The cooperative modes include 'governmental and non-governmental; commercial and non-commercial; global, multilateral, regional or bilateral; and international cooperation among countries in all levels of development'.[44] The determining factors in the cooperative mode are the 'effectiveness' and 'appropriateness' of such modes to the countries concerned.[45]

The UNCOPUOS has long been a major exchange forum for the discussion of space-related issues and the negotiation of legal documents for space activities. This resolution further stresses the important role of the UNCOPUOS in promoting international cooperation in space activities. The States are encouraged to contribute to the UN Programme on Space Applications to realize wider and greater cooperation in space-related issues.[46]

41 Declaration on International Cooperation in the Exploration and Use of Outer Space for the Benefit and in the Interest of All States, Taking into Particular Account the Needs of Developing Countries, UNGA Res 51/122 (13 December 1996) (Benefits Declaration).
42 ibid para 5.
43 ibid para 2.
44 ibid para 4.
45 ibid.
46 ibid para 8.

3.6 The Use of the Geostationary Orbit

The Declaration of the First Meeting of Equatorial Countries (Bogota Declaration), issued in 1976[47] in an attempt to assert sovereignty over the geostationary orbit above the equatorial countries, provided a vivid example testifying that the geostationary orbit is a limited and valuable natural resource. While this declaration proved to be futile in terms of the international responses, it does justify the necessity of a regime to ensure the rational and equitable use of this orbit.

The UNCOPUOS Legal Subcommittee elaborated on the issue and adopted at its thirty-ninth session in 2020 the document entitled 'Some Aspects Concerning the Use of Geostationary Orbit'.[48] This emphasizes the role of the ITU Radio Regulations in ensuring equitable access of the developing countries to the requested orbit/spectrum resources and effective use of the orbit/spectrum resources.

3.7 The Concept of the 'Launching State'

The concept of the 'launching State' was originally defined in the Liability Convention and the Registration Convention as 'a State which launches or procures the launching of a space object' or 'a State from whose territory or facility a space object is launched'.[49] According to the Liability Convention, only a launching State shall be held liable for damages caused by a space object;[50] similarly, the Registration Convention defines a State of registry to be one of the launching States on whose registry a space object is carried.[51] This definition faces serious challenges with the latest technological developments and the emergence of space commercialization and privatization, one typical example being the on-orbit transfer of ownership of space objects.

Against this background, the UNGA adopted a resolution in 2004 regarding the application of the concept of the 'launching State'.[52] The resolution does not resolve the difficulty of applying the concept in situations such as on-orbit transfers; however, the recommendations have to a certain extent helped to close the loophole.

[47] Declaration of the 'First' Meeting of Equatorial Countries (Bogota, Columbia, 3 December 1976) <https://www.jaxa.jp/library/space_law/chapter_2/2-2-1-2_e.html> accessed 22 September 2023.

[48] UNCOPUOS, 'Some Aspects Concerning the Use of the Geostationary Orbit' Thirty-ninth Session of the Legal Subcommittee (Vienna, 27 March–6 April 2000) (20 April 2000) UN Doc A/AC.105/738, Annex III.

[49] Convention on International Liability for Damage Caused by Space Objects (adopted 29 March 1972, entered into force 1 September 1972) 961 UNTS 187, art I(c) (Liability Convention); Convention on Registration of Objects Launched into Outer Space (adopted 14 January 1975, entered into force 15 September 1976) 1023 UNTS 15, art 1(a) (Registration Convention).

[50] Liability Convention (n 49) arts II and III.

[51] Registration Convention (n 49) art I(c).

[52] UNGA Res 59/115 (10 December 2004).

The document calls for the enactment and implementation of national laws on the authorization and continuing supervision of space activities of non-governmental entities within their jurisdictions. It also encourages the conclusion of agreements in joint launches or cooperation programmes regarding the apportionment of liability and compensation as defined in the Liability Convention.[53] In view of the difficulty in adopting a new instrument at the international level, the resolution emphasizes the importance of national space legislation for the solution of issues relating to liability and registration of space objects. The UNCOPUOS shall be in a position to provide necessary information and assistance in the process of enacting national space legislation.

3.8 Enhancing the Practice of Registering Space Objects

As early as 1961, the UNGA called upon the States to register space objects with the UN Secretary-General.[54] The adoption of the 1975 Registration Convention moved the requirement of registration one step further to make it compulsory, which helped in the identification of space objects. However, the registration practice of space objects has not been fully satisfactory so far. There is a need to harmonize the practice of space registration. Accordingly, the UNGA adopted 'Recommendations on Enhancing the Practice of States and International Intergovernmental Organizations in Registering Space Objects' in 2007.[55]

The resolution points out the important role of the Registration Convention and encourages more States to become parties to this Convention. The information required to be provided to the UN Secretary-General includes '1) the Committee on Space Research international designator, where appropriate; 2) Coordinated Universal Time as the time reference for the date of launch; 3) kilometers, minutes and degree as the standard units for basic orbital parameters; 4) any useful information relating to the function of the space object in addition to the general function requested by the Registration Convention'.[56] It also encourages the furnishing of additional information by the State of the registry. Such additional information includes '1) the geostationary orbit location; 2) any change of status in operation; 3) the approximate date of decay or re-entry; 4) the date and physical conditions of moving a space object to a disposal orbit; 5) web links to official information on space objects'.[57] 'The resolution puts forward recommended practices to ensure the most complete registration of space objects.'[58]

[53] Liability Convention (n 49) art IV.
[54] UNGA Res 1721 (XVI) B (20 December 1961) paras 1 and 2.
[55] UNGA Res 62/101 (17 December 2007).
[56] ibid art 2(a).
[57] ibid art 2(b).
[58] ibid art 3.

Under Article 4(a) of the Convention, '[f]ollowing the change of supervision of a space object in orbit, the State of the registry could provide additional information such as the date of change in supervision, the identification of the new owner or operator, any change of orbital position, and any change of function of the space object'.[59] An important recommendation introduced in this resolution is the provision of a model registration form by the Office for Outer Space Affairs to assist the States of registry in their submission of registration information.[60] The Office shall also promote transparency through the provision on its website of contact details of the focal points and web links to the appropriate registries.[61]

3.9 National Space Legislation

In light of the difficulty in concluding new space law conventions, national space legislation becomes increasingly important in the enforcement of obligations under the existing space law treaties and the gap-filling function in the regulation of space activities. In the process of studying space commercialization and privatization, the Space Law Committee of the International Law Association (ILA) picked up the important issue of national space legislation in view of the urgent need for a regulatory regime for commercial space activities. After several rounds of discussion, the ILA in its 75th Conference held in Bulgaria in August 2012 was able to adopt the Sofia Guidelines for a Model Law on National Space Legislation.[62] This also facilitates the discussions in the UNCOPUOS forum, which led to the final adoption of the UNGA resolution one year later, providing recommendations on national legislation relevant to the peaceful exploration and use of outer space.[63]

The resolution recommends the following elements to be included in national space legislation:

> 1) the scope of space activities; 2) national jurisdiction over space activities carried out from territory under its jurisdiction and/or control; 3) licensing and/or authorization of space activities by a competent national authority; 4) the conditions for authorization; 5) appropriate supervision and monitoring procedures; 6) maintenance of a national registry of space objects by an appropriate national authority; 7) ways of seeking recourse from operators or owners of space objects for damage claims, relevant insurance requirements and indemnification procedures; and 8) authorization and continuing supervision of the space activities of

[59] ibid art 4(a).
[60] ibid art 5(a).
[61] ibid art 5(b)–(c).
[62] 'Sofia Guidelines for a Model Law on National Space Legislation' 2012 International Law Association Conference (Bulgaria, 2012) ILA Res 6/2012.
[63] UNGA Res 68/74 (11 December 2013).

non-governmental entities in the event of an on-orbit transfer of ownership or control of a space object.[64]

The resolution recognizes the different approaches taken by the States in the process of national space legislation, in the form of either unified acts or a combination of national legal instruments. Furthermore, the States may come up with specific rules on the above elements taking into account their specific needs and requirements.[65]

4 Selected Soft Law Instruments by the UNCOPUOS and Other non-UN Entities

4.1 UNCOPUOS/IADC Space Debris Mitigation Guidelines

With more and more satellites launched into outer space, space debris mitigation and the protection of the space environment have been major concerns for the international society. The issue of space debris mitigation has been on the agenda of the UNCOPUOS STS since 1994 at its thirty-first session.[66] The Inter-Agency Space Debris Coordination Committee (IADC) presented its proposals on space debris mitigation during the fortieth session in 2003. After several deliberations, the STS adopted the Space Debris Mitigation Guidelines during its forty-fourth session in 2007, followed by endorsements from the UNCOPUOS[67] and the UNGA[68] in the same year.

It is to be noted that Space Debris Mitigation Guidelines were discussed and adopted by the STS, not the Legal Subcommittee. Thus, this document touches more on the technical side of the space debris mitigation measures and not on the legal side, such as the legal definition of 'space debris' and the issues of potential liability and compensation. Nevertheless, the document does indeed define the term 'space debris' in the Background to be 'all man-made objects, including fragments and elements thereof, in Earth orbit or re-entering the atmosphere, that are non-functional'. This definition is further supplemented in the same part by providing the primary sources of space debris in Earth orbit to include '(a) accidental and intentional break-ups which produce long-lived debris; (b) debris released

64 ibid.

65 ibid Preamble.

66 UNCOPUOS, 'Report of the Scientific and Technical Subcommittee on the Work of its Thirty-first Session' (10 March 1994) UN Doc A/AC.105/571, paras 63–74.

67 UNGA, 'Report of the Committee on the Peaceful Uses of Outer Space' UNGAOR 62nd Session Supp No 20, UN Doc A/62/20 (2007) paras 118–19.

68 UNGA Res 62/217 (22 December 2007); UN Doc A/62/20, Annex.

intentionally during the operation of launch vehicle orbital stages and spacecraft; and (c) fragments generated by collisions'. The document further differentiates debris mitigation measures into two categories depending on the short-term and long-term purposes. The short-term measures in curtailing the generation of space debris involve 'the curtailment of the production of mission-related space debris and the avoidance of break-ups'; while the long-term measures concern 'end-of-life procedures that remove decommissioned spacecraft and launch vehicle orbital stages from regions populated by operational spacecraft'.[69]

The document makes it clear that the States take measures voluntarily in ensuring the implementation of the guidelines, which are not legally binding under international law. The guidelines apply to 'the mission planning, design, manufacture and operational (launch, mission and disposal) phases of spacecraft and launch vehicle orbital stages'.[70]

The document lists seven guidelines for the States:

> 1) limit debris released during normal operation; 2) minimize the potential for break-ups during operational phases; 3) limit the probability of accidental collision in orbit; 4) avoid intentional destruction and other harmful activities; 5) minimize the potential for post-mission break-ups resulting from stored energy; 6) limit the long-term presence of spacecraft and launch vehicle orbital stages in the low-Earth orbit (LEO) region after the end of their mission; 7) limit the long-term interference of spacecraft and launch vehicle orbital stages with the geosynchronous Earth orbit (GEO) region after the end of their mission.[71]

The document further calls for international cooperation in researching debris mitigation and possible revision of the document in the future. The document, while adopted by the UNCOPUOS STS, entails legal implications, in particular with regard to the understanding and implementation of Article IX of the Outer Space Treaty and fault liability in the Liability Convention.

4.2 UNCOPUOS Long-Term Space Sustainability Guidelines

Space sustainability is closely related to the issue of space debris mitigation, but with a wider scope in regard to environmental protection and sustainable development. In view of its broad and general nature, the term 'space sustainability' has been coined as 'a vague reference to an ecological ideal'.[72] The adoption of

[69] UNGA, 'Space Debris Mitigation Guidelines' UNGAOR 62nd Session Supp No 20, UN Doc A/62/20 (2007) Annex 1, Background.

[70] ibid Part 4.

[71] ibid.

[72] Christopher J Newman and Mark Williamson, 'Space Sustainability: Reframing the Debate' (2018) 46 SpacePoly 35.

the 'Guidelines for the Long-term Sustainability of Outer Space Activities' by the UNCOPUOS in 2019 provided an opportunity to clarify the concept.[73] For the first time, 'space sustainability' is defined as 'the ability to maintain the conduct of space activities indefinitely into the future in a manner that realizes the objectives of equitable access to the benefits of the exploration and use of outer space for peaceful purposes, in order to meet the needs of the present generation while preserving the outer space environment for future generations'.[74]

The document is comprehensive in substance, covering 'the policy, regulatory, operational, safety, scientific, technical, international cooperation and capacity-building aspects of space activities'.[75] Furthermore, it applies broadly to all types of space activities. It lists four categories of guidelines to ensure the realization of space sustainability: '1) policy and regulatory framework for space activities; 2) safety of space operations; 3) international cooperation, capacity-building and awareness; 4) scientific and technical research and development.'[76]

With regard to policy and regulatory framework for space activities, the following five guidelines are listed:

> 1) adopt, revise and amend, as necessary, national regulatory frameworks for outer space activities; 2) consider a number of elements when developing, revising or amending, as necessary, national regulatory frameworks for outer space activities; 3) supervise national space activities; 4) ensure the equitable, rational and efficient use of the radio frequency spectrum and the various orbital regions used by satellites; 5) enhance the practice of registering space objects.

Ten guidelines are listed to ensure the safety of space operations:

> 1) provide updated contact information and share information on space objects and orbital events; 2) improve accuracy of orbital data on space objects and enhance the practice and utility of sharing orbital information on space objects; 3) promote the collection, sharing and dissemination of space debris monitoring information; 4) perform conjunction assessment during all orbital phases of controlled flight; 5) develop practical approaches for pre-launch conjunction assessment; 6) share operational space weather data and forecasts; 7) develop space weather models and tools and collect established practices on the mitigation of space weather effects; 8) design and operation of space objects regardless of their physical and operational characteristics; 9) take measures to address risks

[73] UN Doc A/AC.105/C.1/L.366 (17 July 2018); UNGA 'Report of the Committee on the Peaceful Uses of Outer Space' UNGAOR 74th Session Supp 20, UN Doc A/74/20 (2019).

[74] ibid para 5.

[75] ibid para 11.

[76] ibid Part II.

> associated with the uncontrolled re-entry of space objects; 10) observe measures of precaution when using sources of laser beams passing through outer space.

Four guidelines are listed to promote international cooperation, capacity-building and awareness: '1) promote and facilitate international cooperation in support of the long-term sustainability of outer space activities; 2) share experience related to the long-term sustainability of outer space activities and develop new procedures, as appropriate, for information exchange; 3) promote and support capacity-building; 4) raise awareness of space activities.'

For the purpose of promoting scientific and technical research and development, the document lists two guidelines: '1) promote and support research into and the development of ways to support sustainable exploration and use of outer space; 2) investigate and consider new measures to manage the space debris population in the long term.'

4.3 European Union (EU) Code of Conduct for Outer Space

The EU first put forward a Code of Conduct for Outer Space in 2008 with the purpose of enhancing 'the safety, security, and sustainability of all outer space activities pertaining to space objects, as well as the space environment'.[77] It was amended several times and then issued in 2014. It was drafted as a non-legally binding, voluntary instrument to facilitate the consensus-building process for norms of responsible behaviours in space activities.[78]

The document includes four guiding principles for space activities:

> 1) the freedom of all States to access, to explore, and to use outer space for peaceful purposes without harmful interference and respecting internationally accepted practices, operating procedures, technical standards and policies associated with the long-term space sustainability; 2) the responsibility of all States to refrain from the threat or use of force in outer space and respect the inherent right of States to individual or collective self-defence; 3) the responsibility of States to take all appropriate measures and cooperate in good faith to avoid harmful interference with space activities; 4) the responsibility of States to promote the peaceful exploration and use of space for the benefit, and in the interest of humankind and to prevent space from becoming an arena of conflict.[79]

[77] Council of the European Union, 'Draft International Code of Conduct for Outer Space Activities' <https://www.eeas.europa.eu/sites/default/files/space_code_conduct_draft_vers_31-march-2014_en.pdf> accessed 22 September 2023 art 1.1 (Code).

[78] ibid art 1.4.

[79] ibid art 2.

It pays great respect to the existing treaties, conventions and other instruments relating to space activities; all other measures that are introduced in this document build on these existing rules and mechanisms in safeguarding the safety, security and sustainability of space activities. It encourages international cooperation and lays down cooperation mechanisms, including:

> 1) notification mechanism (such as pre-notification of risky manoeuvres, predicted conjunctions, launch, collisions, in-orbit break-ups and any other destructions, predicted high-risk re-entry events, malfunctioning of space objects or loss of control);[80] 2) information-sharing mechanism on space strategies and policies, major space research and space application programs, space policies and procedures to prevent and minimize the possibility of accidents, collisions and the creation of space debris, efforts for promoting the adoption of space regulatory instruments;[81] 3) consultation mechanism to prevent or minimize potential risks of damage or harmful interference;[82] 4) regular meetings to define, review and further develop the Code and facilitate its implementation,[83] with the designation of a central point of contact and an electronic database for managing information.[84]

The lukewarm response that has been received from the international community so far makes it clear that the EU Code of Conduct for Outer Space would not be accepted easily within the foreseeable future. However, the negotiation process per se has helped to raise awareness among the international community regarding the importance of space security and sustainability.

4.4 The UN Group of Governmental Experts Report on TCBMs

In view of the threats and risks to space sustainability, safety and security, the Group of Governmental Experts (GGE) was established in 2011 to study Transparency and Confidence Building Measures (TCBMs) in space activities to improve space cooperation and reduce the risks of misunderstanding and mistrust.[85] It has been considered as a pragmatic approach among the States to deal with urgent issues related to space security. The members reached a consensus regarding the voluntary nature of the TCBMs and delivered its final report in July 2013, which was endorsed by the UNGA in the same year.[86]

80 ibid art 5.
81 ibid art 6.
82 ibid art 7.
83 ibid art 8.
84 ibid art 9.
85 UNGA Res 65/68 (5 January 2011).
86 UNGA Res 68/50 (5 December 2013).

Recommended TCBMs for space include:

> 1) information exchange on national space policy and goals, and exchange of information on military space expenditures; 2) information exchange on activities in outer space, including orbital parameters, possible conjunctions, natural space hazards, and planned launches; 3) notifications on risk reductions such as scheduled maneuvers, uncontrolled high-risk re-entries, emergency situations, intentional orbital breakups; and 4) voluntary visits to launch sites and command and control centers, and demonstrations of space and rocket technologies; 5) international cooperation, consultative mechanisms, outreach, and coordination; and 6) voluntary political measures to form the basis for legally binding obligations.[87]

4.5 The COSPAR Planetary Protection Policy

The Committee on Space Research (COSPAR), as a Scientific Committee of the International Council for Science (ICSU), now the International Science Council (ISC), formed its Panel on Planetary Protection in 1999 to consolidate, maintain and update the Planetary Protection Policy for the implementation of Article IX of the Outer Space Treaty. The primary objective is 'to prevent harmful contamination of the Moon and other celestial bodies through the organization of scientific assemblies, publications or any other means'.[88] The COSPAR Planetary Protection Policy was first introduced in October 2002 and then regularly updated to reflect real-time scientific understanding. This recommendatory document seeks to avoid biological contaminations of planetary environments that could compromise scientific exploration.[89] The policy is expected to be based on the most current, peer-reviewed scientific knowledge. It is also expected to enable the exploration of the solar system, rather than prohibiting it.[90]

The policy identified five categories of space mission/target body combinations for planetary protection requirements. No planetary protection requirements are imposed for Category I missions which have no direct interest in understanding the process of chemical evolution or the origin of life. Only simple documentation is required for Category II missions which have only a remote chance that contamination carried by spacecraft might jeopardize future exploration. Category III missions, mostly flyby and orbiter, with a significant chance of contamination jeopardizing future biological experiments, are required, apart from documentation,

[87] ibid.

[88] COSPAR Charter, art I <https://cosparhq.cnes.fr/about/charter/> accessed 22 September 2023.

[89] Tanja Masson-Zwaan and Mahulena Hofmann, *Introduction to Space Law* (4th edn, Kluwer 2019).

[90] 'The COSPAR Panel on Planetary Protection Role, Structure and Activities' (*COSPAR*) <https://cosparhq.cnes.fr/assets/uploads/2019/07/PPP_SRT-Article_Role-Structure_Aug-2019.pdf> accessed 22 September 2023.

to carry out some implementing procedures, including trajectory biasing, the use of cleanrooms during spacecraft assembly and testing, and possibly bioburden reduction.

Category IV missions, mostly probe and lander, require detailed documentation (including a bioassay to enumerate the bioburden, a probability of contamination analysis and an inventory of the bulk constituent organics).[91] Category V missions comprise all Earth-return missions, involving the protection of the terrestrial system, the Earth and the Moon. Requirements for Categories I or II are required only for the outbound phase of Category V missions to solar system bodies with no indigenous life forms according to scientific opinions. Other Category V missions are defined as 'restricted Earth return' with higher requirements prohibiting destructive impact upon return through sterilization procedures and containment techniques.[92]

The COSPAR aims to promote transparency by recommending its members to provide necessary information regarding the procedure and computations used for planetary protection within a reasonable timeframe. It further produces eleven Implementing Guidelines and Category Specifications for Individual Target Bodies:

> 1) Implementation guidelines on the use of clean-room technology for outer-planet missions; 2) Numerical implementation guidelines for forward contamination calculations; 3) Guidelines on the implementation of an organic inventory; 4) Trajectory biasing; 5) Environmental conditions for replication; 6) Implementation guidelines for Category V missions; 7) Category-specific listing of target body/mission types; 8) Category II requirements for missions to the Moon; 9) Category III/IV/V requirements for Mars; 10) Category III/IV/V requirements for Europa and Enceladus; 11) Category requirements for small solar system bodies.

5 Development of Soft Law Rules in Outer Space

In the last few decades, soft law rules have been playing an increasingly critical role in the regulation of space activities. This situation will continue in the near future in view of the stagnation in the enactment of international space law documents. However, as in any other branches of international law, soft law rules in the space law field cannot stand alone without the support of binding international

[91] Category IV focuses on 'an increased number of implementing procedures (including trajectory biasing, cleanrooms, bioload reduction, possible partial sterilization of the direct contact hardware and a bioshield for that hardware)'.

[92] COSPAR, 'COSPAR Policy on Planetary Protection' (3 June 2021) <https://cosparhq.cnes.fr/assets/uploads/2021/07/PPPolicy_2021_3-June.pdf> accessed 22 September 2023.

documents. There is thus a need to systematically examine the development of soft law rules in outer space. The UNCOPUOS Legal Subcommittee timely decided in 2013 to add a new agenda item on 'the general exchange of information on practices in relation to non-legally binding instruments for outer space activities'.[93]

Since the adoption of UNGA resolutions in the early 1980s, three distinct features can be identified for the development of soft laws in outer space. First, the entities involved in formulating soft law rules are increasingly diversified. The UN, in particular the UNGA, is the main entity in creating soft laws in outer space through the adoption of UNGA resolutions. However, we have witnessed entities outside the UN, such as the EU and IADC, taking an increasingly active role in formulating soft law rules. It is expected that more and more international organizations will be involved in this process.

Secondly, soft law rules will expand their regulatory scope to a wider range of space activities, in light of the rapid development of space technologies and the regulatory gap left by the existing space treaties. Before the adoption of any new international treaties, soft law rules are expected to fill the gap for the regulation of commercial space activities (such as the exploitation of natural resources and the development of space tourism), and space sustainability (such as space traffic management and peaceful uses of outer space).

Thirdly, the international society will increasingly strengthen the ethical and legal force of soft law rules through voluntary compliance and national space legislation. The existing space treaties are lagging behind in the regulation of commercial and private space activities. The failure to adopt new binding rules provides an excellent opportunity for soft law rules to take up the alternative role, which is increasingly recognized by the international community. Furthermore, soft law rules, mostly adopted to tackle urgent issues of common concern, can reflect the reality of the needs of the international community in the development of their space activities. The conclusion of such soft law rules can in turn contribute to the healthy and orderly development of space activities.

Broadly speaking, we can expect that soft law will develop in four major directions in the future. First, soft law rules can be transformed into binding documents. One typical example is the transformation of Resolution 1962 (XVIII) in 1963 into the Outer Space Treaty. Secondly, soft laws may develop into international customary law rules through state practice and judicial decisions over time and become legally binding on the States. Thirdly, while not themselves legally binding, soft law rules can help with the interpretation and clarification of relevant provisions in international treaties. For example, the Space Debris Mitigation Guidelines help with the application of Article IX of the Outer Space Treaty and the Liability Convention in the determination of the concept of 'duty

[93] UNCOPUOS, 'Item 13 of the Provisional Agenda' Fifty-second Session of the Legal Subcommittee 8–19 April 2013 (11 April 2013) UN Doc A/AC.105/C.2/L.291.

of care'. Fourthly, soft law rules can be included in national space legislation. For example, the Interim Administrative Measures on Space Debris Mitigation and Protective Management adopted by China in 2009 have taken full account of the UNCOPUOS Space Debris Mitigation Guidelines. The UNCOPUOS published in 2014 a 'Compendium of Space Debris Mitigation Standards Adopted by States and International Organizations' listing relevant instruments, guidelines and measures that have been adopted by States and international organizations.[94]

6 Conclusion

The UN oversaw the development of the five space treaties during the 1960s and 1970s. On the one hand, with more States joining the space club, the conclusion of new space treaties became more difficult; on the other hand, the existing space treaties cannot satisfy the needs of new space activities. This dilemma provides an opportunity for soft law rules to fill the gap in the regulatory regime for space activities. The flexibility in the formulation and operation of soft law rules is particularly attractive to space-faring nations. Soft law, though not legally binding, can be enforced through voluntary compliance or the enactment of similar rules in national laws.

The UNGA has been a major player in issuing resolutions on specific legal issues related to space activities. Other entities, such as the EU, IADC and ILA, have also drafted soft law rules within their jurisdictions in recent years. Nevertheless, it must be noted that soft law is not a law in the real sense, nor can it replace hard law. In practice, some States may even escape their own duties due to the vague wording of soft law rules, which leads to randomness and uncertainty in actual implementation. Despite this, soft law rules for space activities are still an optimal choice in the absence of treaty law and customary law; they are thus the second-best choice for space regulation in view of the difficulty in concluding hard law rules. Soft law remains an important instrument to improve the rule of law in outer space.

[94] 'Compendium of Space Debris Mitigation Standards Adopted by States and International Organizations' (*UNOOSA*) <https://www.unoosa.org/oosa/en/ourwork/topics/space-debris/compendium.html> accessed 22 September 2023.

11

ROLE OF THE INTERNATIONAL TELECOMMUNICATION UNION IN REGULATING SPACE ACTIVITIES

Jennifer A Brobst

1 Introduction

The International Telecommunication Union (ITU) is a large intergovernmental body created in 1865 to regulate cross-border telegraph communications among nations on Earth.[1] The ITU is now an agency of the United Nations (UN), seated in Geneva, Switzerland, with 193 member states and hundreds of other governmental, organisational, and industry sector members and associates.[2] This specialised agency has a role in regulating radio, cellular and internet communications, satellite launches, and other electromagnetic services on Earth and in outer space. Without the capacity for communications across great distances, space travel and exploration would not be possible.

Key aspects of information sharing through telecommunications[3] include observation, mapping, and navigation. For example, the ITU regulates Radio Navigation Satellite Systems (RNSS), which are comprised of various national and

[1] Originally named the International Telegraphic Union, the current, more inclusive title of International Telecommunication Union was adopted in 1932, with inclusion into the United Nations in 1947. See Jens Hinricher, 'The Law-Making of the International Telecommunication Union (ITU)—Providing a New Source of International Law?' (2004) 64 HeidelbergJIntlLaw 489, 489.

[2] 'Status of International Agreements relating to Activities in Outer Space as of 1 January 2022' Sixty-first Session of the UNCOPOUS Legal Subcommittee (Vienna, 28 March–8 April 2022) (28 March 2022) UN Doc A/AC.105/C.2/2022/CRP.10.

[3] In 1947, the International Telecommunication Conference defined 'telecommunication' to include 'any transmission, emission or reception of signs, signals, writings, images and sounds or intelligence of any nature by wire, radio, visual or other electromagnetic systems'. George Arthur Codding, Jr, *The International Telecommunication Union: An Experiment in International Cooperation* (EJ Brill 1952) 1. This remains the ITU's official definition, with the replacement of the term 'visual' with 'optical' in the Radio Regulations. Constitution of the International Telecommunication Union (adopted 22 December 1992, entered into force 1 July 1994) 1825 UNTS 330, Annex para 1012 (ITU Const); Radio Regulations of the International Telecommunications Union (adopted 22 December 1992, entered into force 1 July 1994) 1825 UNTS 330, no 1.3 (2020) (ITU RR). This definition is also adopted by some nation states, such as the United States in 47 USC §702(6) (2018).

Jennifer A Brobst, *Role of the International Telecommunication Union in Regulating Space Activities* In: *International Space Law in the New Space Era*. Edited by: Sandeepa Bhat B., Dilip Ukey, and Adithya Variath, Oxford University Press.
 DOI: 10.1093/9780198909415.003.0011

multinational space satellite navigation (or satnav) systems.[4] Satellite radio installations are also able to surveil and collect a constellation of radio wave signals with the capacity to determine a particular location of persons or objects at a particular time.[5] Such complex communications ensure efficient human and machine communications worldwide,[6] but require administrative coordination to avoid conflict and interference. The ITU provides such coordination, especially prioritising the safety of life[7] and governmental telecommunications.[8] However, the continual increase in spectrum filings for space services has proved to be a regulatory challenge.[9]

Although communications technology is experiencing a renaissance of development, the ITU and international legal structures have remained comparatively static. From the printing press, invented by Johannes Gutenberg in Germany in 1436 enabling public mass communications, to the electronic telegraph, patented by Samuel Morse in the United States in 1844 and offering instant communications across great distances,[10] the expansion of methods of human communication has been profound. The pivotal invention leading to global and space communications is the wireless telegraph, invented by Guglielmo Marconi in Italy in 1901 and demonstrating the promise of radio wave frequencies (RF) transmissions to the world.[11] The ITU has played an important international role throughout most

[4] For a concise overview of the structure and membership of the ITU, as it relates to radiocommunications, see 'ITU-R Radiocommunication Study Groups 2020' (*ITU-R*, 2020) <www.itu.int/dms_pub/itu-r/opb/gen/R-GEN-SGB-2020-PDF-E.pdf> accessed 14 June 2022.

[5] ITU RR no 1.9 (defining radiodetermination as 'the determination of the position, velocity and/or other characteristics of an object, or the obtaining of information relating to these parameters, by means of the propagation properties of radio waves'). See generally 'What is GNSS?' (*European Union Agency for the Space Programme*, 3 December 2021) <www.euspa.europa.eu/european-space/eu-space-programme/what-gnss> accessed 14 June 2022.

[6] See eg Radiocommunications Act 1992 (Australia), 1 September 2021, s 6, No 174, Compilation no 76 (defining radiocommunication as communicating information between persons, things, and parts of things).

[7] ITU Const (n 3) art 40 para 191 ('International telecommunication services must give absolute priority to all telecommunications concerning safety of life at sea, on land, in the air or in outer space, as well as to epidemiological telecommunications of exceptional urgency of the World Health Organization'); ITU RR (n 3) No 4.10 ('Member States recognize that the safety aspects of radionavigation and other safety services require special measures to ensure their freedom from harmful interference; it is necessary therefore to take this factor into account in the assignment and use of frequencies').

[8] ITU Const (n 3) art 41 para 191.

[9] ITU Res 35-1 (WRC-19) ('[F]ilings for frequency assignments to non-geostationary-satellite (non-GSO) systems composed of hundreds to thousands of non-GSO satellites have been received by ITU since 2011, in particular in frequency bands allocated to the fixed-satellite service (FSS) or the mobile-satellite service (MSS)'); ITU REC 36-1 (rev WRC-19) ('[T]he geostationary-satellite orbit and the radio-frequency spectrum are limited natural resources and are being increasingly utilized by space services').

[10] 'Today in History—May 24' (*US Library of Congress*) <www.loc.gov/item/today-in-history/> accessed 19 June 2022 ('As the Civil War progressed, mobile telegraph stations and hastily strung wires began to connect scattered military units').

[11] Donna Coleman Gregg, 'Lessons Learned from the Spectrum Wars: Views on the United States' Effort Going Into and Coming Out of a World Radiocommunication Conference' (2009) 17 CommLConspec 377, 380.

of this technological innovation, and throughout all aspects of the Space Age. As stated by the Director of the ITU Radiocommunications Bureau in 2020:

> Since the launch of the first satellite in 1957, we have witnessed ground-breaking advances in satellite design, manufacturing and launch service capabilities providing immeasurable opportunities for Earth observation, weather forecasting, remote sensing, global positioning systems, satellite television, mobile communication systems and many others.[12]

The ITU's regulation of telecommunications since 1865 has fostered both technological innovation and international cooperation, despite its lack of direct enforcement power and reliance on negotiation in the settlement of disputes.[13] The ITU is neither an enforcement body nor a profitable, bureaucratic agency, but rather provides a comprehensive forum for collaboration.[14] For over 150 years, the ITU has maintained loyal membership with more members than the UN itself. This chapter will address the international legal and policy issues relating to the ITU's role in regulating space activities, particularly satellite operations and the practical coordination of telecommunications.

2 Purpose and Structure of the ITU

The ITU is one of the most mature agencies of the UN, governed since 1865 by the International Telegraph Convention.[15] First incorporating regulation of telephone services, the body then merged with the International Radiotelegraph Union, which was established in 1903 to protect human life from disasters at sea through internationally regulated radio communications.[16] The ITU changed its

[12] Mario Maniewicz, 'Foreword by the Director, Space Science for the Achieving the Sustainable Development Goals' (2020) 6 ITUNewsMag 5.

[13] Stathis Potamitis, 'Competition in International Satellite Telecommunications Services' (1986) 44 UTorontoFacLRev 33, 50 ('[T]he allocation of portion of the frequency spectrum by the ITU depends primarily on the member-states' willingness to comply with the organization's regulations').

[14] In art 1, s 2 of the Constitution of the ITU, the particular roles of the ITU are described in terms of fostering dialogue, coordinating efforts, and facilitating standardisation rather than proscribing specific conduct or imposing sanctions upon members.

[15] Convention télégraphique internationale de Paris (1865) et Règlement de service international (1865) (Paris, 1865) (adopted to improve coordinated, efficient, and rapid telegraphic services).

[16] David Withers, *Radio Spectrum Management: Management of the Spectrum and Regulation of Radio Services* (2nd edn, The Institute of Electrical Engineers 1999) 7. A comprehensive early history of the ITU from the 1800s to the 1950s may be found in Codding (n 3) 461. For a brief and concise history, see 'Coordinating a Limited Resource in Radio Communications to Enable Future Missions' (*NASA*, 3 November 2016) <www.nasa.gov/directorates/heo/scan/news_coordinating_spectrum_to_enable_future_missions.html> accessed 14 June 2022.

name to the International Telecommunication Union in 1934[17] and was adopted as a specialised agency and restructured in 1947 by the newly formed UN.[18] Since the more recent adoption of the ITU Constitution and Convention in 1992, the ITU has repeatedly adjusted its practices,[19] but its primary purpose remains the same, in RF allocation, telecommunication standardisation, and space services. With respect to RF regulation, while the ITU allocates frequency bands for certain purposes, such as meteorological, military, or amateur use, the allotment to specific member Administrations is achieved through negotiation, and each Administration then assigns specific frequency bands to specific radio stations.[20] These procedures seek to avoid harmful interference from other RF, where harmful interference is defined as '[i]nterference which endangers the functioning of a radionavigation service or of other safety services or seriously degrades, obstructs or repeatedly interrupts a radiocommunication service operating in accordance with the Radio Regulations.'[21] It is a concept mirrored in the Outer Space Treaty of 1967, which states in Article IX that nations should seek to avoid harmful interference to the use and exploration of outer space by other nations.[22]

The regulation of telecommunications is not merely market-driven but recognises the integral role of communications as a foundational aspect of a peaceful and cohesive society. As self-described in a Resolution arising from the proceedings of the 2018 ITU Plenipotentiary Conference: 'The Union is committed to connecting all the world's people—wherever they live and whatever their means. ITU's work aims to protect and support everyone's fundamental right to communicate.'[23] The duties and obligations of the ITU and its membership in achieving these aims are set out below.

[17] International Telecommunication Convention (adopted 9 December 1932, entered into force 1 January 1934) US Treaty Series No 867.

[18] Agreement Between the United Nations and the International Telecommunication Union (adopted 2 October 1947, entered into force 1 January 1949) 193 UNTS 3. See generally 'History of ITU Portal' <www.itu.int/en/history/> accessed 14 June 2022.

[19] Convention of the International Telecommunication Union (adopted 22 December 1992, entered into force 1 July 1994) 1825 UNTS 143 (ITU Conv); Gennady Tarasov and Gopinathan Achamkulangare, *Review of Management and Administration in the International Telecommunication Union (ITU)* (UN Joint Inspection Unit, 2016) UN Doc JIU/REP/2016/1 6 (relating that the Constitution of the ITU has been altered by over 274 amendments, and the ITU Convention by over 450 amendments).

[20] See ITU RR (n 3) nos 1.16–1.18, and 5.1. For example, Ofcom, like many nations, allocates and regulates assignments of frequencies within the UK through an auction and licensing scheme. 'Radio Spectrum Regime' (*Practical LawUK*) <https://uk.practicallaw.thomsonreuters.com/5-386-3154?transitionType=Default&contextData=(sc.Default)> accessed 27 September 2023.

[21] ITU Const (n 3) Annex para 1003; ITU Const, art 45 s 2(a) para 11 (harmful interference).

[22] Treaty on Principles Governing the Activities of States in the Exploration and Use of Outer Space, Including the Moon and Other Celestial Bodies (adopted 27 January 1967, entered into force 10 October 1967) 610 UNTS 205, art IX (Outer Space Treaty).

[23] ITU Res 71, Annex 2 s 2.a (rev PP-2018).

2.1 Legal Framework of the ITU

The Constitution of the ITU defines its broader purpose as 'facilitating peaceful relations, international cooperation among peoples and economic and social development by means of efficient telecommunication services'.[24] However, the ITU exempts military radio stations from its purview.[25] Its current infrastructure was established in 1992, with a Secretary-General as legal representative and a Plenipotentiary Conference as the supreme legislative body of the ITU.[26] The Plenipotentiary Conference takes place every four years, electing leaders and setting ITU's strategic and financial priorities,[27] with the implementation of the priorities by the ITU's Council between Plenipotentiary Conferences.[28] Application for membership in the ITU must be approved by at least two-thirds of member states.[29]

The basic instruments of the ITU for radiocommunications are comprised of the Constitution, Convention, Radio Regulations, and the International Telecommunication Regulations, all of which are binding on members.[30] The member states, in turn, are bound to enforce the ITU provisions upon operating Administrations in their jurisdiction, including private ITU members, when they engage in either international activities or activities that could cause harmful interference to the radio services of other nations.[31] The ITU also provides standardisation recommendations for non-radio communications, such as optical communications using fibre optic cable to transmit voice, images, and data through the reflection of light.[32] Optical or laser communications provide useful analogues for radio transmissions in space activities, and both forms of communications work in tandem for new solutions in deep space. To their advantage, optical communications do not radiate signals or cause interference,[33] and thus require

[24] ITU Const (n 3) Preamble para 1.

[25] ibid (n 3) art 48 s 1 para 202 ('Member States retain their entire freedom with regard to military radio installations').

[26] ibid (n 3) art 11 s 1(1) para 73*bis*. See Joint Inspection (n 19) 5.

[27] ibid (n 3) art 8 para 47. See 'ITU PP Bucharest 2022' (*ITU*) <www.itu.int/pp22/en/> accessed 19 June 2022.

[28] ibid (n 3) art 10 s 3 para 68.

[29] ibid (n 3) art 2 s c para 23.

[30] ibid (n 3) art 4 s 3 para 31; art 54 s 1 para 215 (regulations as binding).

[31] ibid (n 3) art 6 s 2 para 38.

[32] See eg 'Submarine Telecom Cables Enhance climate Monitoring and Tsunami Forecasts' (*ITU News*, 8 June 2022) <www.itu.int/hub/2022/06/submarine-cables-telecom-climate-monitoring-tsunami-forecasts/> accessed 18 June 2022; Hamid Hemmati and Abhijit Biswas, 'Future Prospects and Applications' in Hami Hemmati (ed), *Deep Space Optical Communications* (John Wiley 2006) 643–58.

[33] For examples of current uses of optical communications in space, see 'Fiber Optic Sensing System Readied for Space Use' (*NASA*, 22 June 2020) <www.nasa.gov/centers/armstrong/features/Fiber-Optic-Sensing-System-Readied-for-Space-Use.html> accessed 18 June 2022; '22.LCRD [Laser Communication Relay Demonstration]—The Future, NASA's Invisible Network Podcast' (*NASA*) <www.nasa.gov/mediacast/goddard/2021/22-lcrd-the-future-nasas-the-invisible-network-podcast> accessed 18 June 2022 (discussing heightened security offered by fibre optics and the Laser-Enhanced Mission Communications Navigation and Operational Services pipeline project (LEMNOS), which is

standardisation coordination but not the complex negotiations that the ITU oversees for wireless RF and its inherent risk of interference.

In 2016, in its most recent review, the UN Joint Inspection Unit described the ITU as unparalleled in the complexity of governance, based on adopting a 'federal' structure, 'rather than the "unitary" model that is commonly found in the rest of the United Nations system'.[34] This assessment relates in part to the ITU relying on self-monitoring by members, but also by delegating work through three sectors: the Radiocommunication Sector (ITU-R),[35] the Telecommunication Standardization Sector (ITU-T),[36] and the Telecommunication Development Sector (ITU-D).[37]

In addition, the Constitution of the ITU expressly states that it will engage in co-operation with intergovernmental organisations and non-governmental organisations in matters relating to telecommunications.[38] The work of the ITU intersects with that of many other international bodies, such as the World Meteorological Organization, a specialised body of the UN which precisely monitors weather, climate, and water on Earth through the use of weather satellites.[39] The World Trade Organization addresses controversial cybersecurity and privacy concerns in telecommunications.[40] Membership in the ITU as a member state also requires recognition of political sovereignty, which may require interpretations of UN General Assembly resolutions, the UN Charter, and the Universal Declaration of Human Rights.[41] Finally, some international organisations are also designated to manage specific frequency assignments, such as the UN's International Civil Aviation Organization (ICAO).[42]

With respect to space activities, the ITU-R facilitates negotiations at World Radiocommunication Conferences (WRC) every four years,[43] as well as Regional Radiocommunication Conferences,[44] in order to determine efficient radio

developing the Orion Artemis II Optical Communications System (O2O) that will enable lunar astronauts to send ultra-high-definition video to Earth from the Moon).

[34] Tarasov and Achamkulangare (n 19) 6. The UN Joint Inspection Unit has reviewed the ITU on two prior occasions in 2001 (UN Doc JIU/REP/2001/3) and 2009 (UN Doc JIU/REP/2009/3).

[35] ITU Const (n 3) art 7 paras 39–46. See ITU-R Study Groups (n 4) 5.

[36] ITU Const (n 3) art 17 s 1(1) para 104 ('studying technical, operating and tariff questions and adopting recommendations on them with a view to standardizing telecommunications on a worldwide basis').

[37] ibid art 21 s 2 paras 120–29 (supporting the UN's economic and social development priorities).

[38] ibid art 1 s 2(g) para 9.

[39] Petteri Tallas, 'The Importance of Progress in Earth Observation from Space to Protect Our Planet' (2020) 6 ITU NewsMag 13.

[40] Scott J Shackelford, 'The Law of Cyber Peace' (2017) 18 ChiJIntlL 1, 24 (providing the example of commercial concerns with US National Security Administration (NSA) surveillance and its impact on trade).

[41] See eg ITU Res 12 (WRC-19) (promoting the ability of Palestine to modernise its radio spectrum use and telecommunication networks).

[42] ICAO Handbook on Radio Frequency Spectrum Requirements for Civil Aviation, ICAO Doc 9718 (2018).

[43] ITU Const (n 3) art 13.

[44] ibid art 43 para 194; ITU Conv (n 19) art 9.

spectrum use on Earth and in space. At the WRC and Plenipotentiary Conferences, only member states have voting rights and each member state has one vote.[45] In 2019, Resolution 811 of the WRC-19 presented the agenda for the next WRC in 2023, including five chapters: (1) fixed, mobile and broadcasting issues; (2) aeronautical and maritime issues; (3) science issues; (4) satellite issues; and (5) general issues.[46] Such negotiations are essential to the safety and success of national and international space programmes:

> NASA's [National Aeronautics and Space Administration] participation and collaboration with the NTIA [National Telecommunications and Information Administration], FCC [Federal Communications Commission], and ITU ensure that existing missions are protected from interference and that sufficient bandwidth is available for future missions. Today and in the coming years, NASA's missions demand highly reliable communications to explore the solar system and beyond and to support technology development for scientific and aeronautical research as well as human exploration on the Journey to Mars.[47]

Multiple study groups within the ITU conduct extensive research to further develop access and quality of radiocommunications technology, and to promote technological innovations. The Science Services Study Group specifically addresses a number of space-related subjects, including: meteorological satellite services for weather forecasting and climate change; environmental monitoring (land mass, oceanic, biomass, and atmospheric); disaster detection and relief operations (earthquakes, tsunamis, oil spills); radioastronomy; human and robotic spacecraft and exploration; research satellites; and standardisation of frequency and time signals.[48]

Finally, the ITU-R relies in part on governance through a Bureau and a Board, which address registration and conflict resolution. The Radiocommunication Bureau assigns RF allocations and maintains a Master International Frequency Register, currently including approximately 3.14 million terrestrial radio station records and 2,200 satellite station records.[49] The Radio Regulations Board has 12 elected members who are independent advisors in the allocation and use of RF, serving as 'custodians of an international public trust'.[50] The Board implements the regulations, reviews appeals from the Bureau's frequency assignments

[45] ITU Conv (n 19) art 32A s 1 para 340A; ITU Res 14 (rev PP-2006) (recognising that non-member-state sector members 'may participate in all activities of the Sector concerned, with the exception of formal votes and some treaty-making conferences').

[46] See 'ITU-R Preparatory Studies for WRC-23, Chapters of the Draft CPM Report to WRC-23' (*ITU-R*) <www.itu.int/en/ITU-R/study-groups/rcpm/Pages/wrc-23-studies.aspx> accessed 15 June 2022.

[47] Coordinating a Limited Resource (n 16).

[48] ITU-R Study Groups (n 4) 47.

[49] ibid 8. See ITU Const, art 16.

[50] ITU Const (n 3) art 14. See ITU-R Study Groups (n 4) 12.

and unresolved claims of interference, and provides advice to the WRC.[51] While ITU member states and sector members must agree to abide by the provisions of the ITU Constitution and Convention, as well as the various Resolutions adopted in Conferences, the Radio Regulations also serve as binding international authority and provide the detailed guidance necessary for the implementation of radiocommunications.

2.2 ITU Radio Regulations as Binding International Authority

The ITU Radio Regulations, last revised in 2019 by the WRC-19, govern radiocommunication activities through a detailed categorisation and allocation process.[52] For example, the regulations categorise RNSS that provide radionavigation services on land and sea, as well as through the Earth's atmosphere and into outer space.[53] The regulations also distinguish radiocommunication safety services which specifically safeguard human life and property,[54] such as aeronautical radio navigation services (ARNS) which are particularly important for safety-of-life applications in air traffic control and congested flight plans.[55] Protection of all of these services requires freedom from harmful interference and careful assignment and use of frequencies for compatibility.[56] Efficiency also requires that frequency allocations be used, or the designed member may forfeit access.[57] While the ITU's allocation to members considers broad categories of use for certain frequencies on the spectrum, the assignment of specific frequencies occurs at the national level and is negotiated among member states at WRCs.[58]

The Radio Regulations demand that members ensure proper use of the RF that have been allotted to them.[59] That is, they must 'endeavour to limit the number of frequencies and the spectrum used to the minimum essential to provide in a satisfactory manner the necessary services. To that end, they shall endeavour to apply

[51] ITU-R Study Groups (n 4) 12.

[52] See generally Fabio Tronchetti, *Fundamentals of Space Law & Policy* (Springer/International Space University 2013) 40.

[53] ITU RR (n 3) no 1.43.

[54] ibid no 1.59.

[55] 'GNSS Signal' (*ESA Navipedia*) <https://gssc.esa.int/navipedia/index.php/GNSS_signal> accessed 12 June 2022.

[56] ITU RR (n 3) no 4.10.

[57] Delbert D Smith and Christopher Stott, 'Private Sector Utilization of the Moon: A Right of Use' (2022) 34(3) Air&SpaceLaw ABA 12, 13 (noting a requirement of continued justification of spectrum rights to the ITU once allocated).

[58] David M Leive, *International Telecommunications and International Law: The Regulation of the Radio Spectrum* (AW Sijthoff/Oceana Publications 1970) 20.

[59] ITU RR (n 3) no 3.14 ('To ensure compliance with these Regulations, administrations shall arrange for frequent checks to be made of the emissions of stations under their jurisdiction').

the latest technical advances as soon as possible.'[60] This structure and practice of self-regulation have remained fairly consistent within the ITU during its relatively long history as a specialised agency of the UN.

With respect to satellite monitoring, ITU radio regulations do provide a complex licensing and planning scheme for launch and orbit by member states. Application to the ITU is made by a nation on behalf of those within its jurisdiction, without a need to prove justification for acquisition of rights to orbital and frequency designations.[61] This liberal process incentivises financial speculation in the application of rights not yet needed, known as spectrum warehousing,[62] and also permits domination by the few nations that emerge first in the field.[63] By 1986, nearly half of all geostationary satellites were controlled by the United States, but in the next decade Intelsat, a commercial cooperative of 130 states providing an international public satellite telecom network, would emerge to bring interconnectivity across the planet.[64]

As of 2021, 4,550 satellites orbited the Earth, including at least one satellite from 75 different nations.[65] Of these, 565 satellites were geostationary, 3,790 were in Low Earth Orbit, 159 in Medium Earth Orbit, and 56 in a Highly Elliptical Orbit.[66] It is predicted that the American company SpaceX will soon own more satellites in orbit than any other company or nation combined, as it currently operates one-third (1,655) of all satellites in space, compared to ownership by the governments of China (218), the United States (193), Russia (125), Europe (51), and India (47).[67] The Radio Regulations do not constrain the competitive success of those members that are able to utilise new satellite technology first.

However, the ITU only grants voting rights to member states, a right not granted to private sector members, but it is widely apparent that private investment has become crucial to space development. At the national level, with the early development of satellites in the 1960s, a number of nations created public-private partnerships, such as the Communications Satellite Corporation (COMSAT), federally

[60] ITU Const (n 3) art 44 s 1 para 195; ITU RR (n 3) Preamble no 0.2; ITU RR (n 3) no 3.8 ('every effort should be made to keep frequency tolerances and levels of unwanted emissions at the lowest values which the state of the technique and the nature of the service permit').

[61] See generally Albert N Delzeit and Robert F Beal, 'The Vulnerability of the Pacific Rim Orbital Spectrum Under International Space Law' (1996) 9 NYIntlLRev 69.

[62] ITU Res 35 (WRC-19) (identifying the need to ensure actual use of NGSS frequencies, particularly considering restrictions due to congestion of satellites potentially impacting the Space Stations). See also ITU REC 36-1 (rev WRC-19) (promoting the Master International Frequency Register to 'reflect more accurately the actual use being made of these resources').

[63] See eg Delzeit and Beal (n 61) 70–73 (discussing Tonga's unsuccessful application for 16 orbital satellite positions over the lucrative Pacific Rim).

[64] ibid 76.

[65] Kelly Kizer Whitt, 'Who Owns All the Satellites?' (*EarthSky.org*, 8 February 2022) <https://earthsky.org/space/who-owns-satellites-company-country/> accessed 14 June 2022.

[66] ibid.

[67] ibid. Note that the listed figures are meant to reflect 'government' ownership, but the proliferation of public-private partnerships makes such designations less accurate as they vary in structure and are not consistently defined among nations.

funded but privately run to provide international satellite communications for the United States.[68] The International Telecommunication Satellite Organization (INTELSAT) was created in 1964 as a single quasi-commercial global satellite system and public utilities company.[69] Not only has commercialisation influenced the work of the ITU in space activities, but the addition of regional satellite networks has also significantly changed the regulatory landscape and opportunities for all nations to participate.[70]

2.3 Intersection with International and National Space Law

For space activities involving the ITU, the member states must also comply with several international space treaties and customary international law, while the private sector members are bound by their own nation's space legislation. These binding treaties include the Outer Space Treaty of 1967,[71] the Rescue Treaty of 1968,[72] the Liability Convention of 1972,[73] the Registration Convention of 1975,[74] and the Moon Treaty of 1979, the last of which has relatively fewer signatories than the other space treaties.[75] The binding instruments of the ITU and the various space treaties do not cross-reference each other, but they are often harmonised in substance. For example, Articles 46 and 47 of the ITU Constitution, which require urgent response to distress calls but prohibit false or deceptive safety signals, intrinsically support the provisions of the Rescue Treaty.[76] Notably, the ITU

[68] For an overview of the historical development of satellite communications regulation in the United States, see *National Association of Broadcasters v FCC* (DC Cir 1984) 740 F2d 1190, 1195–96 (47 USC §§701–44).

[69] 'About Us' (*Intelsat*) <www.intelsat.com/about-us/> accessed 27 September 2023; Lorenza Sebesta, 'U.S.-European Relations and the Decision to Build Ariane, the European Launch Vehicle' in Andrew J Butrica (ed), *SP-1247 Beyond the Ionosphere Fifty Years of Satellite Communication* (The NASA History Series 1997) ch 11 para 140 <https://history.nasa.gov/SP-4217/ch11.htm> accessed 14 June 2022; Potamitis (n 13) 33.

[70] 'What is Ofcom?' (*Ofcom*) <www.ofcom.org.uk/about-ofcom/what-is-ofcom> accessed 15 June 2022 (describing the United Kingdom regulator of telecommunications as 'independent' but taking 'duties from Parliament'); Attila Matas, 'Radio Navigation Satellite Service and the ITU Radio Regulations, ICG-8 (Dubai, United Arab Emirates)' (*ITU Radiocommunication Bureau*, 9–14 November 2013) (noting that, before 2000, there were two GPSS (Navstar-GPS and GLONASS), but that by 2013 there were satellite filings representing 238 RNSS networks from 22 national administrations).

[71] Outer Space Treaty (n 22).

[72] Agreement on the Rescue of Astronauts, the Return of Astronauts and the Return of Objects Launched into Outer Space (adopted 22 April 1968, entered into force 3 December 1968) 672 UNTS 119 (Rescue Agreement).

[73] Convention on International Liability for Damage Caused by Space Objects (adopted 29 March 1972, entered into force 1 September 1972) 961 UNTS 187 (Liability Convention).

[74] Convention on Registration of Objects Launched into Outer Space (adopted 14 January 1975, entered into force 15 September 1976) 1023 UNTS 15 (Registration Convention).

[75] Agreement Governing the Activities of States on the Moon and Other Celestial Bodies (adopted 18 December 1979, entered into force 11 July 1984) 1363 UNTS 3 (Moon Treaty). For the status of the five international space treaties, see Status (n 2).

[76] ITU Const (n 3) arts 46 and 47 paras 200–01.

Convention is older and has been ratified by more member states than any of the other space treaties.[77] With this established history, the ITU has become mutually reliant on the established space treaties when considering ongoing questions as to the meaning of sovereignty and jurisdiction of nations in space, space debris mitigation for satellite systems, and the resolution of claims of liability associated with information and communications systems in space.

For example, whether cyberspace (interconnected digital technology) is part of the global commons and therefore immune to sovereign appropriation is still in question.[78] In the sharing of Earth observation data to address climate change, some assert that open access is critical[79] and supported by the Outer Space Treaty which promotes sharing research findings with the public and the international scientific community.[80] Customary international law would define global commons to include outer space, the Earth's atmosphere, the high seas, and Antarctica,[81] and all of these make use of satellite-based radio wave telecommunications and cyberspace. With respect to geostationary orbits, the United States Congress has interpreted Article 44 of the ITU Constitution as declining to grant any sovereign rights over segments of the geostationary orbit, 'nor does any country have preferential rights over any such segment'.[82] Indeed, as satellites are positioned around the Earth, they do not inhabit the jurisdictional space of the nearest land on Earth, unlike civil aviation laws applicable to the Earth's atmosphere.

The clearest foundational support for a global commons framework resides in the Outer Space Treaty, which declares in Article I: 'The exploration and use of outer space, including the moon and other celestial bodies, shall be carried out for the benefit and in the interests of all countries, irrespective of their degree of economic or scientific development, and shall be the province of all mankind.'[83]

However, if strictly applied, this common heritage principle could require consent from all nations to extract minerals or water from celestial bodies, which is the approach of the Moon Treaty,[84] but contrary to both the rights of exclusivity agreed upon by other nations in the Artemis Accords and the interests of investors.[85]

[77] Status (n 2) 10.

[78] Shackelford (n 40) 30.

[79] Marco Borghi, 'Toward Full and Open Access: Challenges and Opportunities for the Legal Interoperability of Earth Observation Data' (2021) 45 JSpaceL 142, 174 (discussing the challenges facing the Global Earth Observation System of Systems and the European Union's Earth Observation Copernicus programme).

[80] Outer Space Treaty (n 22) art XI.

[81] Shackelford (n 40) 30.

[82] Committee on Foreign Relations, 'Report on International Telecommunication Union Constitution and Convention' (27 October 1997) 105th Congress Senate Exec Rep 105(3) 14.

[83] Outer Space Treaty (n 22) art I.

[84] Joseph A Bosco, 'International Law Regarding Outer Space—An Overview' (1990) 55 JAirL&Com 609, 620; Moon Treaty (n 75).

[85] Artemis Accords: Principles for Cooperation in the Civil Exploration and Use of The Moon, Mars, Comets, and Asteroids for Peaceful Purposes (adopted 13 October 2023) s 10.2 (Artemis Accords) ('The Signatories affirm that the extraction of space resources does not inherently constitute national appropriation under Article II of the Outer Space Treaty, and that contracts and other legal instruments

The lack of more directly applicable binding space treaties has begun to place pressure on the work of the ITU. If substantial questions of interpretation of the older space treaties persist and the ITU does not have the necessary authority to rectify emerging concerns, then competitive efforts by those outside the ITU leadership will inevitably arise. For example, the risk of satellite collision with space debris[86] could be addressed through space insurance regimes,[87] which would not necessarily prevent further debris from being created nor properly mitigated. According to a 2016 report by the UN's Joint Inspection Unit:

> ITU operates in a fast-changing environment and faces competition, especially from the private sector, in several domains of its core activities. Therefore, it is crucial that ITU ensure its relevance and secure its place in the rapidly evolving information and communications technology (ICT) universe.[88]

Some scholars assert that this lack of a rigid framework creates a useful polycentric approach to governance of telecommunications and cyberspace, where custom and the process of trial-and-error between numerous public and private partners and their respective legal bodies effects a market-like form of governance.[89] In contrast, during the 1990s, a newly restructured and formalised ITU was deemed a stabilising force in the face of rapid technological development. As one member of the industry stated at a Congressional hearing with respect to the United States signing of the ITU Convention: 'Mr. Chairman, frankly, if the plenipotentiary is left to private industry, it will be a cat fight. I think we really need an organizational hand in this that will bring in the nations of the world together to make some decisions that industry will fully support and that they had a voice in setting.'[90]

In recent decades, non-binding space law agreements continue to be adopted, affecting the work of the ITU, such as the 2007 Space Debris Mitigation Guidelines following the Chinese anti-satellite test (ASAT) that had created a significant

relating to space resources should be consistent with that Treaty'). See Fengna Xu and others, 'A Re-Examination of the Fundamental Principles of International Space Law at the Dawn of Space Mining' (2020) 44 JSpaceL 1, 9; Mark J Sundahl and Jeffrey A Murphy, 'Set the Controls for the Heart of the Moon: Is Existing Law Sufficient' (2020) 48 GaJIntlL&CompL 683, 685 (arguing that investors need assurances of property-related rights to be willing to fund and engage in space exploration and settlement).

[86] ITU REC no 1003.2 (12/2010) (providing expressly non-binding guidelines on space debris mitigation).

[87] See Christopher TW Kunstadter. 'What Keeps Space Insurers Up at Night?' (2022) 34(3) Air&SpaceLaw ABA 10,10 (positing that improved collision detection warning systems have increased willingness to insure satellite operations, despite the increase in space debris).

[88] Joint Inspection (n 19) iii.

[89] Shackelford (n 40) 41.

[90] Committee on Foreign Relations (n 82) 51 (David F Fisher, Vice President, General Counsel, and Corporate Secretary, ADC Telecommunications, Inc).

amount of hazardous debris.[91] In 2022, the United States became the first spacefaring nation to expressly declare its intention not to conduct ASAT weapons testing, which would commonly use ground or air-launched missiles to destroy satellites.[92] Early historians, such as George Codding, attributed the ITU's characteristic 'disregard for legal formalities' to the fact that most of the work was conducted by telecommunications experts, rather than politicians,[93] but they may not have foreseen that rapid and successful technological development would eventually create a more politicised, competitive, and dangerous space arena. In the soft law approach to space debris is the understanding that innovations that enable satellite debris removal inherently present the risk of dual use by also creating a military weapon to destroy the satellites of other nations.[94]

A looser and more open approach to telecommunications regulation has galvanised public-private partnerships and commercialisation in space at the international and national levels. In the United Kingdom, for example, privatising government-owned telecommunications companies and opening the market to commercial competition required a transition to expanded access to licensure, a strategy that arose in part to deter the unlicensed use of radio equipment causing substantial interference to authorised systems in the 1960s.[95] The United States FCC has repeatedly chosen a relaxed regulatory approach in the face of new technology in order to allow for telecommunications efficiencies to develop naturally.[96] However, this approach required careful oversight to avoid monopolies contrary to the public interest.[97] In 1975, as the FCC faced a rapidly evolving satellite communications industry, the federal judiciary in the United States wrote that 'regulatory practices and policies that will serve the "public interest" today may be quite different from those that were adequate to that purpose in 1910, 1927, or 1934, or that may further the public interest in the future'.[98] This need for flexibility arguably

[91] For a discussion of these legal developments in historical context, see Jack M Beard, 'Soft Law's Failure on the Horizon: The International Code of Conduct for Outer Space Activities' (2017) 38 UPaJIntlL 335421; Scott J Shackelford, 'Governing the Final Frontier: A Polycentric Approach to Managing Space Weaponization and Debris' (2014) 51 AmBusLJ 429.

[92] Daryl G Kimball, 'U.S. Commits to ASAT Ban' (*Arms Control Today*, *Arms Control Association*, May 2022) <www.armscontrol.org/act/2022-05/news/us-commits-asat-ban#:~:text=In%20a%20statement%20on%20April,for%20responsible%20behavior%20in%20space.%E2%80%9D&text=The%20announcement%20follows%20the%20Russian%20launch%20on%20Nov> accessed 15 June 2022.

[93] Codding (n 3) 461.

[94] Major Adam G Mudge, 'Incentivizing "Active Debris Removal" Following the Failure of Mitigation Measures to Solve the Space Debris Problem: Current Challenges and Future Strategies' (2022) 82 AFLRev 88, 138.

[95] See Withers (n 16) 3–4 (discussing variation in national administration and regulation after ITU allocation).

[96] *National Ass'n of Broadcasters v FCC*, 740 F2d 1190, 1207 (DC Cir 1984) ('[I]n the early stages of the development of cable television, the FCC refused to restrict the number of channels of origination that a cable owner could operate; such restrictions were thought to constrain desirable experimentation within a new industry').

[97] ibid 1208.

[98] *Washington Utilities & Transportation Commission v FCC* 513 F2d 1142, 1157 (9th Cir 1975), *cert denied*, 423 US 836 (1975).

remains true and the development of telecommunications has only quickened in pace,[99] but now requires difficult decisions regarding when to abandon traditional approaches to spectrum use for new applications and technology.[100]

The non-binding soft law relied on by the ITU, such as internationally agreed upon principles, charters, codes of conduct, and recommendations, is more flexible but less protective than treaties.[101] Two decades ago, space law scholars were more sanguine:

> Restrained by the laws of physics, not by the laws of politics, co-operation is vital and unavoidable, thus turning simple recommendations and other legally non-binding decisions into powerful and authoritative 'law', which is, to use the words of a former ITU Secretary-General, almost universally followed.[102]

Some remain so.[103] Whether this level of cooperation is sustainable without supplementary binding international law and enforcement power, or whether custom in ITU regulatory actions affecting outer space crystallises into binding customary law, are open questions today. The reality of negotiations at ITU WRCs, however, is that they involve strongly conflicting views.[104] If it is cooperative conduct over time that eventually generates customary international law, the business practices of the growing commercialised space industry and their influence on governments in promoting economic development will certainly be a factor.[105] The potential for customary international space law is there, given the ITU's long history and collaborative framework, but without the benefit of a clear record of international dispute resolution or arbitral interpretations from the ITU.

[99] Thomas W Hazlett and Brent Skorup, 'Tragedy of the Regulatory Commons: LightSquared and the Missing Spectrum Rights' (2014) 13 DukeL&TechRev 1, 34 (asserting that without a free market in telecommunications, 'top-down [governmental] allocations . . . effectively force bureaucrats to make trade-offs uninformed by market data and heavily influenced by political pressure').

[100] 'Technical Compatibility Challenges Between Fixed Satellite Service and 5G in C-Band, 21/2/8628-C-band' (*Intelsat*, February 2021) <www.intelsat.com/wp-content/uploads/2021/02/intelsat-C-band-whitepaper.pdf> accessed 15 June 2022.

[101] See Beard (n 91) 345.

[102] Hinricher (n 1) 495.

[103] Steven Freeland and Yun Zhao, 'Rules of the "Space Road:" How Soft Law Principles Interact With Customary International Law for the Regulation of Space Activities' (2020) 44(2) JSpaceL 405, 417 (supporting the 'long-established understanding that soft law is a well-accepted approach—and, perhaps, a more practical and achievable approach than hard law—for advancing humankind's endeavors in outer space').

[104] Gregg (n 11) 386; Jeffrey A Eisenach, 'Spectrum Reallocation and the National Broadband Plan' (2011) 64 FedCommLJ 87, 126 (addressing the costly risks of spectrum interference due to lack of regulation and increasing demand).

[105] Melissa J Durkee, 'Interstitial Space Law' (2019) 97 WashULRev 423, 471. See eg Ranjana Kaul, 'Recent Space Reforms in India: Perspectives on Policy and Law' (2020) 44 JSpaceL 450, 478 (calling for national legislation to harmonise the activities of a growing private space sector in India with the ratified space treaties); Steven Semararo, 'Regulating Information Platforms: The Convergence to Antitrust' (2002) 1 Telecom&HighTechLaw 143, 144 (asserting that antitrust law will be key in a less regulated telecommunications environment).

3 The ITU's Limited Regulation of International Disputes

The risk of harmful interference to RF use in space-related telecommunications could jeopardise the safety and feasibility of space operations, both in deep space, such as a launch from the Moon to Mars, or in Low Earth Orbit, where myriad satellite uses have become vital to the lives of those on Earth. Regulation through the published allocation of RF by the ITU is a means of preventing such interference, but when conflicts do occur, the ITU also has a limited legal framework to address them.

3.1 Competition for Scarce Resources

The ITU Constitution describes the basic tasks of the ITU on Earth and in space in Article I:

> [to] effect allocation of bands of the radio-frequency spectrum, the allotment of radio frequencies and the registration of radio-frequency assignments and, for space services, of any associated orbital position in the geostationary-satellite orbit or of any associated characteristics of satellites in other orbits, in order to avoid harmful interference between radio stations of different countries.[106]

The radio frequencies are resources allocated by the ITU in the form of electromagnetic (radio) waves. Specifically, the ITU allocates which of nine identified RF bands are designated for certain purposes, such as maritime navigation or meteorological aids. The agency also facilitates negotiations among member states for the acquisition of frequency use and geostationary orbital satellite positions.[107] The allocation of frequency bands for services is divided into three geographical regions around the Earth, which may be uniform for all or distinguished based on need or environmental factors.[108] Such a structured system requires cooperation to work effectively.

The risk of frequency interference has grown with the increase in the number of satellites and participating nations. As an a priori system, the ITU registration of member orbital locations affords each member state at least one orbital allocation,[109] but as an a posteriori system a first registrant has some priority over adjacent satellite operations in different frequencies on a 'first come, first served'

[106] ITU Const (n 3) art 1 s 2(a) para 11.

[107] ITU RR (n 3) no 2.1 and Section IV—Table of Frequency Allocations.

[108] ITU RR (n 3) no 5.2.

[109] Artemis Accords (n 85) 17 (addressing the ITU's modified 'first-come, first served' policy to space access); Francis Lyall, 'The International Telecommunication Union and Development' (1994) 22(1–2) JSpaceL 23, 30.

basis.[110] However, the relationship between radio stations in the same or adjacent regions using the same RF bands for different services applies the basic principle of 'equality of right to operate' with respect to claims of harmful interference.[111]

In space activities, the national and international satnav systems, such as Galileo (the European Union's global navigation satellite system), GLONASS (Russia's global navigation satellite system), IRNSS (the Indian Regional Navigation Satellite System), and NAVSTAR GPS (the United States Global Positioning System), all vie for RF access. Nevertheless, the ITU fosters an ethic of mutual technical assistance among its members, particularly for the benefit of developing nations.[112] For example, although not all nations have equal satellite access, 'Earth observations are performed for the benefit of the whole international community and the data are generally made available at no cost'.[113] This should reflect a cooperative spirit in the face of conflict, but commercial interests primarily seek financial profit. For example, as of 2022, in lieu of large geostationary satellites, constellations of smaller commercially owned Low Earth Orbit satellite networks have now emerged, exacerbating the risk of interference as more satellites are placed into orbit.[114]

Competition is heightened by the fact that the radio spectrum is a limited resource. Electromagnetic radiation is harnessed through technology by directing RF from a transmitter to a receiver through antennae that operate over a limited range of RF,[115] whether RF occurs naturally from radiation emitted by stars or lightning, or by human activity such as from industrial or medical technology.[116] Statutory definitions further circumscribe the legal meaning of the term 'radio wave' for regulatory purposes,[117] but the inherent properties of RF are consistent.

Unlike other natural resources, the radio spectrum is occupied but not consumed during use.[118] 'Electromagnetic waves do not need a medium in order to propagate, which means they can travel through air, solid materials, and also

[110] ITU RR no 8.3; ITU Res 2 (rev WRC-03) (calling for equitable use, by all countries, with equal rights of the geostationary-satellite and other satellite orbits and of frequency bands for radiocommunication services). See Jannat C Thompson, 'Space for Rent: The International Telecommunications Union, Space Law, and Orbit/Spectrum Leasing' (1996) 62 JAirL&Com 279, 290.

[111] ITU RR (n 3) no 4.8.

[112] ITU Res 15 (rev WRC-03).

[113] ITU Res 673-1, s (h) (rev WRC-12).

[114] Rachel Jewett, 'Will LEO Change the Paradigm of IFC Service?' (*Avionics International*, June/July 2021) <https://interactive.aviationtoday.com/avionicsmagazine/june-july-2021/will-leo-change-the-paradigm-of-ifc-service/> accessed 14 June 2022.

[115] See ITU Res 63 (rev WRC-12) (addressing harmful interference with radiocommunications from industrial, scientific, and medical equipment).

[116] 'What Are Radio Waves?'(*NASA*, 31 August 2018) <www.nasa.gov/directorates/heo/scan/communications/outreach/funfacts/what_are_radio_waves> accessed 14 June 2022 (explaining that radio waves are naturally occurring waves of electromagnetic radiation produced from the intersection of oscillating electric and magnetic fields).

[117] See eg New Zealand Radiocommunications Act 1989, s 2 (28 October 2021) ('radio waves means electromagnetic waves of frequencies lower than 3 000 gigahertz, propagated in space without artificial guide').

[118] Withers (n 16) 1.

through the vacuum of space.'[119] The repeated cycles of the wave are measured in frequency and size by velocity per second in units called hertz (Hz), which are categorised into RF bands for certain uses by the ITU, such as military or meteorological use or communications by certain space programmes.[120] Although multiple users can coexist in the same frequency range,[121] the RF spectrum is a scarce resource because a single frequency cannot be used at the same time and place and in the same direction by more than one user.[122] At higher frequencies, RF 'punch through' the atmosphere into outer space, rather than being reflected back to Earth from the ionosphere like other high frequency (HF) bands.[123] While communications between the Earth and the International Space Station have used the S-band (2–4 GHz),[124] the ITU-R is now adapting regulations for deep space explorations to Mars and beyond.[125] Currently, the Radiocommunication Sector Study Group is conducting research in preparation for WRC-27 to develop uses of higher bands up to 700 GHz.[126]

Thus far, the ITU's space services efforts remain primarily in the regulation of Low Earth Orbit where most satellites reside.[127] A number of scholars have argued that the ITU allocation process of RF as a scarce resource reflects a tragedy of the commons, where new players and demands encounter resistance from incumbents, and diminish the potential of a more diverse market.[128] Thus, owners of new

[119] 'Introduction to Electromagnetic Spectrum' (*NASA*, 27 June 2018) <www.nasa.gov/directorates/heo/scan/spectrum/overview/index.html> accessed 14 June 2022.

[120] Philip Dearman and Paula Pyburne, 'Radiocommunications Legislation Amendment (Reform and Modernisation Bill 2020) [and Associated Bills]' (*Parliament of Australia*, 18 November 2020) <www.aph.gov.au/Parliamentary_Business/Bills_Legislation/bd/bd2021a/21bd025#_Toc56599893> accessed 27 September 2023 ('Like waves in the ocean, radio waves have physical properties: they have a wavelength (a distance between peaks) and a frequency (the number of wave peaks passing any given point over a period of time)').

[121] 'What is GNSS?' (n 5).

[122] WRC-79, Res 3 (identifying the orbit-spectrum as a scarce resource which requires equitable access for all nations); 'Introduction to the Electromagnetic Spectrum' (*NASA*, 27 June 2018) <www.nasa.gov/directorates/heo/scan/spectrum/overview/index.html> accessed 14 June 2022.

[123] 'HF Makes a Comeback' (*IWCE's Urgent Communications*, 1 March 2011) <https://urgentcomm.com/2011/03/01/hf-makes-a-comeback/> accessed 14 June 2022.

[124] See 'Satellite Frequency Bands' (*The European Space Agency*) <www.esa.int/Applications/Telecommunications_Integrated_Applications/Satellite_frequency_bands> accessed 14 June 2022.

[125] ITU RR (n 3) no 4.24 (WRC-15) ('Space research systems intended to operate in deep space may also use the space research service (deep space) allocations, with the same status as those allocations, when the spacecraft is near the Earth, such as during launch, early orbit, flying by the Earth and returning to the Earth').

[126] ITU-R Study Groups (n 4) 18. See also Jianxin Li and others, 'Satellite Communication on the Non-geostationary System and the Geostationary System in the Fixed Satellite Service' (2019) 28 WOCC978-1-7281-0660-1/19 1 ('[T]he world has been paying attention to the applications of the Q/V frequency bands and even higher bands for the next-generation satellite communications').

[127] Whitt (n 65) (explaining that communications satellites currently comprise about half of the satellites in space, whether global and orbiting the Earth, or geostationary and fixed in orbit to serve a single region on Earth).

[128] Hazlett and Skorup (n 99) 22 (addressing the regulatory domination of GPS through a successful complaint to the FCC against LightSquared LLC, regarding potential RF adjacent interference from proposed 4G long-term evolution (LTE) communications).

technology strive to take advantage of higher RF bands where spectrum licensing faces less competition, such as NASA's Deep Space Optical Communications programme for interplanetary communications.[129] Whether the allocation process suppresses innovation or results in favoritism, waste, and inefficiency are important policy questions for WRCs, but when actual unresolved disputes occur the ITU also has a role.

3.2 Arbitration and the Negotiation of Disputes

The ITU-R assists in averting international conflict by laying the foundation for member states and sector members to coordinate and negotiate disputes related to RF use. A large number of ITU members, both governmental and non-governmental, have assented to cooperate with each other, and specifically to avoid harmful interference.[130] They rely on the ITU's notice and publication of frequency allotments to protect against such interference, a publication which may be revoked for non-compliance.[131] As one scholar noted regarding the early formation of the ITU, '[i]ndeed, the ITU regulatory regime offers an excellent example of how national self-interest can be harnessed to promote compliance with international law.'[132] Essentially, the restrictions of the laws of physics constrain all members 'despite the uncertain sanctions of International Law'.[133]

3.2.1 ITU Dispute Procedures

When disputes do occur, the binding instruments of the ITU call for resolution through negotiation, diplomatic channels, and arbitration, but not as a tribunal.[134] In 1994, the ITU adopted an Optional Protocol on the Compulsory Settlement of Disputes.[135] However, nations have been reticent to engage in the ITU's arbitration

[129] Meltem Civas and Ozgur B Akan, 'Terahertz Wireless Communications in Space' (2021) 2(7) ITUJFut&EvolTech ('THz space links can pave the way for services including live video feeding, high-resolution imagery, and virtual reality from space').

[130] ITU Const (n 3) art 45 s 1 para 197 ('All stations, whatever their purpose, must be established and operated in such a manner as not to cause harmful interference to the radio services or communications of other Member States or of recognized operating agencies, or of other duly authorized operating agencies which carry on a radio service, and which operate in accordance with the provisions of the Radio Regulations').

[131] See ITU RR (n 3) no 9.2 B.1, note 11 (providing a right of cancellation of publication of frequency registration upon non-compliance with payment requirements, in which the Bureau 'shall inform all administrations of such action, and that the network specified in the publication in question no longer has to be taken into consideration by the Bureau and other administrations').

[132] Leive (n 58) 24.

[133] Lyall (n 109) 30.

[134] ITU Const (n 3) art 56 paras 233–35; ITU Conv (n 19) art 41 para 5; ITU RR (n 3) no 9.3 (resolution of harmful interference to satellite systems).

[135] ITU Optional Protocol on the Compulsory Settlement of Disputes Relating to the Constitution of the International Telecommunication Union, to the Convention of the International Telecommunication Union and to the Administrative Regulations (adopted 22 December 1992, entered into force 1 July 1994) 1825 UNTS 330.

process, and only one-third of member states have ratified, accepted, or acceded to the Optional Protocol.[136] The Radio Regulations Board of the ITU-R has independent authority, upon request, to consider appeals of frequency assignments made by the Radiocommunication Bureau.[137] The ITU Convention provides that the Bureau shall assist in the resolution of harmful interference claims upon request by member states, including investigative services and production of a report of recommendations 'for consideration' by the Board.[138] Ultimately, if negotiation is ineffective and an unresolved dispute over harmful interference is referred to the Bureau, the final action by ITU is to assist and facilitate, not to impose a binding determination.[139] However, if the ITU's Board is called upon to assist in a dispute related to harmful interference, it seeks to find a technical solution if possible,[140] and 'administrations are urged to exercise the utmost goodwill and mutual cooperation'.[141]

The ITU's limited role in the resolution of disputes has garnered both criticism and praise. According to one proponent, '[b]y not excluding any method of conflict resolution, the ITU Convention makes available the benefits of them all: the control of bilateral negotiations, the flexibility of mediation, the objectivity of arbitrations, or the certainty of adjudication.'[142] However, with this approach regulation at the national level is critical. Investors are less likely to be engaged in spectrum acquisition if they are uncertain as to the ITU's ability to ensure protection against interference,[143] therefore national sanctions by member states ensure compliance with ITU regulations. For example, in the United States, the FCC restrictions upon licensure include a prohibition on 'harmful interference' from the usage of RF,[144] and those operating without a licence to use RF are also sanctioned by revocation of

[136] 'List of Countries for an Agreement' (*ITU*) <www.itu.int/online/mm/scripts/gensel25?agrmtid=0000925245> accessed 18 June 2022 (listing, as of 18 June 2022, 64 countries that have ratified the 1992 ITU Optional Protocol on the Compulsory Settlement of Disputes, out of 193 countries that have ratified or acceded to the ITU Constitution, Convention, and Radio Regulations). See Robert R Bruce and others, 'Dispute Resolution in the Telecommunications Sector: Current Practices and Future Directions'(October 2004) Discussion Paper for ITU & World Bank 45 (noting that the ITU's involvement in formal dispute resolution regarding radio frequencies is rare); Donna C Gregg, 'Capitalizing on the National Self-Interest: The Management of International Telecommunication Conflict by the International Telecommunication Union' (1982) 45(1) Law&ContempProbs 37, 50 ('In the past, the arbitration procedures have been largely ignored, with members preferring to resolve their disputes through bilateral negotiation').

[137] ITU Conv (n 19) art 10 s 2 para 140.

[138] ITU Conv (n 19) art 12 s 2(f) para 173.

[139] See ITU RR (n 3) no 9.65; ITU Res 169-4 (WRC-19) (instructing the Director of the Radiocommunication Bureau 'to take all necessary actions to facilitate the implementation of this Resolution, together with providing any assistance for the resolution of interference, when required').

[140] Leive (n 58) 129.

[141] ITU RR (n 3) no 7.8.

[142] Gregg (n 136) 50–51.

[143] Ellen P Goodman, 'Progress Toward Rational Spectrum Rights: Are We Getting Anywhere?' in J Pierre de Vries and Kaleb A Sieh (eds), 'The Unfinished Radio Revolution: Eight Perspectives on Wireless Interference' (2011) 9 JTelecom&HighTechL 501, 509.

[144] Hazlett and Skorup (n 99) 6 (addressing wireless services interference regulations under 47 CFR §15.5).

licence, fine, or criminal prosecution.[145] The New Zealand Radiocommunications Act 1989 provides that 'the arbitral tribunal has the power to determine the cause of the harmful interference and has the power to direct the parties to the dispute or the Secretary to take action to reduce or eliminate harmful interference'.[146] Notably, the Radio Regulations and ITU-R reports and recommendations are relevant but not determining factors for consideration by the tribunal in New Zealand.[147] The ITU also leaves to national law, international arbitration, and other international treaties the determination of individual or joint civil liability in contract and tort law which may arise from space-related telecommunication activities.[148]

3.2.2 Inadvertent or Aggressive Harmful Interference

Inadvertent interference with the function of RF use is often the primary concern for ITU members and is usually managed through negotiation without the need for formal dispute resolution.[149] Also, sources beyond the actions of ITU members may cause interference, such as atmospheric conditions from rain and fog,[150] or unexpected solar radiation changes affecting orbital paths and the positioning of satellites, necessitating corrections rather than dispute resolution.[151] But aggressive conduct remains a concern, such as the need to prevent cyberattacks that could disrupt satellite orbits and functions[152] or privacy concerns related to geotracking and aggression against individuals.[153] The privacy provisions of the ITU's Constitution

[145] ibid 23 (noting that unauthorised minor radio transmissions are usually dealt with at the local government level).

[146] New Zealand Radiocommunications Act (n 117) s 109C(1).

[147] ibid s 109A(1).

[148] ITU Const (n 3) art 36 para 183 ('Member States accept no responsibility towards users of the international telecommunication services, particularly as regards claims for damages'). See also Artemis Accords (n 85), s 2(1)(b) <www.nasa.gov/specials/artemis-accords/img/Artemis-Accords-signed-13Oct2020.pdf> (asserting the need for cooperation and transparency in applying the principles of the Accords to bilateral instruments, including those addressing liability); Dejian Kong, 'Civil Liability for Damage Caused by Global Navigation Satellite Systems: A Conceptual Analysis' (2016) 41 AnnAir&SpaceL 313, 320.

[149] See Adrian Copiz, 'Scarcity in Space: The International Regulation of Satellites' (2002) 10 ComLConspec 207, 208.

[150] Withers (n 16) 301–04.

[151] Copiz (n 149) 210. See also *Space Weather*, 42 USC §18388 (2018) (repealed 21 October 2020) (finding that '[t]he effects of severe space weather events on the electric power grid, telecommunications and entertainment satellites, airline communications during polar routes, and space-based position, navigation and timing systems could have significant societal, economic, national security, and health impacts'); Novah Maras Nurul Kamsah, 'Star Sensor Availability of Equatorial Satellite based on LAPAN-A2 Satellite Observation' (2021) ICARES978-1-6654-0179-1/21 1, 5 (evaluating the heightened solar interference to Indonesia's first small satellite in the equatorial regions).

[152] Shackelford (n 40) 31.

[153] Andrew Guthrie Ferguson, 'The Internet of Things and the Fourth Amendment of Effects' (2016) 104 CalLRev 805, 822 (highlighting the individual privacy implications of cheaply built smart home devices using radio-frequency identification devices connected to the Internet and sharing data widely without much data protection); A Michael Froomkin, 'The Death of Privacy?' (2000) 52 StanLRev 1461, 1498 (addressing the various law enforcement intrusions upon individual privacy through telecommunications surveillance, including thermal imaging from remote-sensing satellites).

are fairly opaque, requiring agreement to take 'all possible measures' to ensure the 'secrecy of international correspondence'.[154]

Radiocommunications create a conduit for aggressive action, including space piracy, commercial sabotage, and military action in space, much of which remains relatively unregulated.[155] However, radiocommunications also enhance protection from armed conflicts, such as the ITU Resolution setting forth procedures for ships and aircraft to be able to identify with urgency that they are not parties to armed conflict.[156] Since the 1950s, satellites have always been capable of diverse use, where the governments of the first space-faring nations initially designed satellites for both weather observation[157] and commercial telecommunications.[158] The ITU asserts the goal 'to promote the extension of the benefits of the new telecommunication technologies to all the world's inhabitants', including technology for individual, personal mobile communications[159] and amateur satellite stations.[160] Other treaties also declare that space activities must be used for peaceful purposes, such as the multilateral Limited Nuclear Test Ban Treaty which predates the Outer Space Treaty.[161] Also, the Committee on the Peaceful Uses of Outer Space (COPUOS) became a permanent body of the UN in 1959,[162] actively involved in many aspects of the regulation of activities in space to reduce conflict.

Government military influence continues to direct many space activities, although international diplomacy fostered by the ITU's coordination has shown that restraint is possible. In the United States, for example, the US Space Force now operates GPS.[163] When concern arose as to a potential violation of the International

[154] ITU Const (n 3) art 37 paras 184–85.

[155] Sarah M Mountain, 'The Legality and Implications of Intentional Interference with Commercial Communication Satellite Signals' (2014) 90 IntlLStud 101, 196 (identifying a degree of permissive lawlessness where 'States are unwilling to vest regulatory authorities with enforcement powers, often fail to report incidents of satellite signal interference and do not pursue reparations under international law'). See Jorge Ciccorossi, 'Harmful Interference to Satellite Systems and the Current Challenge to GNSS' (*Eurocontrol Stakeholder Forum on GNSS*, 4 March 2021) (reporting that 99.94 per cent of the radio spectrum is 'free' of harmful interference but noting that 'reporting harmful interference is key').

[156] ITU Res 18 (rev WRC-15).

[157] The first artificial satellite in space, the Soviet Union's Sputnik 1, was launched on 4 October 1957 and was designed to obtain atmospheric data and information on the propagation of radio waves in the upper levels of the atmosphere. See 'Sputnik 1' (*NASA Space Science Data Coordinated Archive, 1957-001B*) <https://nssdc.gsfc.nasa.gov/nmc/spacecraft/display.action?id=1957-001B> accessed 15 June 2022.

[158] Samuel D Estep and Amalya L Kearse, 'Space Communications and the Law: Adequate International Control After 1963' (1962) 60 MichLRev 873, 873 (describing the first commercial satellite launch by AT&T using a NASA launching site to provide communications services to the United States, England, France, and Germany in 1962).

[159] ITU Res 25 (rev WRC-03) (Operation of Global Satellite Systems for personal communications).

[160] ITU Res 642.

[161] Treaty Banning Nuclear Weapon Tests in the Atmosphere, in Outer Space, and Under Water (adopted 5 August 1963, entered into force 10 October 1963) 480 UNTS 43. See Bosco (n 84) 629–30.

[162] UNGA Res 1348 (XII) (13 December 1958). See 'Committee on the Peaceful Uses of Outer Space' (*UN Office for Outer Space Affairs*) <www.unoosa.org/oosa/en/ourwork/copuos/index.html> accessed 27 September 2023.

[163] Brooke Wise, 'The Magic of GPS Simplified' (*Space Operations Command (SPOC)*, 14 December 2021) <www.spoc.spaceforce.mil/News/Article-Display/Article/2874327/> accessed 15 June 2022

Telecommunication Convention and ITU Radio Regulations based on the military satellite jamming capabilities of the United States and other nations, the international community vigorously objected.[164] In May 2000, at the direction of President Bill Clinton, the US government discontinued its use of Selective Availability of satellites in order to make GPS more responsive to civil and commercial users worldwide.[165] The US Space-Based Positioning, Navigation, and Timing Policy of 2021 directs:

> Engage with international GNSS providers to ensure compatibility, encourage interoperability with likeminded nations, promote transparency in civil service provision, and enable market access for United States industry. Encourage foreign development of PNT services and systems based on GPS and the inclusion of GPS as an essential element in systems that integrate multiple PNT services. At a minimum, seek to ensure that all foreign systems are compatible with GPS and its augmentations, that they do not interfere with GPS military and civil signals, and that mutual security concerns are addressed to prevent hostile use of United States space-based PNT services.[166]

As a related issue, national export control measures seek to protect national security and prevent missile proliferation, but also the international distribution of satellite equipment capable of interfering with these weapons systems.[167] Cyberattacks may operate indiscriminately through RF use which can disrupt critical infrastructures and risk the life and health of civilian populations as a component of armed conflict and violation of the principle of distinction in international humanitarian law (the law of armed conflict).[168] There is also some debate as to

('Space Delta 8—Satellite Communications and Navigational Warfare, 2nd SOPS, is responsible for GPS operations . . . A total of 170 Guardians, Airman and civilians make sure that the satellite constellation is operating properly 24/7').

[164] ITU RR (n 3) no 3.15 ('The use of damped wave emissions is forbidden in all stations'). See Christine M. Schenone, 'Jamming the Stations: Is There an International Free Flow of Information?' (1984) 14(3) CalWIntlLJ 501, 502.

[165] 'Selective Availability' (*GPS*) <www.gps.gov/systems/gps/modernization/sa/> accessed 15 June 2022 ('In September 2007, the U.S. government announced its decision to procure the future generation of GPS satellites, known as GPS III, without the SA feature').

[166] 'Memorandum on Space Policy Directive 7, The United States Space-Based Positioning, Navigation, and Timing Policy' (*GPS*, 15 January 2021) s 4(i) <www.gps.gov/policy/docs/2021/> accessed 15 June 2022.

[167] See eg 'Receiving Equipment for "Navigation Satellite Systems"', Cont Cas Fed (CCH US) P EAR 7A105 (CCH) (2021), 2020 WL 5099136 (addressing US regulations for export control relating to satellite equipment for use in missiles, or designed to gain access to secure data, or to employ anti-jam features, but not applicable to commercial, civil aviation, or safety-of-life satellite equipment).

[168] See 'Cyber Warfare: Does International Humanitarian Law Apply?' (*International Committee of the Red Cross*, 25 February 2021) <www.icrc.org/en/document/cyber-warfare-and-international-humanitarian-law#:~:text=In%20our%20views%2C%20the%20law,the%20International%20Court%20of%20Justice> accessed 15 June 2022; International Committee of the Red Cross, 'International Humanitarian Law and Cyber Operations During Armed Conflicts' (November 2019) (ICRC Position Paper) 5.c.

whether the International Court of Justice has defined the legality of RF-enabled cyberweapons,[169] based on its opinion that international humanitarian law applies to 'all forms of warfare and to all kinds of weapons'.[170]

While the ITU has a role to play in internet governance,[171] and its members continue to face serious cybersecurity threats,[172] the ITU has never been an adjudicatory body and is not structured to resolve conflicts or to enforce action. The ITU must also take into account that at this time member states continue to share data from scientific exploration and innovative radiocommunications technology that benefit all nations, such as the Mars Rover discovery of saltwater remnants on the planet's surface.[173]

4 ITU Accountability and the Future of Space Telecommunications

Even without an enforcement mechanism, the ITU as an international body has acknowledged its dedication to social cohesion and fair access against the risks of national, economic, and military domination, particularly through its International Development Sector or ITU-D.[174] The Constitution of the ITU states:

> In using frequency bands for radio services, Member States shall bear in mind that radio frequencies and any associated orbits, including the geostationary-satellite orbit, are limited natural resources and that they must be used rationally, efficiently and economically, in conformity with the provisions of the Radio Regulations, so that countries or groups of countries may have equitable access to those orbits and frequencies, taking into account the special needs of the developing countries and the geographical situation of particular countries.[175]

[169] Shackelford (n 40) 28.

[170] ICRC Position Paper (n 168) 4 (citing International Court of Justice, *Legality of the Threat or the Use of Nuclear Weapons* (Advisory Opinion) 8 July 1996, para 86).

[171] ITU REC 207 (rev WRC-19) (promoting the study and future development by ITU-R of international mobile telecommunications, including broadband technologies).

[172] Shackelford (n 40); US-China Economic and Security Review Commission, 'China's Position on a Code of Conduct' (8 September 2017) Staff Research Report 1, 3 (discussing recent research that explores technology that would employ intentional satellite interference capabilities in the course of cyber-electronic warfare).

[173] Jim Taylor and others, 'Mars Exploration Rover Telecommunications' in Jim Taylor (ed), *Deep Space Communications* (1st edn, John Wiley 2016) 251 (assessing communications between the first two Mars Rovers launched in 2003, Spirit (now silent) and Opportunity (active until 2019) and their X-band telecommunications links); Joe Palca, 'NASA's Mars Rover Opportunity is Officially Declared Dead' (*NPR*, 13 February 2019) (reporting that the Rover Opportunity was intended to operate for only 90 days, but operated for 14 years, returning thousands of images of Mars, including proof that saltwater had covered Mars surface areas).

[174] ITU Const (n 3) art 21 s 2 paras 120–29 (supporting the UN's economic and social development priorities). For an early history of the ITU-D, see Liability Convention (n 73) 23–32.

[175] ITU Const (n 3) art 44 s 2 para 196; ITU RR, Preamble no 0.3.

While three nations (the Soviet Union/Russia, United States, and China) have launched their own crews of astronauts into space, citizens of approximately 42 nations have flown in space, 12 nations have launched their own satellites, at least 90 nations and organisations currently have satellites in orbit, and all nations now rely on space-based communications technology.[176]

To manage this varied representation, while still ensuring economic benefit and opportunity to developing nations, the ITU has shifted to supporting more national and regional cooperation and infrastructure.[177] The ITU is also currently engaged in contributing to the 2015 UN Sustainable Development Goals to use space science for the monitoring of global challenges such as 'providing food security, reducing the risk of disasters, preventing humanitarian crises, monitoring natural resources, and reducing poverty'.[178] The WRC-19 produced several resolutions focusing specifically on the use of satellites for disaster relief, such as using space weather monitoring and global warning systems to protect the Earth.[179] These efforts are supported in tandem with the commercialisation of space telecommunications, and the balance of competing interests is inevitably affected by the elected leadership of the ITU.

The ITU's lack of diversity of leadership and perspective is facing calls for reform. In March 2022, for example, Australia announced a leadership initiative to support women's empowerment and gender equality at the next ITU Plenipotentiary Conference, given the historical lack of women's representation in the telecommunications industry generally and in ITU decision-making.[180] The absence of women, particularly in the highest leadership positions of the ITU, has been officially noted by the UN in reviews of ITU structure and functions.[181] Representation within the ITU and at WRCs must directly affect the agency's understanding and ability to provide proper guidance in world economic, military, and space activities and the needs of developing nations.[182]

How new technologies and space activities affect existing inequities may be at the forefront of many ITU agendas. Broadband technology, for example, which

[176] 'Space Exploration' (*US National Archives and Records Admin*) <www.archives.gov/research/alic/reference/space-timeline.html> accessed 19 June 2022; 'Satellites by Countries and Organizations' (*N2YO.com*) <www.n2yo.com/satellites/?c=&t=country> accessed 19 June 2022.

[177] ITU Res 25 (rev PP-2018) (Strengthening the regional presence).

[178] Maniewicz (n 12) 4. See also UN Department of Economic and Social Affairs, 'Transforming Our World: The 2030 Agenda for Sustainable Development' (2015) UN Doc A/RES/70/1 .

[179] ITU Res 646, 647, and 657 (WRC-19).

[180] 'Advancing Women's Decisionmaking at the ITU Plenipotentiary' (*ITU*, 8 March 2022) <www.itu.int/en/mediacentre/Pages/PR-2022-03-08-Australia-Women-Empowerment.aspx> accessed 15 June 2022.

[181] Joint Inspection (n 19) v (providing, in Recommendation 8, that the Secretary-General of the ITU should present a plan to improve gender balance, 'especially at senior management level, within each component of the Union, and report annually to the Council on its implementation').

[182] See eg ITU Res 5-2 (WRC-15) (instructing the Secretary-General to offer ITU assistance to developing countries in the tropical areas to study, improve, and develop their radiocommunications).

enables the use and transmission of a wide range of frequencies and a large number of messages to be communicated simultaneously, is an advancement deeply impacting national economies and social inequality. The World Bank reports that '[b]roadband (or high-speed) internet access is not a luxury, but a basic necessity for economic and human development in both developed and developing countries', but that only about 35 per cent of the population in developing nations has access to the internet in any form.[183] In the United States, despite its being a major economic power where broadband technology is promoted by Congress, 60 per cent of healthcare centres outside metropolitan areas and 40 per cent of schools in the United States continue to lack broadband access.[184]

Information and communication technology is increasingly based in space, implicating numerous emerging policy concerns. With respect to space activities, the agenda for the ITU's WRC-23 focuses on upgrading technology and climate change-related concerns, such as implementing spaceborne radar to monitor Earth's ice sheets and obtaining regulatory recognition of space weather sensors.[185] Much promise lies in future space exploration,[186] but the ITU will need to ensure its accountability and sustainability as an agency of the UN committed to enhancing the peaceful use of space. As one of the UN's most long-standing specialised agencies, the ITU has an impressive track record of bringing the world together, harnessing the collaborative skills of experts, and fostering a forum for negotiations among both public and private members.

[183] 'Connecting for Inclusion: Broadband Access for All' (*The World Bank*) <www.worldbank.org/en/topic/digitaldevelopment/brief/connecting-for-inclusion-broadband-access-for-all> accessed 12 June 2022.

[184] Compare USC §1301(1) (2018) ('The deployment and adoption of broadband technology has resulted in enhanced economic development and public safety for communities across the Nation, improved health care and educational opportunities, and a better quality of life for all Americans') with '21 Million Americans Still Lack Broadband Connectivity' (*Pew Charitable Trust*, 10 July 2019) <www.pewtrusts.org/en/research-and-analysis/fact-sheets/2019/07/21-million-americans-still-lack-broadband-connectivity> accessed 15 June 2022 (based on FCC Reports from 2016 and 2019).

[185] John Zuzek, Chairman, 'Spectrum Management for Science Services After WRC-19' (2020) 6 ITUNewsMag 11; ITU Res 812 (WRC-19) (setting a similar agenda for the WRC-27).

[186] See eg ITU REC 208 (WRC-19) (intelligent transportation systems); Stephanie L Booth and Michael A Marsden, 'Radio Frequency and Optical Communication Link Trade Studies Between Earth, Deep Space Gateway, and Lunar Surface' (December 2019) NASAGlennResCtr 2019-220203 (higher bandwidth applications for Earth to Moon communications, including via NASA's Deep Space Gateway Communications link for an Artemis program lunar orbital outpost); 'NASA Moon Mission Set to Break Record in Navigation Signal Test' (*NASA*, 2 June 2022) <www.nasa.gov/feature/goddard/2022/nasa-moon-mission-set-to-break-record-in-navigation-signal-test> accessed 19 June 2022 (testing GNSS on the Moon for the first time); 'Safeguarding Radioastronomy on the Moon' (*ITU News*, 25 May 2022) <www.itu.int/hub/2022/05/moon-based-radio-astronomy-spectrum/> accessed 19 June 2022 (preparing the Moon for service as an outpost for further exploration to Mars and other bodies); Charlie Wood, 'The New History of the Milky Way' (*Quantamagazine*, 15 December 2020) <www.quantamagazine.org/the-new-history-of-the-milky-way-20201215/> accessed 15 June 2022 (utilising a massive set of data from the European Gaia satellite in 2018 to discover that the Milky Way galaxy continues to be reformed from its lineage of galactic mergers).

12
THE ARTEMIS ACCORDS AND SUBSEQUENT DEVELOPMENTS

Melissa de Zwart and Joel Lisk

1 Introduction

As of August 2023, twenty-eight countries have signed the United States-led Artemis Accords.[1] Officially announced on 14 October 2020, with eight foundation partners (Australia, Canada, Italy, Japan, Luxembourg, United Arab Emirates, United Kingdom, and United States of America), the Artemis Accords embody both a programme to return humans to the Moon and then travel to Mars, and principles to, in the words of the US National Aeronautics and Space Administration (NASA), 'guide space exploration cooperation among nations participating' in the US-led Artemis Program.[2] It is these principles which will set the tone for how each partner engages in the Artemis Program. These Accords will shape the evolution of space governance and norm development as more partners sign on to the Accords and subsequently adopt practices and approaches to increased human uses of outer space that reflect the content of the Accords. Whilst the Artemis Accords do not meet the minimum requirements to be a formal treaty instrument, primarily due to the variable nature of the entities signing the document (such as national space agencies and government departments) rather than the State ratifying the document, their nature will inevitably influence the evolution of international law through State practice and through the soft power of diplomacy, embodiment in domestic laws and creation of international standards and norms. This approach has been criticised by some States as an attempt by the US to drive unilateral law-making.[3] Others have heralded this as a major step forward in an impasse to develop practical rules for how humans will cohabit on the Moon.

[1] Artemis Accords: Principles for Cooperation in the Civil Exploration and Use of The Moon, Mars, Comets, and Asteroids for Peaceful Purposes (adopted 13 October 2023) (Artemis Accords).

[2] 'NASA, International Partners Advance Cooperation with First Signings of Artemis Accords' (*NASA*, 14 October 2020) <www.nasa.gov/press-release/nasa-international-partners-advance-cooperation-with-first-signings-of-artemis-accords> accessed 13 July 2022.

[3] Loren Grush, 'Head of Russian Space Program Calls for More International Cooperation in NASA's Moon Plans' *The Verge* (13 October 2020) <www.theverge.com/2020/10/12/21512712/nasa-roscosmos-russia-dmitry-rogozin-artemis-moon-interntational-cooperation> accessed 22 August 2023; Sa'id Mosteshar, 'Artemis: The Discordant Accords' (2020) 44 JSpaceL 591; Alexander Stirn, 'Do NASA's Lunar Exploration Rules Violate Space Law?' *Scientific American* (12 November 2020) <www.

Melissa de Zwart and Joel Lisk, *The Artemis Accords and Subsequent Developments* In: *International Space Law in the New Space Era.* Edited by: Sandeepa Bhat B., Dilip Ukey, and Adithya Variath, Oxford University Press.
 DOI: 10.1093/9780198909415.003.0012

This chapter will identify and examine the principles of the Artemis Accords and how they fit within current international law. This analysis will be accompanied by a discussion of the Artemis missions to explain how these principles may operate in practice. It will then address how the development of these principles may influence and modify the expansion of international space law. In particular, it will consider the importance of influence, diplomacy and soft law. We conclude that the Artemis Accords will perform a major role in influencing the future governance of space activities. They represent an inevitable evolution of international practice which is required to facilitate extended human uses of outer space.

2 The Artemis Accords

The Artemis Accords prescribe ten principles which are expressed to 'to enhance the governance of the civil exploration and use of outer space' and 'to increase the safety of operations, reduce uncertainty, and promote the sustainable and beneficial use of space for all humankind'.[4] It is acknowledged that these principles are not intended to be an exhaustive articulation of the principles that should apply to activities undertaken under the Accords. Participants are expected to enter into specific agreements which adopt the principles of the Accords and prescribe other necessary arrangements such as ownership and use of intellectual property, liability and exchange of goods and data.[5]

The principles of the Accords are expressed to 'represent a *political commitment* to the principles described herein, many of which provide for operational implementation of important obligations contained in the Outer Space Treaty and other instruments'.[6] The Preamble to the Accords affirms the importance of compliance with four of the United Nations (UN) Space Treaties, namely the Treaty on Principles Governing the Activities of States in the Exploration and Use of Outer Space, Including the Moon and Other Celestial Bodies (Outer Space Treaty); the Agreement on the Rescue of Astronauts, the Return of Astronauts and the Return of Objects Launched into Outer Space (Rescue and Return Agreement); the Convention on International Liability for Damage Caused by Space Objects (Liability Convention); and the Convention on Registration of Objects Launched into Outer Space (Registration Convention).[7] Whilst the Preamble to the Accords

scientificamerican.com/article/do-nasas-lunar-exploration-rules-violate-space-law/> accessed 13 July 2023; Matthew Gross, 'The Artemis Accords: International Cooperation in the Era of Space Exploration' *Harvard International Review* (27 January 2023) <hir.harvard.edu/the-artemis-accords/> accessed 13 July 2022.

[4] Artemis Accords (n 1) s 1.
[5] Artemis Accords (n 1) s 2.
[6] Artemis Accords (n 1) s 1 (emphasis added).
[7] See Artemis Accords (n 1) preamble para 9; Treaty on Principles Governing the Activities of States in the Exploration and Use of Outer Space, Including the Moon and Other Celestial Bodies (adopted

also affirms 'the benefits of coordination via multilateral forums, such as the UN Committee on the Peaceful Uses of Outer Space (UNCOPUOS), to further efforts toward a global consensus on critical issues regarding space exploration and use',[8] the fifth UN space treaty, the Agreement Governing the Activities of States on the Moon and Other Celestial Bodies (Moon Agreement), is missing from this list.[9] Importantly in this context, the US is not a party to the Moon Agreement, which has only eighteen signatories.[10] None of the traditional major space-faring States are ratified States Parties to the Moon Agreement and it is apparent that the Moon Agreement is not reflective of customary international law.[11] This view has been further affirmed most recently by the US in the context of the Artemis Accords.[12] The 2020 US 'Executive Order on Encouraging International Support for the Recovery and Use of Space Resources' clearly stated:

> [T]he United States does not consider the Moon Agreement to be an effective or necessary instrument to guide nation states regarding the promotion of commercial participation in the long-term exploration, scientific discovery, and use of the Moon, Mars, or other celestial bodies. Accordingly, the Secretary of State shall object to any attempt by any other state or international organization to treat the Moon Agreement as reflecting or otherwise expressing customary international law.[13]

Whilst the accompanying White House statement affirms the position that this Executive Order is addressed primarily to US policy regarding resource recovery

27 January 1967, entered into force 10 October 1967) 610 UNTS 205 (Outer Space Treaty); Agreement on the Rescue of Astronauts, the Return of Astronauts and the Return of Objects Launched into Outer Space (adopted 22 April 1968, entered into force 3 December 1968) 672 UNTS 119 (Rescue and Return Agreement); Convention on International Liability for Damage Caused by Space Objects (adopted 29 March 1972, entered into force 1 September 1972) 961 UNTS 187 (Liability Convention); Convention on the Registration of Objects Launched into Outer Space (adopted 14 January 1975, entered into force 15 September 1976) 1023 UNTS 15 (Registration Convention).

[8] See Artemis Accords (n1) Preamble para 9.

[9] Agreement Governing the Activities of States on the Moon and Other Celestial Bodies (adopted 18 December 1979, entered into force 11 July 1984) 1363 UNTS 3 (Moon Agreement).

[10] As of 1 January 2022, the *Moon Agreement* had eighteen ratified States Parties and an additional four signatories, see 'Status of International Agreements relating to activities in outer space as of 1 January 2022' Sixty-first Session of the UNCOPOUS Legal Subcommittee (Vienna, 28 March–8 April 2022) (28 March 2022) UN Doc A/AC.105/C.2/2022/CRP.10. On 5 January 2023, Saudi Arabia notified the United Nations Secretary-General that it was withdrawing from the Moon Agreement effective 5 January 2024, see Depositary Notification C.N.4.2023.TREATIES-XXIV.2 of 5 January 2023 (Withdrawal: Saudi Arabia) C.N.4.2023.TREATIES-XXIV.2 of 5 January 2023 (Depositary Notification) available at <https://treaties.un.org/doc/Publication/CN/2023/CN.4.2023-Eng.pdf> accessed 21 July 2023.

[11] While there are no traditional space-faring nations party to the Moon Agreement, France and India are both signatories.

[12] See President Donald J Trump, 'Executive Order 13914 of April 6, 2020: Encouraging International Support for the Recovery and Use of Space Resources' (2020) 85 Federal Register 20381.

[13] ibid.

and utilisation in outer space, the Order explicitly reaffirms the position that the US will object to 'any attempt to treat the 1979 Moon Agreement as expressing customary international law'.[14] Several of the sections of the Accords are essentially a restatement of existing principles of space law under the Outer Space Treaty, whilst some reflect principles or norms being addressed in current working groups, and others seek to push the boundaries of current consensus positions on uses of space.[15] In this way, the Accords have been said to promulgate and reinforce US interpretations of international space law.[16]

Each principle will be addressed in turn in the context of its operation, purpose and fit within current international space law.

2.1 Preamble, Sections 1 and 2

The Artemis Accords, much like traditional international agreements, contain a broad preamble referencing the background of the Accords and acknowledging a range of matters linked to both the Accords themselves and the Artemis Program, including (but not limited to) matters such as:[17]

- 'Noting the benefit for all humankind to be gained from cooperating in the peaceful use of outer space';
- '[u]shering in a new era of exploration, more than 50 years after the historic Apollo 11 Moon landing and more than 20 years after the establishment of a continuous human presence aboard the International Space Station';
- 'Recognizing the global benefits of space exploration and commerce'; and
- 'Acknowledging a collective interest in preserving outer space heritage'.

While important to providing context to the Accords, the Preamble is ultimately only background. Sections 1 and 2 of the Accords are purely administrative provisions, affirming the purpose of the Accords and confirming that further

[14] 'Fact Sheet: Encouraging International Support for the Recovery and Use of Space Resources' (*The White House*, 6 April 2020) <https://trumpwhitehouse.archives.gov/wp-content/uploads/2020/04/Fact-Sheet-on-EO-Encouraging-International-Support-for-the-Recovery-and-Use-of-Space-Resources.pdf> accessed 5 July 2023.

[15] Jeff Foust, 'What's in a Name When it Comes to an "Accord"?' *The Space Review* (13 July 2020) <www.thespacereview.com/article/3987/1> accessed 5 July 2023; Chris Johnson, 'SpaceWatchGL Feature: The Space Law Context of the Artemis Accords (Part 2)' (*SpaceWatch.Global*, 28 May 2020) <spacewatch.global/2020/05/spacewatchgl-feature-the-space-law-context-of-the-artemis-accords-part-2/> accessed 5 July 2023; Rossana Deplano, 'The Artemis Accords: Evolution or Revolution in International Space Law?' (2021) 70 ICLQ 799–819.

[16] Joshua Lee, Elsbeth Magilton, and Amelia Ruffolo, 'Diplomatic Impact in the Stars? A Review of the Impact of the Artemis Accords on Global Relationships' (2022) 30 CathUJL&Tech 1, 2.

[17] Artemis Accords (n 1) Preamble.

agreements may be necessary to fully implement the provisions of the Accords or formally codify cooperative arrangements between parties to the Accords.

2.2 Peaceful Purposes (Section 3)

Section 3 (Peaceful Purposes) states 'that cooperative activities under these Accords should be exclusively for peaceful purposes and in accordance with relevant international law'. This requirement echoes the wording of the Outer Space Treaty, in which the term 'peaceful purposes' appears four times: twice in the preamble and twice in Article IV. Article IV articulates the agreement by States Parties to the Outer Space Treaty to refrain from placing nuclear weapons or other weapons of mass destruction in orbit around the Earth, on celestial bodies or stationing 'such weapons in space in any other manner'. The second paragraph of this Article states:

> The Moon and other celestial bodies shall be used by all States Parties to the Treaty exclusively for peaceful purposes. The establishment of military bases, installations and fortifications, the testing of any type of weapons and the conduct of military manoeuvres on celestial bodies shall be forbidden. The use of military personnel for scientific research or for any other peaceful purposes shall not be prohibited. The use of any equipment or facility necessary for the peaceful exploration of the Moon and other celestial bodies shall also not be prohibited.[18]

The explicit incorporation of general principles of international law is contained in Article III of the Outer Space Treaty, which provides:

> States Parties to the Treaty shall carry on activities in the exploration and use of outer space, including the Moon and other celestial bodies, in accordance with international law, including the Charter of the United Nations, in the interest of maintaining international peace and security and promoting international co-operation and understanding.[19]

This provision of the Outer Space Treaty is seen as essential to the preservation and maintenance of the international legal order and ensured that the existing body of international law (and the international law that was yet to come) did not simply 'stop' once humanity reached outer space.[20] The concept of 'peaceful purposes' is fundamental to international space law, being the language used in the first UN

[18] Outer Space Treaty (n 7) art IV.
[19] ibid art III.
[20] Olivier Ribbelink, 'Article III' in Stephan Hobe, Bernhard Schmidt-Tedd, Kai-Uwe Schrogl, and Gerardine Meishan Goh (eds), *Cologne Commentary on Space Law*, vol I (Carl Heymanns Verlag 2009) 64–65.

General Assembly resolution directed at the use of outer space in 1957.[21] It was repeated in a further UNGA resolution in 1958 addressing 'Questions of the Peaceful Uses of Outer Space' and establishing the UN Committee on the Peaceful Uses of Outer Space (UNCOPUOS) as an ad hoc committee of the UNGA.[22] UNCOPUOS was established as a permanent committee of the UNGA in December 1959.[23] UNCOPUOS deals only with issues relating to the peaceful uses of outer space, with matters relating to military uses of space referred to the UN Conference on Disarmament.

Artemis is a project to be conducted by civilian space agencies and commercial partners, thus the 'peaceful purposes' article is both a fundamental and obvious foundational principle, affirming both the peaceful nature of the proposed activities and founding the Artemis Accords clearly within the history of international space law and scope of the Outer Space Treaty.

2.3 Transparency (Section 4)

Section 4 (Transparency) requires signatories to commit 'to transparency in the broad dissemination of information regarding their national space policies and space exploration plans in accordance with their national rules and regulations'. Further, they agree to 'plan to share scientific information resulting from their activities pursuant to these Accords with the public and the international scientific community on a good-faith basis, and consistent with Article XI of the Outer Space Treaty'.[24]

Transparency, information sharing and verification are all vital to multi-partner cooperation. Further, overtures to information sharing have been present since the first UN Declaration in 1957, with the underlying concept of cooperation in what was then a 'new' domain being key.[25] The Outer Space Treaty includes several provisions that deal with transparency and information sharing, including Article V (information regarding phenomena which could constitute a danger to the life or health of astronauts); Article IX (requests for consultation regarding an activity or experiment potentially causing harmful interference); Article X (requests for an opportunity to observe the flight of objects launched by another state); Article XI (obligation to inform the UN, the public and international scientific community regarding the nature, conduct, location and results of activities in outer space); and Article XII (requiring all stations, installations, equipment and space vehicles on the Moon and other celestial bodies to be open to representatives of other States

[21] UNGA Res 1148 (XII) (14 November 1957) para [1](f).
[22] UNGA Res 1348 (XIII) (13 December 1958).
[23] UNGA Res 1472 (XIV) (12 December 1959).
[24] Artemis Accords (n 1) s 4.
[25] UNGA Res 1148 (n 21).

Parties on the basis of reciprocity, subject to reasonable notice and appropriate consultations).

Obligations of transparency and information sharing with respect to hazards and risks are vital for safety reasons in the harsh environment of space. Transparency can also reduce the risk of hostility being created from misunderstandings and false attribution. However, it is noted that whilst these provisions exist there is little evidence of their application in practice.[26]

2.4 Interoperability (Section 5)

Section 5 introduces the key concept of interoperability, recognising 'that the development of interoperable and common exploration infrastructure and standards, including but not limited to fuel storage and delivery systems, landing structures, communications systems, and power systems, will enhance space-based exploration, scientific discovery, and commercial utilization'. Section 5 of the Accords does not mandate interoperability, potentially an inherent limitation in the nature of the Accords, but requires signatories to 'commit to use reasonable efforts to utilize current interoperability standards for space-based infrastructure, to establish such standards when current standards do not exist or are inadequate, and to follow such standards'.[27]

The exploration and use of space have witnessed a surprising combination of great power rivalry and cooperation, with one of the major impediments to cooperation being the capability for interoperability. Naturally, this has been largely fuelled by the competition between the original space superpowers for technological superiority and Cold War competition. Whilst that rivalry spurred the rapid advancement of space technology by the competing States, each seeking to achieve the important 'firsts'—the first to orbit the Earth, the first human on the Moon and so on—it also generated a culture of secrecy which necessitated the rapid development of technologies addressing the same capability gap in isolation. The 1975 Apollo-Soyuz mission represented a significant element in the Nixon-Ford administration's détente initiatives, building upon visions of space as a new domain for cooperation and 'brotherhood'.[28] Regardless of whether it was a successful political mission, it could not have occurred but for the technical interoperability of the 'androgynous' docking system which had to be developed to link the American Apollo capsule with the Soviet Soyuz, resulting in an embrace between the commanders of the two modules. This unique docking system, jointly developed by

[26] Francis Lyall and Paul B Larsen, *Space Law: A Treatise* (2nd edn, Routledge 2018) 63.
[27] Artemis Accord*s* (n 1) s 5.
[28] Thomas Ellis '"Howdy Partner!" Space Brotherhood, Detente and the Symbolism of the Apollo–Soyuz Test Project' (2019) 53(3) JAmStud 744, 764–69.

the American and Soviet space programmes, was essential as the two spacecraft used incompatible docking systems. It was also vital to develop a system whereby neither spacecraft was seen as 'submissive' to the other. Thus, the issue of interoperability became a matter of politics as much as engineering.[29] The successful docking of the rival powers' spacecraft using a neutral docking mechanism opened up the possibility of further cooperation in space, including the construction of an orbital space station. This process led in due course to the most successful example of international cooperation in space: the International Space Station (ISS).

The ISS is operated pursuant to the International Space Station Intergovernmental Agreement (IGA),[30] supported by four Memoranda of Understanding between NASA, the European Space Agency, the Canadian Space Agency, ROSCOSMOS and the Japanese Space Agency.[31] As Stewart and Dittmer observe, the ISS, has served as an 'assemblage' which operates as a site for international scientific cooperation.[32] The IGA, like the Artemis Accords, prescribes the terms for engagement in the project, to which all partners must adhere. This has the dual effect of creating a formal channel for engagement between the parties (and their host States) outside of other (political) venues and necessitates a certain level of reliance and trust.[33] They argue that adopting a '"more-than-human" approach to space diplomacy highlights the role of materials in fostering cooperation and peaceful negotiations' which may prevail even in the face of terrestrial conflict.[34] This suggests that a similar role is possible for the Artemis Accords in advancing the development of projects and hence norms of engagement through the conduct of space projects with like-minded partners, which has the potential to attract other participants.[35]

In more recent years, the emergence of commercial space start-ups has seen disputes regarding intellectual property and confidential information.[36] Concurrently,

[29] William Stewart and Jason Dittmer 'More-than-Human Space Diplomacy: Assembling Internationalism in Orbit' (2023) 18 HagueJDipl 219, 232–33.

[30] Agreement among the Government of Canada, Governments of Member States of the European Space Agency, the Government of Japan, the Government of the Russian Federation, and the Government of the United States of America Concerning Cooperation on the Civil International Space Station (adopted 29 January 1998, entered into force 27 March 2001) TIAS 12927 (1998) (IGA).

[31] Agreement among the Government of the United States of America, Governments of Member States of the European Space Agency, the Government of Japan, and the Government of Canada on Cooperation in the Detailed Design, Development, Operation, and Utilization of the Permanently Manned Civil Space Station (adopted 29 September 1988, entered into force 30 January 1992) 37 ZLW 341; Alexander Yakovenko, 'The Intergovernmental Agreement on the International Space Station' (1999) 15(2) SpacePoly 79–86.

[32] William Stewart and Jason Dittmer, 'More-than-Human Space Diplomacy: Assembling Internationalism in Orbit' (2023) 18 HagueJDipl 219–52.

[33] ibid 239.

[34] ibid 244.

[35] Nancy Riordan, Miloslav Machon and Lucia Csajkova, 'Space Diplomacy and the Artemis Accords' (2023) 18 HagueJDipl 380–408.

[36] Joshua T Smith, 'The Future of Space Exploration: SpaceX's Petitions for Inter Partes Review against Blue Origin's Rocket-Landing Technology' (2017) 19 SouthMethodUSci&TechLRev 65.

while disputes may occur between competitors on access to technology, it is important to recognise that international standards have emerged over time to ensure interoperability. For example, the 'CubeSat' and its underlying unit-style structure are now one of the most common platforms for small satellite development across numerous (and competing) manufacturers.

Interoperability ensures greater capability for cooperation and lowers the cost of working across different providers. Commercial off-the-shelf technology has revolutionised the speed and lower cost of development of space technologies.[37] NASA's contracts with SpaceX to develop a crewed lunar lander as part of the Artemis project are a clear example of the central importance of commercial operators to the success of the programme.[38]

2.5 Emergency Assistance (Section 6)

Section 6 (Emergency Assistance) requires all signatories to 'commit to taking all reasonable efforts to render necessary assistance to personnel in outer space who are in distress, and to acknowledge their obligations under the Rescue and Return Agreement'.

The Outer Space Treaty provides a special status for astronauts, designating them in Article V as 'envoys of [hu]mankind' and requiring States Parties to 'render them all possible assistance in the event of accident, distress, or emergency landing on the territory of another State Party or on the high seas'. Astronauts must be 'safely and promptly returned to the State of registry of their space vehicle'. Astronauts of one State must render 'all possible assistance' to astronauts of another State. These obligations are further elaborated on in the Rescue and Return Agreement.

Notably, the Artemis Accords create an obligation with respect to 'personnel in outer space' rather than limiting this to astronauts. The Rescue and Return Agreement, whilst using the word 'astronaut' in its full title, uses the phrase 'personnel of a spacecraft' throughout, echoing the use of the word 'personnel' in Article VIII of the Outer Space Treaty. In this way the Artemis Accords avoid any complexity created by the issue of whether a person is an 'astronaut' or not. With the growth of commercial spaceflight, space tourism and the need for a broader range of crew skills, there has been considerable debate regarding whether an individual spaceflight participant qualifies as an astronaut. The bestowal of 'wings'

[37] 'The NESC 2014 Technical Update' (*NASA*, 2014) <www.nasa.gov/sites/default/files/atoms/files/techup2014_pdf9-1pageview.pdf> accessed 15 July 2023; 'Robotics and Automation on Earth and in Space Roadmap 2021–2030' (*Australian Space Agency*, 2022) <www.industry.gov.au/sites/default/files/January%202022/document/robotics_and_automation_on_earth_and_in_space_roadmap_2021-2030.pdf> accessed 15 July 2023.

[38] 'NASA Awards SpaceX Second Contract Option for Artemis Moon Landing' (*NASA*, 16 November 2022) <www.nasa.gov/press-release/nasa-awards-spacex-second-contract-option-for-artemis-moon-landing-0> accessed 15 July 2023.

has traditionally been regarded as one of the clearest indicia of a person being an astronaut, in the absence of a more formal definition. The various branches of the US military apply differing requirements according to service branch with respect to qualification for the award of wings. For the Air Force, Navy and Marines, the requirement is the completion of specified training and having travelled beyond 50 miles (approximately 80 km) above sea level; for the Army, the qualification is the completion of specified training and undertaking orbit of the Earth.[39] For NASA astronauts, silver pins are awarded to those who have completed astronaut training and gold pins to those who have flown in space.[40] Since 2004, the Federal Aviation Administration (FAA) has awarded Commercial Space Astronaut Wings to individuals who reach 50 miles above the surface of the Earth. The purpose of the FAA programme was to bring attention to space activities and celebrate the achievements of those brave enough to venture into space and 'promote the development of vehicles designed to carry humans into space'.[41] Notably in December 2021, the FAA announced that, with the significant expansion of space tourism, it would no longer be awarding wings to private spaceflight participants but would rather recognise those who launch on an FAA-licensed or permitted launch and reach 50 statute miles above the surface of the Earth by listing them on a website.[42]

The activities proposed to be undertaken as part of the Artemis Program will result in a much larger number of people going into space and undertaking various activities. Some may remain off Earth for extended periods of time, effectively dwelling in space or on the Moon. The clarification of terminology necessitated both by the inconsistent wording of the UN Space Treaties and the evolution of space activities, that obligations of assistance apply regardless of the individual status of the person, reflects the fact that space will remain a hostile environment for humanity.

2.6 Registration of Space Objects (Section 7)

Section 7 deals with the registration of space objects. This provision reflects and reinforces the obligations of the Registration Convention:

[39] 'Jeff Bezos and Sir Richard Branson May not be Astronauts, US Says' *BBC News* (23 July 2021) <www.bbc.com/news/world-us-canada-57950149> accessed 15 July 2023.

[40] Robert Z Pearlman, '"Pin"-nacle Achievement: The Story Behind NASA's Astronaut Pin' (*Space.com*, 7 January 2020) <www.space.com/nasa-astronaut-pin-history.html> accessed 15 July 2023.

[41] 'Commercial Human Spaceflight Recognition' (*Federal Aviation Administration*, 9 June 2022) <www.faa.gov/space/human_spaceflight/recognition/> accessed 15 July 2023.

[42] 'FAA Ends Commercial Space Astronaut Wings Program, Will Recognise Individuals Reaching Space on Website' (*Federal Aviation Administration*, 10 December 2021) <www.faa.gov/newsroom/faa-ends-commercial-space-astronaut-wings-program-will-recognize-individuals-reaching> accessed 15 July 2023.

> For cooperative activities under these Accords, the Signatories commit to determine which of them should register any relevant space object in accordance with the Registration Convention.
>
> For activities involving a non-Party to the Registration Convention, the Signatories intend to cooperate to consult with that non-Party to determine the appropriate means of registration.

Registration is regarded as a principle underpinning the responsible use of space and is embodied in Article VIII of the Outer Space Treaty, as well as the Registration Convention.

Relevantly, registration also plays an essential role in determining the prevailing jurisdiction that applies to particular space objects by virtue of the Outer Space Treaty, with deemed 'jurisdiction and control' arising as a consequence of registration.[43] The Artemis Program will inevitably see some components (and potentially, spacecraft) manufactured by different States as contributions to the program. This mixed nationality of space objects and componentry causes registration to become important. Section 7 essentially calls for agreements like those contained in the IGA that saw components of the ISS assigned to particular States from the outset. This ultimately resulted in States bearing the responsibility to register those components in their name and retain jurisdiction as a consequence.[44]

Importantly, aspects of the IGA are being carried directly into Artemis, with NASA entering into agreements with the European Space Agency, Canada and Japan for the Lunar Gateway using the 'Evolution' provision of the IGA.[45] Practically, this places the Lunar Gateway, an essential component of the Artemis Program, under a legal arrangement nearly identical to the ISS, which ultimately necessitates compatibility of the Accords with the content of existing international cooperation practices.[46]

[43] Outer Space Treaty (n 7) art VIII.

[44] IGA (n 30) art 5 Annex.

[45] ibid art 14; 'NASA, European Space Agency Formalize Artemis Gateway Partnership' (*NASA*, 28 October 2020) <www.nasa.gov/press-release/nasa-european-space-agency-formalize-artemis-gateway-partnership> accessed 15 July 2023; 'NASA, Canadian Space Agency Formalize Gateway Partnership for Artemis Program' (*NASA*, 17 December 2020) <www.nasa.gov/press-release/nasa-canadian-space-agency-formalize-gateway-partnership-for-artemis-program> accessed 15 July 2023; 'NASA, Government of Japan Formalize Gateway Partnership for Artemis Program' (*NASA*, 13 January 2021) <www.nasa.gov/press-release/nasa-government-of-japan-formalize-gateway-partnership-for-artemis-program> accessed 15 July 2023.

[46] Although the NASA Office of Inspector General noted in a 2023 report that this arrangement cannot continue indefinitely as it can only be used for collaborations between States that are already party to the International Space Station IGA; 'NASA's Partnerships with International Space Agencies for the Artemis Campaign' (*NASA Office of Inspector General*, 17 January 2023) 23–24 <oig.nasa.gov/docs/IG-23-004.pdf> accessed 15 July 2023.

2.7 Release of Scientific Data (Section 8)

As noted above, the UN Space Treaties have a focus on information sharing and transparency. Section 8 establishes obligations with respect to the release of scientific data. This reflects Article XI of the Outer Space Treaty which provides:

> In order to promote international cooperation in the peaceful exploration and use of outer space, States Parties to the Treaty conducting activities in outer space, including the Moon and other celestial bodies, agree to inform the Secretary-General of the United Nations as well as the public and the international and scientific community, to the greatest extent feasible and practicable, of the nature, conduct, locations and results of such activities. On receiving the said information, the Secretary-General of the United Nations should be prepared to disseminate it immediately and effectively.

Artemis signatories, through section 8, commit to 'the open sharing of scientific data'.[47] Scientific results obtained from cooperative Artemis activities will be made 'available to the public and the international scientific community, as appropriate, in a timely manner'.[48] Whilst there is an emphasis on the desirability for signatories to 'communicate and release information to the public regarding their own activities' in a timely fashion, the release of information is also necessary for coordination and consultation to ensure 'appropriate protection for any proprietary and/ or export-controlled information'. Notably, the 'commitment to openly share scientific data is not intended to apply to private sector operations unless such operations are being conducted on behalf of a Signatory to the Accords'.[49] This position goes some of the way to protecting the interests of commercial partners, especially in connection with the generation of new intellectual property that may be of value.

2.8 Preserving Outer Space Heritage (Section 9)

Section 9 represents a new and potentially controversial principle: preserving outer space heritage. Signatories are obliged 'to preserve outer space heritage' defined as 'historically significant human or robotic landing sites, artifacts, spacecraft, and other evidence of activity on celestial bodies in accordance with mutually developed standards and practices'.[50]

[47] Artemis Accords (n 1) s 8.
[48] ibid s 8.
[49] ibid s 8.
[50] ibid s 9.

Further, the experiences under the Accords with respect to space heritage will be used 'to contribute to multilateral efforts to further develop international practices and rules applicable to preserving outer space heritage'. Whilst there is a growing recognition of the importance of space heritage,[51] there has been some concern that the focus of the Artemis Accords is only on the US Apollo landing sites and lunar missions and that it would be preferable for an international body (such as the UN) to oversee preservation of unique sites according to a broader cultural remit.[52] It is worth noting the cultural and diplomatic significance that lunar memorials have already served. Small memorials to cosmonauts were placed on the Moon by Apollo 11 (a pair of medals gifted by the widows of Yuri Gagarin and Vladimir Komarov to astronaut Frank Borman) and Apollo 15 (a miniature sculpture, titled 'The Fallen Astronaut', accompanied by a plaque listing the names of fourteen deceased astronauts and cosmonauts).[53]

This particular section continues to build upon the work that has been taking place within the US to protect what are seen as sites with heritage value. In 2020, the Congress expressed the view that the 'lunar landing sites of the Apollo 11 spacecraft [. . .] and the crewed and robotic spacecraft that followed, are of outstanding universal value of humanity'.[54] The One Small Step to Protect Human Heritage in Space Act[55] was passed by the US Congress on 31 December 2020 and requires NASA to take steps to protect sites with heritage value through its contractual arrangements associated with lunar activities. This includes promoting recognition and respect for NASA's 2011 'Recommendations to Space-Faring Entities' on the protection of sites of 'historic and scientific value' to the US.[56] The 2020 Law does not directly impose obligations or prohibitions in connection with historical sites, rather it requires NASA to act in a certain way. The findings at the beginning of the Law are also informative, purporting to couch protection of US sites within Articles I, VIII and IX of the Outer Space Treaty.[57] In many respects, the inclusion of section 9 in the Accords is a consequence of the 2020 Law and bolsters the US policy positions with respect to its existing historic lunar sites.

[51] 'NASA's Recommendations to Space-faring Entities: How to Protect and Preserve the Historic and Scientific Value of US Government Lunar Artifacts' (*NASA*, 28 October 2011) <www.nasa.gov/pdf/617743main_NASA-USG_LUNAR_HISTORIC_SITES_RevA-508.pdf> accessed 15 July 2023.

[52] Alice Gorman, 'Who is Responsible for Heritage in Outer Space?' (*Apollo*, 23 November 2020) <www.apollo-magazine.com/artemis-accords-heritage-outer-space/> accessed 15 July 2023.

[53] Ellis (n 28) 756–57.

[54] One Small Step to Protect Human Heritage in Space Act (2020). Public L No 116-275, §2(a)(6), 134 Stat 3358, 3359.

[55] ibid §3, 3360–61.

[56] ibid §3(b), 3360.

[57] ibid §§2(a)(10)–(12), 3359–60.

2.9 Space Resources (Section 10)

Underpinning the establishment of a long-term human habitat on the Moon, as well as obtaining the fuel and other resources necessary for the journey to Mars, is the extraction of resources from the Moon and other celestial bodies. Section 10—Space Resources—articulates the common position on space resource utilisation to be adopted by all parties to the Artemis Accords. The key elements of this section are as follows:

- Signatories acknowledge that the use of space resources 'can benefit humankind by providing critical support for safe and sustainable operations';
- Extraction and utilisation of space resources should be undertaken in compliance 'with the Outer Space Treaty and in support of safe and sustainable space activities';
- Signatories affirm the position 'that the extraction of space resources does not inherently constitute national appropriation under Article II of the Outer Space Treaty';
- All 'contracts and other legal instruments relating to space resources should be consistent with' the Outer Space Treaty;
- Signatories will inform 'the Secretary-General of the United Nations as well as the public and the international scientific community of their space resource extraction activities in accordance with the Outer Space Treaty'; and
- Experience gained from participation in the Accords will be used 'to contribute to multilateral efforts to further develop international practices and rules applicable to the extraction and utilization of space resources, including through' UNCOPUOS.

Article II of the Outer Space Treaty provides: 'Outer space, including the Moon and other celestial bodies, is not subject to national appropriation by claim of sovereignty, by means of use or occupation, or by any other means.' This principle of 'non-appropriation' has been cited as an impediment to the development of resource extraction as the scope of the Article II prohibition remains unsettled.[58] The question remains: does the non-appropriation principle apply only to the purported appropriation of the entire celestial body (eg the Moon) or territory thereon, or does it limit any and all appropriation including the extraction and use of mineral and regolith?[59]

[58] See, Board of Directors, Position Paper on Space Resource Mining (*International Institute of Space Law*, 20 December 2015).

[59] Melissa de Zwart, Stacey Henderson and Michelle Neumann, 'Space Resource Activities and the Evolution of International Space Law' (2023) 211 ActaAstro 155.

The US has continued to assert its position that the 'utilisation of space-based resources—including commercial utilisation—is consistent with the four main United Nations space treaties'.[60] Maintaining that the Outer Space Treaty does not provide a 'comprehensive international regime for space resource utilisation activities', the US position is that the Artemis Accords 'form a starting point for the US Government's future work on space resources'.[61] Several proposed models for management of resource extraction have been developed, including the Hague International Space Resources Group's Building Blocks for the Development of an International Framework on Space Resource Activities[62] and the Moon Village Association Best Practices for Sustainable Lunar Activities.[63] The UNCOPUOS Legal Subcommittee established a Working Group on Space Resource Activities in 2021 which will produce its final report in 2027.[64]

For States such as Australia that are party to the Moon Agreement and also the Artemis Accords, this may result in a need to reconcile the extent to which Australia can be involved in Artemis missions that include space resource activities, in the absence of any international regime governing the exploitation of space resources. Australia has, in a statement in June 2021 to the Legal Subcommittee of the UNCOPUOS, expressed that its position is 'that the Artemis Accords are consistent with Australia's international legal obligations'.[65] The statement goes on to confirm that Australia supports the 'further development of norms that could guide the orderly and safe development, as well as rational management of the Moon's natural resources'.[66] Saudi Arabia, the only other signatory to the Artemis Accords also party to the Moon Agreement, notified the UN of its intention to withdraw from the latter agreement, effective from January 2024.[67] No reason has been publicly provided for this withdrawal, making it difficult to authoritatively conclude that such a withdrawal was due to any incompatibility.

[60] Gabriel Swiney, 'US Statement Agenda Item 14: Potential Legal Models for Activities in the Exploration, Exploitation and Utilization of Space Resources' Sixtieth Session of the UNCOPUOS Legal Subcommittee (2021) <www.unoosa.org/documents/pdf/copuos/lsc/2021/statements/item_14_USA_ver.1_1_June_AM.pdf> accessed 13 July 2023.

[61] ibid.

[62] Hague International Space Resources Governance Working Group, 'Building Blocks for the Development of an International Framework on Space Resources Activities' (*International Institute of Air and Space Law*, 12 November 2019) <www.universiteitleiden.nl/en/law/institute-of-public-law/institute-of-air-space-law/the-hague-space-resources-governance-working-group> accessed 15 July 2023.

[63] Moon Village Association, 'Global Expert Group on Sustainable Lunar Activities Recommended Framework and Key Elements for Peaceful and Sustainable Lunar Activities' (2022) <moonvillageassociation.org/gegsla/documents/gegsla-recommended-framework/> accessed 15 July 2023.

[64] UNGA, 'Report of the Committee on the Peaceful Uses of Outer Space' (25 August–3 September 2021) UNGAOR 76th Session Supp No 20, UN Doc A/76/20.

[65] Australian Space Agency, 'Statement-Australia, Agenda Item 14, 60th COPOUS Legal Subcommittee' <www.unoosa.org/documents/pdf/copuos/lsc/2021/statements/item_14_Australia_ver.1_4_June_PM.pdf> accessed 13 July 2023.

[66] ibid.

[67] Status of International Agreements (n 10).

2.10 Deconfliction of Space Activities (Section 11)

Section 11 presents another issue which pushes the current boundaries of international space law—deconfliction of space activities—including the creation and recognition of safety zones.

Section 11 establishes its foundation in the concepts of 'due regard' and 'harmful interference' recognised in Article IX of the Outer Space Treaty, and Signatories are obliged to share information and refrain from any activities that may create harmful interference or 'pose a safety hazard to . . . space-based activities' of other Signatories. Further, the Signatories 'affirm that the exploration and use of outer space should be conducted with due consideration to the United Nations Guidelines for the Long-term Sustainability of Outer Space Activities adopted by the [UN]COPUOS in 2019, with appropriate changes to reflect the nature of operations beyond low-Earth orbit'.[68]

In order to give effect to these principles, Signatories agree to 'use their experience under the Accords to contribute to multilateral efforts to further develop international practices, criteria, and rules applicable to the definition and determination of safety zones and harmful interference'. It is significant that these safety zones are expressed to be implementing obligations under the Outer Space Treaty, including the avoidance of harmful interference.

To avoid the risk of harmful interference created by a certain activity or 'anomalous event', Signatories are required to designate particular areas as 'safety zones', with notification to be provided to others of the existence of that safety zone. The provision does not specify any particular dimensions or size for a safety zone, only that the 'size and scope of the safety zone, as well as the notice and coordination, should reflect the nature of the operations being conducted and the environment that such operations are conducted in' and 'leveraging commonly accepted scientific and engineering principles'.[69] Further, the expected duration of such zones is left undetermined, with only a statement that they are temporary, ending when the relevant operation ceases.[70] Whilst this provides the necessary latitude to encompass activities conducted both in orbit and on the surface of a celestial body, such as the Moon, the very difference of such environments leaves the potential coverage of such zones in both time and size undetermined.

Recognising both the contentious nature of the safety zones concept and cognisant of evolving understanding of the space environment, the Artemis Accords state that

[68] Artemis Accords (n 1) s 11(2).
[69] ibid s 11(7).
[70] ibid s 11(11).

> the nature and existence of safety zones is expected to change over time reflecting the status of the relevant operation. If the nature of an operation changes, the operating Signatory should alter the size and scope of the corresponding safety zone as appropriate.[71]

Provisions also refer to making information available regarding the nature of operations being undertaken within the safety zone, with reservations for proprietary and export-controlled information, as well as 'scientific discovery and technology demonstration'.[72] Specific reference is also made to 'the safe and efficient extraction and utilization of space resources in support of sustainable space exploration and other operations'.[73]

Notably, '[t]he Signatories commit to respect the principle of free access to all areas of celestial bodies and all other provisions of the Outer Space Treaty in their use of safety zones.' This provision appears to tread an important line between a safety zone being a purely safety matter and a safety zone becoming a de facto claim over a part of space for 'safety' reasons. Temporally, section 11 of the Accords requires Signatories 'to adjust their usage of safety zones over time based on mutual experiences and consultations with each other and the international community'.[74]

Safety zones are not formally recognised in international space law, although the concept has been recognised as providing potential for giving practical effect to the need to avoid collision and conflict. The ISS has a 200 km 'keep out' zone.[75] Commercial operators have been required to actively consider the ISS and growing congestion in Low Earth Orbit, with a prominent example being restrictions on SpaceX's Starlink orbits.[76] Safety zones have also been suggested in the context of resource extraction and military uses of space.

An example of a safety zone in practice might relate to any significant movement on the surface of the Moon, such as a landing of a spacecraft or even an equipment drop, which creates 'plume-surface interactions' resulting in regolith, dust, rocks and other surface material being distributed across the relevant area creating serious hazards to surrounding activities or habitats.[77] The exact risk created by such plumes remains unknown due to the limited experience of interaction with the Moon's surface. The size of the affected area is also unknown, with landings having

[71] ibid s 11(7)(c).

[72] ibid s 11(11).

[73] ibid s 11(11).

[74] ibid s 11(11).

[75] Jessy Kate Schingler, 'Imaging Safety Zones: Implications and Open Questions' *The Space Review* (8 June 2020) <www.thespacereview.com/article/3962/1> accessed 15 July 2023.

[76] See eg *In the Matter of Space Exploration Holdings, LLC, Request for Orbital Deployment and Operating Authority for the SpaceX Gen2 NGSO Satellite System*, Federal Communications Commission (FCC 22-91, 1 December 2022) 72[cc].

[77] Gabriel Swiney and Amanda Hernandez, 'Lunar Landing and Operations Policy Analysis' (*NASA Office of Technology, Policy and Strategy*, 30 September 2022) 22.

'the potential to transport material across the entire surface of the Moon and even into outer space'.[78]

The Artemis Accords are not the first to propose the use of safety zones. The Hague Space Resources Working Group recommended that the international framework should permit the creation of safety zones in the context of resource utilisation:

> Taking into account the principle of non-appropriation under Article II [Outer Space Treaty], the international framework should permit States and international organizations responsible for space resource activities to establish a safety zone, or other area based safety measure, around an area identified for a space resource activity as necessary to assure safety and to avoid any harmful interference with that space resource activity. Such safety measure shall not impede the free access, in accordance with international law, to any area of outer space by personnel, vehicles and equipment of another operator. In accordance with the area-based safety measure, a State or international organization may restrict access for a limited period of time, provided that timely public notice has been given setting out the reasons for such restriction.[79]

Notably, this proposal places the concept of the safety zone firmly within existing international space law and the scope of Article II of the Outer Space Treaty.

2.11 Space Debris (Section 12)

Section 12 of the Accords contemplates the important and ever-relevant matter of orbital debris. Broadly, 'debris' can refer to a wide range of objects including defunct satellites that have not yet re-entered the atmosphere, flecks of paint that may fall off space objects, used rocket bodies, stage separation rings, other components that are left to remain in outer space and fragments of spacecraft that have been kinetically destroyed. Due to the characteristics of the space domain, this debris presents a serious threat to space operations. Section 12, while not being couched in mandatory language, commits signatories to the Accords to:

- Planning for the mitigation of orbital debris—specifically the safe, timely and efficient passivation and disposal of the debris—as part of the mission planning process; and

[78] Michelle Munk, 'Plume Surface Interaction: Surface Excavation and Construction Impacts' (Surface Construction and Excavation Workshop, NASA, as cited in Gabriel Swiney and Amanda Hernandez (n 77) 23).

[79] The Hague International Space Resources Governance Working Group (n 62).

- Limiting, as far as practicable, the generation of new long-lived harmful orbital debris.

In the modern space era, these commitments are not novel or surprising and align with existing international practices associated with orbital debris.[80] The commitment to consider orbital debris in the planning stage is important, codifying the position that debris should be mitigated *by design* as opposed to debris mitigation plans being fitted to a mission in an ad hoc manner after it has been designed.

While the international law applicable to space activities does not actively address the issue of space debris, there are provisions of relevance. Article IX of the Outer Space Treaty is the prime candidate, with its focus on due regard and potentially harmful interference.[81] State practices are also changing, with a large number of States (including Artemis Accords Signatories) codifying debris mitigation requirements for those operating within their jurisdictions.[82]

What this section of the Accords does not do is consider the unique circumstances of the Artemis Program. The most common manifestation of debris mitigation is re-entry through Earth's atmosphere.[83] This causes the debris to disintegrate as it passes through the atmosphere. Given that the Artemis Program will eventually involve the placement of space objects in lunar orbit and on the Moon's surface, the traditional disposal mechanism is not achievable. This leaves an ongoing lacuna in the area of debris mitigation and end-of-mission-life measures associated with debris in lunar orbit and on the lunar surface. De-orbit of debris in the orbit of the Moon, could potentially result in debris 'crash landing' on the Moon's surface due to the lack of atmosphere which could destroy or decelerate the debris. While this strategy would ensure that lunar orbit remains clear of debris, it does present an ongoing risk to installations and space objects (including heritage sites) on the lunar surface.

Despite the potential gap in connection with the mitigation of debris around the Moon, section 12 codifies the growing need to actively consider debris mitigation as part of mission planning. The commitment that this section of the Accords represents is an important signal for the future of space activities and the need to consider the sustainability of operations.

[80] eg 'Space Debris Mitigation Guidelines' UNGAOR 62nd Session Supp No 20, UN Doc A/62/20 (2007) Annex.

[81] Outer Space Treaty (n 7) art IX.

[82] See eg Space (Launches and Returns) Act 2018 (Australia) s 34(2)–(4).

[83] See 'Space Debris Mitigation Guidelines' (n 80) Guideline 6.

2.12 Final Provisions (Section 13)

Section 13 outlines the process for the administration of the Accords. Signatories 'commit to periodically consult to review the implementation of the principles' and 'exchange views on potential areas of future cooperation'. No bespoke mechanism for consultation and review is established by the Accords, rather discussions should build upon 'any consultative mechanisms in pre-existing arrangements'. The US government 'will maintain the original text of these Accords'. A copy of the Accords has been transmitted to the Secretary-General of the UN, for the purpose of circulating to all members of the UN as an official document of the UN. However, the Accords are 'not eligible for registration under Article 102 of the Charter of the United Nations'.[84] The effect of non-registration is that parties to the Accords may not invoke the Accords before any organ of the UN.[85]

3 The Artemis Program and Missions

The Artemis Program emerged out of restructuring to NASA's programme to return to the Moon coinciding with the commencement of the Trump Administration in 2017. Aspects of the programme can be traced to historic programmes including the George W Bush and Barack Obama era Constellation Program. The Artemis Program gets its name from Greek mythology, with Artemis being the goddess of the Moon and the twin sister of Apollo.[86]

Artemis differs from traditional NASA programmes. Rather than having a single focus, the programme is led by NASA with a 'broad articulation of a unified purpose across missions, funding lines, directorate, and partnerships'.[87] As noted above, it will involve commercial and government partnerships, with various elements potentially involving a different set of partners. For example, Australia has entered into an agreement with NASA to develop a Moon rover. Pursuant to the Australian Moon to Mars Initiative, the Australian government has awarded $8 million to a commercial consortium to design early-stage prototypes of a semi-autonomous rover.[88] The rover is intended to collect lunar soil samples, which will

[84] Artemis Accords (n 1) s 13(2).

[85] Art 102(2) of the Charter of the United Nations provides: 'No party to any such treaty or international agreement which has not been registered in accordance with the provisions of paragraph 1 of this Article may invoke that treaty or agreement before any organ of the United Nations.' Charter of the United Nations (adopted 26 June 1945, entered into force 24 October 1945) 1 UNTS XVI, art 102.

[86] 'What is Artemis?' (*NASA*, 26 July 2019) <www.nasa.gov/what-is-artemis> accessed 15 July 2023.

[87] Gabriel Swiney and Amanda Hernandez (n 77).

[88] Malcolm Davis, 'Australia on a Path to the Moon with NASA's Artemis Program' (*The Strategist*, Australian Strategic Policy Institute, 17 November 2022) <www.aspistrategist.org.au/australia-on-a-path-to-the-moon-with-nasas-artemis-program/> accessed 15 July 2023.

then be used by NASA to study the viability of oxygen extraction from those samples. Launch of the Australian rover is proposed for 2026.[89]

The Artemis Program currently includes the Human Landing System, Orion, Space Launch System and Commercial Lunar Payload systems, with the planned construction of the Lunar Gateway and a permanent human habitat on the Moon.[90]

NASA also released its 'Moon to Mars Objectives' which further detail its approach to human exploration of deep space, encompassing the Artemis missions.[91] Developed by NASA in consultation with US industry, academia and international partners, it is anticipated to evolve as the Artemis missions are conducted. This document reflects a further detailed examination of how the Artemis principles may be developed and articulated in a practical framework.

The first theme, 'Recurring Tenets: Common Themes Across Objectives', reinforces the focus on collaboration with international partners and with industry. It also re-articulates the underpinning Artemis principles of responsibility, requiring the 'conduct [of] all activities for the exploration and use of outer space for peaceful purposes consistent with international obligations, and principles for responsible behaviour in space' and interoperability, to ensure commonality of 'technical, operations and process standards' across both systems and crews.[92] In this way, the Artemis principles are embedded in programme design from the outset, operationalising their role in both practice and norm development. This document is expressed to be a living document that will be updated as the Artemis missions and accomplishments evolve. The purpose of this 'objectives-based approach' is to ensure that the basic approach remains the same, regardless of the technology or mechanism that is deployed to achieve that objective. This again echoes the claim that the Artemis approach will be one of evolution and reflection. It is not a single mission or technology; it is being developed by NASA with its partners, and it is envisaged that different partners will fulfil different roles.

4 The Impact of the Artemis Accords on International Lawmaking

As has already been observed, there are many who assert that the Artemis Accords are being used to control and shape the future development of international space law.[93] Some argue that the Artemis Accords will lead to a fracturing of the order

[89] Department of Industry, Science and Resources, 'Space Industry Partners Chosen for Australia's First Moon Mission' (*Department of Industry, Science and Resources, Australian Government*, 20 March 2023) <www.industry.gov.au/news/space-industry-partners-chosen-australias-first-moon-mission> accessed 15 July 2023.

[90] Gabriel Swiney and Amanda Hernandez (n 77) 8.

[91] Moon to Mars Objectives (*NASA*, September 2022) <www.nasa.gov/sites/default/files/atoms/files/m2m-objectives-exec-summary.pdf> accessed 15 July 2023.

[92] ibid.

[93] Christopher Newman, 'Artemis Accords: Why Many Countries are Refusing to Sign Moon Exploration Agreement' *The Conversation* (19 October 2020) <theconversation.com/

established by the Outer Space Treaty and will result in 'fragmentation and competition'.[94] In particular, this criticism focuses on the fact that the Artemis Accords process has taken place entirely outside of the UN, which has nurtured the growth of international space law since the 1950s.

Russia, via ROSCOSMOS, was critical of both the Artemis Accords and President Trump's Executive Order on Space Resources, levelling accusations of the US privatising and colonising space.[95] China did not provide any official response to the Accords, but the Deputy Director of the China National Space Administration Space Law Centre, Guoyu Wang, asserted that the Artemis Accords seek 'to build a community of interests led by the US in the field of lunar and deep space exploration, as well as to establish its dominant position in space resources exploitation and recovery, and, finally, to consolidate its leading position in space'.[96]

Others view the Accords as both a transparent and legitimate process, with Fidler opining that, '[t]hrough the [A]ccords, the United States is using its influence as the predominant spacefaring nation to strengthen international space law'.[97]

International law can of course only be made by States, through the creation of multilateral treaties and by development of new customary international law through persistent and dispersed State practice. Ud Din argues that the Accords both 'replicate the design of the existing international space law' and 'espouse a definitive interpretation of issues that are contentious and uncertain in existing space law'.[98] Thus, the Artemis Accords provide an alternative and hence competing view of several key issues of international space law, which threaten to undermine multilateral approaches to space activities and increase the potential for conflict.[99]

Deplano also undertook an analysis of several provisions of the Artemis Accords to assess their compliance with existing international law. Deplano concluded that, whilst the provisions of the Artemis Accords do not appear to deviate from principles of international law, particularly the Outer Space Treaty, she too formed the view that they do reflect a clear departure from the traditional practice of negotiating space law in the UN.[100] It is necessary to recognise that this approach

artemis-accords-why-many-countries-are-refusing-to-sign-moon-exploration-agreement-148134> accessed 15 July 2023.

[94] Athar ud Din, 'The Artemis Accords: The End of Multilateralism in the Management of Outer Space?' (2022) 20(2–3) Astropol 136.

[95] Namrata Goswami, 'The Strategic Implications of the China-Russia Lunar Base Cooperation Agreement' *The Diplomat* (19 March 2021) <thediplomat.com/2021/03/the-strategic-implications-of-the-china-russia-lunar-base-cooperation-agreement/> accessed 15 July 2023.

[96] Guoyu Wang, 'NASA's Artemis Accords: The Path to a United Space Law or a Divided One?' *The Space Review* (24 August 2020) <www.thespacereview.com/article/4009/1> accessed 15 July 2023.

[97] David Fidler, 'The Artemis Accords and the Next Generation of Outer Space Governance' (*Council on Foreign Relations*, 2 June 2020) <www.cfr.org/blog/artemis-accords-and-next-generation-outer-space-governance> accessed 15 July 2023.

[98] ud Din (n 94) 146.

[99] ibid.

[100] Deplano (n 15) 812.

may be entirely intentional and represents an attempt at shifting the interpretation of the UN Space Treaties through subsequent State practice and agreement of States Parties—approaches that are recognised through the Vienna Convention on the Law of Treaties.[101]

Operating below international law is the practice of diplomacy and the creation of soft law and norms of behaviour. The Artemis Accords initially drew together a group of like-minded States, with the initial signatures being executed virtually due to the Covid-19 pandemic.[102] As the number of signatories has grown, there is an almost inevitable tug towards the interpretations of complex international space law principles articulated in the Accords.

It should be noted that Artemis is not the only multistate project to establish a human presence on the Moon. China and Russia announced plans in 2021 to establish a permanent human base on the Moon, inviting other States to become involved in the International Lunar Research Station (ILRS).[103] As with Artemis, the ILRS will involve extractive and utilisation activities, as well as scientific and human studies.[104]

5 Conclusion

This chapter has analysed the provisions of the Artemis Accords and their place within international space law. It has examined claims that the Artemis Accords are an attempt by the US to forge a unilateral pathway for the development and evolution of international law with respect to the use of cislunar space and beyond.

Rather than reflecting a complete break with the UN Space Treaties, many of the Accords' provisions are explicitly grounded in concepts such as 'peaceful purposes', 'emergency assistance' and 'registration'. Newer concepts, such as 'deconfliction' and 'space debris' are grounded in a practical need to ensure that space remains accessible and usable. More contentious are issues such as the utilisation of space resources, but the Artemis Accords are not the first or only international project to attempt to deal with the rules applying to the extraction and use of space resources. These matters have all been raised in other fora.

It also should be kept in mind that, whilst Artemis is a specific programme of NASA-led activities, the Accords are structured as a set of principles which may

[101] Vienna Convention on the Law of Treaties (adopted 23 May 1969, entered into force 27 January 1980) 1155 UNTS 331, art 31(3)(a)–(b).

[102] Riordan (n 35).

[103] Andrew Jones, 'China, Russia Enter MOU on International Lunar Research Station' *Space News* (9 March 2021) <spacenews.com/china-russia-enter-mou-on-international-lunar-research-station/> accessed 15 July 2023.

[104] Tereza Pultarova, 'Russia, China Reveal Moon Base Roadmap but no Plans for Astronaut Trips' *Space* (17 June 2021) <www.space.com/china-russia-international-lunar-research-station> accessed 15 July 2023.

evolve as the partners develop activities and agreements which give effect to the vision of the Artemis Accords. The Accords also reflect the practical need to move faster than the UN has been able to with respect to evolving needs for rules regarding resource extraction, safety zones and large-scale human and robotic activities in outer space. The failure of the Moon Agreement to attract a critical mass of signatories evidences the likelihood that the early success of multilateral space treaty development is over. The exercise of soft power and the diplomatic success of bringing governments and space agencies together to work on a large-scale, multifaceted project which will return humans to the Moon demonstrates new ways of working together. Practices, policies and agreements will start to shape State practice, which in time will evidence the evolution of international space law.

There remains much formal work still to be done to ensure that such activities do not fragment space nor lead to conflict, and formal space law has a vital role to play in this regard. However, as humans venture back to the Moon and on to Mars, the law will need to evolve with them, and Artemis is one important element of this process.

13

INTERNATIONAL SECURED TRANSACTIONS LAW IN OUTER SPACE

*Hamza Hameed**

1 Introduction

The space industry has grown exponentially in the past decade. The value of the global space economy rose to US$546 billion in 2022, which was more than 55 per cent higher than a decade ago.[1] Based on findings, by 2040 the industry is anticipated to achieve an annual revenue of $1 trillion. A significant 95 per cent reduction in launch expenses is predicted, paving the way for an expanded range of services from space.[2] Advancements in technology, heightened investment from the private sector, and a growing need for space-related data are transforming the space industry.

In order to continue to support this growth, international law needs to be present, given the inherent cross-border nature of this industry and the involvement of parties from all parts of the world. The space sector's primary international law-related guidelines stem from the Outer Space Treaty of 1967, which was developed by the United Nations Office for Outer Space Affairs. This was further supported by four additional treaties developed in the 1970s and 1980s, a majority of which now serve as the cornerstone of public international space law, as well as the foundation of most national space legislation.

One area of international law that is particularly relevant to the space sector is secured transactions law. Secured transactions law deals with the creation and enforcement of security interests in personal property, which can be used as collateral

* Hamza Hameed is a Senior Policy Manager for Satellite and Telecommunications at Access Partnership in Singapore. Prior to this, Hamza worked as a Legal Consultant at the International Institute for Unification of Private Law (UNIDROIT) in Rome, Italy.

[1] Space Foundation Editorial, 'Space Foundation Releases the Space Report 2023 Q2, Showing Annual Growth of Global Space Economy to $546B' (*Space Foundation*, 7 September 2023)<www.spacefoundation.org/2023/07/25/the-space-report-2023-q2> accessed 10 October 2023.

[2] Morgan Stanley projects that by 2040, the worldwide space industry's revenue will surge to US$1.1 trillion or higher, a significant rise from $350 billion recorded in 2016. See 'Space: Investing in the Final Frontier' (*Morgan Stanley*) <www.morganstanley.com/ideas/investing-in-spacet> accessed 10 October 2023.

Hamza Hameed, *International Secured Transactions Law in Outer Space* In: *International Space Law in the New Space Era*. Edited by: Sandeepa Bhat B., Dilip Ukey, and Adithya Variath, Oxford University Press. © Hamza Hameed 2024.
DOI: 10.1093/9780198909415.003.0013

to secure the repayment of a debt or other obligation. In industries with high-value assets, this is often used as an effective financing tool.

The rapid expansion of the space industry over the last ten years underscores the importance of a strong and efficient international framework for secure transactions within the space realm. The Space Protocol to the Cape Town Convention provides a framework for this, but it is important that it be implemented and enforced effectively in order to continue to support the growth of the space industry. This chapter will explore these issues in more detail, examining the role of international secured transactions law in supporting the growth of the space sector, and the specific provisions of the Space Protocol that can enable this. The chapter will also touch upon new developments in the industry that make the creation, recognition, and enforcement of security rights in space assets even more important.

2 Asset-Based Financing in the Space Sector

Asset-based financing is a form of raising capital in which a company uses its tangible assets as collateral to secure a loan or line of credit. In asset-based financing, the borrower pledges specific assets as collateral to secure the loan. These assets serve as a guarantee to the lender that they will recover their funds if the borrower defaults on the loan.[3]

Asset-based financing is often used to provide working capital to a company. It allows the business to access cash tied up in assets, which can be used for various purposes, such as covering operating expenses, expanding operations, or financing growth initiatives. The loan amount is typically based on the appraised value of the collateralised assets. Lenders often require borrowers to provide regular reports on the status and value of the collateralised assets. This monitoring ensures that the collateral remains sufficient to cover the loan.[4]

From the lender's perspective, asset-based financing reduces risk because the collateral provides a means of recovery in case of default. As a result, asset-based loans may be accessible to companies with lower credit ratings. The terms and interest rates of asset-backed loans can fluctuate based on the lending institution, the nature of the assets, and the credit standing of the borrower.

Asset-based financing is a versatile financial tool that allows businesses to leverage their tangible assets to secure funding. It can be an effective way to access

[3] For a general definition of asset-based lending, see 'What Is Asset-Based Lending?' (*British Business Bank*) <www.british-business-bank.co.uk/finance-hub/asset-based-lending/> accessed 10 October 2023; for a detailed discussion of asset-based financing in the context of international secured transactions law, see, in general, Roy Goode, *Transnational Commercial Law: Text, Cases, and Materials* (OUP 2015).

[4] For Cape Town Convention specifics, see Roy Goode, 'The Cape Town Convention on International Interests in Mobile Equipment: A Driving Force for International Asset-Based Financing' (2002) 7 UnifLRev 3.

capital for growth, manage cash flow, or navigate financial challenges, especially when traditional financing options may be limited. Asset-based financing can become complicated for space assets as they are movable and hard for creditors to take possession of while on-orbit. Given the significant capital required to finance space assets, it is vital for lenders to trust that, should a debtor fail to meet payment or other obligations, their contractual and ownership rights will be upheld and executed. Furthermore, when multiple parties lay claim to the same space asset, creditors will want assurance of their claim's priority.[5]

So far, secured loan obligations have generally been secured not on space assets but on more easily accessible, immobile, and marketable terrestrial collateral in order to provide lenders with some protection in the event of default. Financing without collateral, or unsecured financing, relies on the borrower's credit standing or their available cash. If the borrower does not fulfil repayment obligations, lenders cannot lay claim to the borrower's assets. This type of financing is more friendly to triple-A rated companies with their creditworthiness and terrestrial assets. However, for smaller space companies, a creditor will not have enough confidence and will be reluctant to extend credit or will charge a higher interest rate.

3 The Space Protocol of the Cape Town Convention

3.1 The Cape Town Convention

The Convention on International Interests in Mobile Equipment (hereinafter the 'Cape Town Convention' or 'Convention')[6] was adopted on 16 November 2001 in Cape Town, South Africa. It was negotiated under the auspices of the International Institute for the Unification of Private Law (UNIDROIT) and the International Civil Aviation Organization (ICAO). The Convention is considered one of the most successful international treaties in the field of commercial and private law,[7] with 85 Contracting States representing over 80 per cent of the global population and over 75 per cent of global nominal gross domestic product.[8] The European Union has also approved the Convention.[9]

[5] Anna Veneziano, Hamza Hameed, Oliver Heinrich, and Erik Pellander, 'Asset-Based Financing for the Space Sector' (71st International Astronautical Congress, October 2020).

[6] Convention on International Interests in Mobile Equipment (adopted 16 November 2001, entered into force 1 April 2004) 2307 UNTS 285 (Cape Town Convention).

[7] For additional information on the Cape Town Convention, see Roy Goode, *Official Commentary on the Convention on International Interests in Mobile Equipment and Protocol thereto on Matters Specific to Aircraft Equipment* (5th edn, UNIDROIT 2022).

[8] 'Convention on International Interests in Mobile Equipment (Cape Town, 2001)—States Parties' (*UNIDROIT*) <www.unidroit.org/instruments/security-interests/cape-town-convention/states-parties/> accessed 10 October 2023.

[9] As of 25 July 2023. For the most up-to-date status, see: <www.unidroit.org/instruments/security-interests/cape-town-convention> accessed 10 October 2023.

The Treaty establishes a global framework for secured transactions law that encourages asset-based financing and leasing activities for select categories of high-value mobile equipment. This offers enhanced legal assurance to those holding security interests in mobile assets, especially when these assets traverse international boundaries. The Convention is complemented by Protocols focusing on high-value equipment in particular sectors.[10]

Currently, there are four Protocols to the Convention: the Protocol on Matters Specific to Aircraft Equipment (2001);[11] the Luxembourg Protocol on Matters Specific to Railway Rolling Stock (2007);[12] the Protocol on Matters Specific to Space Assets (2012);[13] and the Protocol on Matters Specific to Mining, Agriculture and Construction Equipment (2019).[14]

The need for a global secured transactions law structure for distinct industry sectors, such as the Cape Town Convention, emerged in the 1980s and 1990s. This was due to the rising private sector financing requirements for certain high-value assets—notably in aviation—in the absence of standardised rules to aid this process. The driving force behind this demand was the swift growth of air travel combined with diminished public funding for the industry.[15] Yet, the growing demand for private funding faced challenges due to legal obstacles associated with asset-based financing and leasing, especially concerning international transactions. The situation was characterised by a lack of uniformity in legal standards, with traditional conflict-of-laws approaches often ill-suited to movable assets. Furthermore, legal systems varied in their treatment of the recognition of security interests, procedures for enforcement, and the remedies available to creditors seeking to assert their rights over an asset.[16]

[10] Roy Goode, 'From Acorn to Oak Tree: The Development of the Cape Town Convention and Protocols' (2012) 17(4) UnifLRev 599.

[11] Protocol to the Convention on International Interests in Mobile Equipment (adopted 16 November 2001, entered into force 1 March 2006) <www.unidroit.org/instruments/security-interests/aircraft-protocol/> accessed 10 October 2023 (Aircraft Protocol).

[12] Luxembourg Protocol to the Convention on International Interests in Mobile Equipment on Matters Specific to Railway Rolling Stock (adopted 23 February 2007) <www.unidroit.org/instruments/security-interests/rail-protocol/> accessed 10 October 2023.

[13] Protocol to the Convention on International Interests in Mobile Equipment on Matters Specific to Space Assets (adopted 9 March 2012) <www.unidroit.org/instruments/security-interests/space-protocol/> accessed 10 October 2023 (Space Protocol).

[14] Protocol to the Convention on International Interests in Mobile Equipment on Matters Specific to Mining, Agricultural and Construction Equipment (22 November 2019) <www.unidroit.org/instruments/security-interests/mac-protocol/> accessed 10 October 2023.

[15] For a detailed history of the Cape Town Convention, see Anton N Didenko, *The Cape Town Convention. A Documentary History* (Bloomsbury Publishing 2021); see also Ron Cuming, 'Cape Town Convention and AEP Canadian Ratification Meeting' (*Transport Canada and Department of Justice Canada, Ottawa*, January 2008).

[16] See eg Iwan Davies, 'The New Lex Mercatoria: International Interests in Mobile Equipment' (2003) 52 ICLQ 151.

In response to these challenges, the Cape Town Convention was established as a treaty-based mechanism to institute a standardised legal framework. This framework was designed to facilitate the financing and leasing of specific categories of uniquely identifiable, high-value, mobile equipment used in distinct industry sectors. The main goal was to offer enhanced assurance and clear legal guidance to private financiers and creditors involved in secured lending practices within the industries covered by the Treaty.

3.2 The Space Protocol

The Space Protocol, which is part of the Cape Town Convention system, was adopted at the conclusion of a Diplomatic Conference convened by UNIDROIT in Berlin in 2012 with over 40 States and multiple intergovernmental organisations participating. The Space Protocol extends the application of the Cape Town Convention system to space assets and addresses specific challenges relevant to asset-based financing of space equipment.[17] The space sector formed part of the original set of industries intended to be covered by the Cape Town Convention system and is the subject of its third Protocol.

Generally, the Cape Town Convention and all its Protocols have the following key features:[18]

1. International Interest: Parties have the ability to establish an autonomous 'international interest' over equipment, which can originate from a security agreement, a title reservation agreement, or a lease agreement.
2. International Registry: Each Protocol has a dedicated international asset-based and fully electronic registry that ensures transparency, predictability, and the effectiveness of creditors' rights against competing creditors and in insolvency.[19]
3. Clear Priority Rules: The Registry establishes a definitive guideline for deciding priorities, known as the 'first-to-file rule', which stands even against interests defined by domestic law, subject to few and clearly specified exceptions.[20]
4. Effective Enforcement Measures Applicable also in Insolvency: The Cape Town Convention presents efficient and prompt enforcement actions,

[17] Roy Goode, *Official Commentary to Convention on International Interests in Mobile Equipment and Protocol thereto on Matters Specific to Space Assets* (1st edn, UNIDROIT 2013) 11.

[18] Ludwig Weber, 'Public and Private Features of the Cape Town Convention' (2015) 4(1) CapeTownConvJ 53.

[19] See Rory McPhillips and others, 'Comparative Analysis of Aircraft, Rail and Space International Registries and their Regulatory Provisions' (2016) 5(1) CapeTownConvJ 29.

[20] Roy Goode, 'The Priority Rules under the Cape Town Convention and Protocols' (2012) 1(1) CapeTownConvJ 95.

encompassing out-of-court solutions if Contracting States concur, along with preliminary relief awaiting a conclusive decision in court proceedings.[21]

5. Flexibility through Declarations: States can make policy choices, to a limited extent, through declarations to the Convention and to the Protocols. These allow Contracting States to apply the provisions of the Convention in a manner that continues to respect their own domestic legal systems.[22]
6. Enhanced Creditor Protection through Protocols: Protocols enable Contracting States to enhance the rights of creditors during enforcement and bankruptcy proceedings via declarations.

In addition, the Space Protocol, similar to other Protocols of the Cape Town Convention, prioritises the following key principles:

- Practicality: Reflecting the essential elements of asset-based financing and leasing transactions within the space sector.
- Party Autonomy: Acknowledging that all entities engaged in high-value international space transactions possess expertise and experience in these dealings, and they are adequately represented to guarantee that their agreements receive respect and execution.
- Predictability: Ensuring predictability in the application of the legal regime, as referred to in the interpretation provisions of Article 5(1) and the priority rules of the Convention itself, which prioritise certainty and simplicity.
- Transparency: Promoting transparency through rules within the Convention that provide for publicity of an international interest through registration, giving notice to third parties and ensuring the application of a transparent priority system.
- Sensitivity to National Legal Cultures: Permitting a Contracting State to evaluate the economic advantages of the standardised legal framework against existing national laws. This is achieved through declarations that can omit certain specific treaty provisions if deemed incompatible or choose specific provisions that bolster creditors' rights under the Treaty.[23]

The Space Protocol extends the benefits of the Cape Town Convention system to the space industry. Its contents are divided into five parts: scope of application, technical requirements for creating interests, specific remedies, priority rules, and

[21] Giles Cuniberti, 'Enforcement of Court Decisions under the Cape Town Convention' (2018) 7 CapeTownConvJ 93.

[22] Thomas Traschler, 'The Significance of the Qualifying Declarations under the Cape Town Convention' (2019) 24 UnifLRev 42.

[23] See Roy Goode, *Convention on International Interests in Mobile Equipment and Protocol thereto on matters specific to Mining, Agricultural and Construction equipment: Official Commentary* (1st edn, UNIDROIT 2021) 2.23.

registration procedure. It protects an assignment of a debtor's rights to a creditor by recording such assignments against the registered international interest. The Protocol also aids creditors in capturing additional collateral sums payable by third parties for the lease or access to the space asset.[24] The rest of this chapter covers the specific provisions of the Space Protocol and how they enable the use of asset-based financing in the space sector.

3.3 Scope of the Protocol

The scope of the Space Protocol can be found in its section on definitions. The Space Protocol defines the term 'space asset' as:

> Any man-made uniquely identifiable asset in space or designed to be launched into space, and comprising:
> (i) a spacecraft, such as a satellite, space station, space module, space capsule, space vehicle or reusable launch vehicle, whether or not including a space asset falling within (ii) or (iii) below;
> (ii) a payload (whether telecommunications, navigation, observation, scientific or otherwise) in respect of which a separate registration may be effected in accordance with regulations; or
> (iii) a part of a spacecraft or payload such as a transponder, in respect of which a separate registration may be effected in accordance with the regulations, together with all installed, incorporated or attached accessories, parts and equipment and all data, manuals and records relating thereto.[25]

The definition mentioned above encompasses all space objects typically employed by entities within the space industry. The definition recognises the considerable potential for technological advancements and progress within the space sector. Consequently, it delegates authority to the Space Registry Regulations established under the Space Protocol, allowing them to determine the inclusion of additional categories of space assets, such as payloads and parts, as deemed

[24] Mark J Sundahl, *The Cape Town Convention, Its Application to Space Assets and Relation to the Law of Outer Space* (Martinus Nijhoff 2013) 29.

[25] See Space Protocol (n 13) art II; for a detailed discussion on the definition of a 'space asset' under the Space Protocol, see Sundahl (n 24) 35; Pai Zheng, 'Space Asset under the Space Protocol of the Cape Town Convention' (GlobaLex 2014) <www.nyulawglobal.org/globalex/Space_Asset_Protocol_Cape_Town_Convention.html> accessed 10 October 2023; for a discussion of the early evolution of the term 'space asset' in the Cape Town Convention (n 6), see Sylvia Ospina, 'The Concepts of Assets and Property: Similarities and Differences, and Their Applicability to Undertakings in Outer Space' (2003) 45 ProceColloqLOuterSpace 12; Olivier M Ribbelink, 'The Protocol on Matters Specific to Space Assets' (2004) 12 EuRevPrivL 37, 39.

appropriate. This determination is to be made following consultation with the industry, should a recognised need for independent financing of such assets arise.[26]

The present version of the Space Registry Regulations was concluded at the fourth meeting of the Space Preparatory Commission.[27] This version distinguishes between 'transponders or other communication equipment' and other categories of payloads and parts, including 'observation payloads', 'navigation payloads', 'scientific payloads', and 'other components of a spacecraft or payload'. For the latter category, Annex 1 to the Space Registry Regulations includes an Explanatory Note signifying that the bankability of these specific asset types has yet to be thoroughly tested. Consequently, the Space Registry, at present, exclusively accepts registrations related to spacecraft and transponders or other communication equipment, pending revisions to the draft Space Registry Regulations.

This is in line with the current financial leasing practices within the space sector, chiefly concerning satellites and satellite transponders. However, it is vital to acknowledge that this specific approach may undergo adjustments in the future, necessitated by alterations to the Space Registry Regulations to meet the evolving demands of the industry.[28] The definition in itself has a broad scope and is designed to capture any assets which may be the subject of individual financing.

3.4 Identification of Space Assets

The success of any Protocol of the Cape Town Convention system rests on the proper functioning of the International Registry which that Protocol mandates, which operates on an asset-based registration model.[29] Naturally, in order for security rights to be created against an asset, the assets used as collateral must be uniquely identifiable. As such, identification criteria for assets are a critical part

[26] For the Regulations, see UNIDROIT, 'Summary Report of the Fourth Session' (Rome, 10–11 December 2015) Preparatory Commission for the Establishment of the International Registry for Space Assets Pursuant to the Space Protocol, Appendix III <www.unidroit.org/english/documents/2015/depositary/ctc-sp/pcs-04-07rev-e.pdf> accessed 3 August 2022.

[27] The Preparatory Commission was instituted as per Res. 1 of the Diplomatic Conference held in Berlin. The members of the Space Preparatory Commission include Brazil, the People's Republic of China, the Czech Republic, France, Germany, India, Italy, the Russian Federation, Saudi Arabia, South Africa, and the United States of America. Additional industry and regulatory bodies such as the ITU are also actively involved in the meetings of this Commission. For more, see <www.unidroit.org/work-in-progress/space-prepcom> accessed 3 August 2022.

[28] See McPhillips and others (n 19).

[29] Importantly, while the Protocol mandates for the creation of an asset-based registry, this does not have any implications with regard to asset tracing or tracking. The purpose of the 'asset-based' structure of the Registry is to enable the creation of interests against identifiable assets. Additionally, the Registry under the Cape Town Convention (n 6) has no association with, impact on, or correlation with the Registry of space assets maintained by the United Nations, which is closely associated to the Registration Convention 1974. These two registries have fundamentally different purposes, whereby the one under the Cape Town Convention is a secured transactions registry, and the one under the UN space treaties seeks to keep a record of all spacecraft launched.

of every Protocol of the Cape Town Convention. The Space Protocol, alongside providing an extensive definition of space assets, delineates the procedures for including these assets in the Registry. These procedures are governed by the Space Registry Regulations and will be binding upon the Registry's establishment.

Under the current Regulations, all assets must be linked to a 'unique identification number', with the issuance process for these numbers detailed in Annex 2 of these Regulations. To obtain this number, the owner of a space asset is required to furnish the Registrar with specific details, including their name, the manufacturer's name, the manufacturer's contract reference number, and the category of the space asset. If an identical unique identification number already exists for a particular asset, the Registrar is obligated to utilise it.

Additionally, the Registrar is responsible for creating a 'unique identification file' for each space asset assigned a unique identification number, where registered international interests are recorded. This file also includes supplementary information for reference to facilitate collateral identification, without affecting priority determination. This supplementary data may consist of details such as the UTC time of the object's launch, its allocated frequency, or any other Committee on Space Research unique identifier associated with it.[30]

3.5 Registry, Priority, and Insolvency Mechanism

The efficacy of the Cape Town Convention system is largely dependent upon the registration of international interests in dedicated international registries set up under each Protocol.[31] Within the Cape Town framework, an interest that is registered takes precedence over interests registered later and interests that have not been registered in the event of the debtor's insolvency. Consequently, registration significantly influences the priority of the interest, creating a strong incentive for parties to protect their rights through registration in an international registry. In the instance of default, creditors can utilise the default and insolvency solutions offered by the Cape Town Convention system to assert their rights concerning the asset. These remedies encompass interim measures, such as asset preservation during the process of exercising full remedies, taking control of the charged object, selling or leasing it, and collecting any income or profit arising from its management or use.

Under the Space Protocol, Contracting States have the flexibility to select one of two alternatives for the enforcement of their right to repossess the asset in cases

[30] For the Regulations, see UNIDROIT (n 26); see also McPhillips and others (n 19).

[31] For a discussion on the Priority Regime, see Nuria de la Peña, 'Reforming the Legal Framework for Security Interests in Mobile Property' (1999) 4 UnifLRev 347; Michel Deschamps, 'The Perfection and Priority Rules of the Cape Town Convention and the Aircraft Protocol: A Comparative Law Analysis' (2013) 2(1) CapeTownConvJ 51.

of debtor insolvency. These alternatives differ in their timelines and procedural mechanisms for enforcing repossession or rectifying the default.[32]

3.6 TT&C Enforcement Mechanism

The Space Protocol was drafted with the needs of the space industry in mind. Repossession under the other Protocols of the Cape Town Convention entails the physical takeover of an asset by the creditor. However, the Space Protocol took into account the current impracticality of physically repossessing space assets. To address this challenge, the drafters focused on Tracking, Telemetry, and Control (TT&C) systems inherent in space assets, specifically tied to their command codes. These satellite command codes, functioning as encryption keys, confer control over the satellites.

As such, the Space Protocol uses the concept of functional control as the equivalent of physical repossession. In the case of the insolvency of a debtor, the creditor would take control over the space asset through its TT&C. In fact, Article XIX of the Protocol permits parties to an agreement to mutually decide on the placement of command codes and related data and materials with a third party. This arrangement provides the creditor with the means to establish control over or operate the space asset effectively, thereby facilitating the exercise of their rights as conferred by the Space Protocol. Notably, Article XXVI(2)(c) grants Contracting States the authority to restrict the application of Article XIX through laws or regulations, potentially prohibiting, restricting, or attaching conditions to the placement of command codes with third parties. However, naturally, as a Contracting State diminishes creditor protection, the effectiveness of the Cape Town Convention in attracting investment and reducing the cost of credit diminishes accordingly.[33]

Enforcement by the transfer of TT&C is one way in which the Space Protocol gives confidence to creditors for the recovery of their investment in case a debtor becomes insolvent. The Space Protocol also includes other such tools. For example, the Space Protocol contains provisions that allow debtors to easily reassign the rights to revenue collected from services provided by a spacecraft to a creditor, which can be used as an alternative means of payment in case the debtor is experiencing financial difficulties.[34]

[32] See Kristin Van Zwieten, 'The Insolvency Provisions of the Cape Town Convention and Protocols: Historical and Economic Perspectives' (2012) 1 CapeTownConvJ 53.

[33] See Traschler (n 22).

[34] See UNIDROIT, 'Asset Baset Financing in the Space Industry—Facilitating Space Financing through the Space Protocol of the Cape Town Convention' <www.unidroit.org/english/conventions/mobile-equipment/conference2012/useful%20documentation/summary-e.pdf> accessed 10 October 2023.

3.7 Public Service Exception

Acknowledging the pivotal role played by space assets in delivering crucial public services, the drafters of the Space Protocol took deliberate measures to avert the abrupt cessation of such services in the event of debtor defaults. Article XXVII of the Space Protocol imposes constraints on the remedies available to a creditor in relation to space assets providing public services, in cases where a specific public service notice is registered against the particular asset.[35]

The underlying principle rests on the premise that States have a natural interest in ensuring that creditor actions do not precipitate the sudden termination of services of paramount public significance, such as weather monitoring or Global Positioning System services. Article XXVII is invoked upon the registration of a 'public service notice', a process necessitating consensus between the parties involved in the public service contract and the Contracting State. While creditor consent is not obligatory for the registration of a public service notice, the debtor's participation in the public service contract empowers the creditor to include a provision that restricts the debtor's authority to consent to such registration.

Upon the registration of a public service notice, throughout the 'suspension period', which commences with the creditor's registration of a default notice signifying potential remedies, the creditor is precluded from actions that render the space asset unavailable. The duration of this 'suspension period', ranging from three to six months, is determined by the Contracting States through a mandatory declaration under Article XXVII(4).

Article XXVII(9) sets forth an exception to the halting of remedies under a public service announcement. This exception is valid when: (1) the international interest is logged before the issuance of the public service announcement; (2) the creditor is unaware of the public service agreement or its announcement; and (3) the public service announcement has not been registered within six months of the space asset's initial launch.[36]

3.8 Application to Physically Connected Space Assets

Spacecraft are often composed of distinct modules interconnected physically, as exemplified by the International Space Station. Article XVII(3) of the Space Protocol imposes a constraint on the exercise of remedies concerning physically

[35] See Anna Veneziano and Hamza Hameed, 'The Space Protocol of the Cape Town Convention: An International Secured Transactions Regime for Space Assets' (69th International Aeronautical Congress, 2018).

[36] See Tugrul Cakir, 'The Public Service Exemption in the Space Protocol in light of that found in the Luxembourg Protocol' (67th International Aeronautical Congress 2016); Howard Rosen, 'Public Service and the Cape Town Convention' (2013) 2 CapeTownConvJ 131.

linked space assets. Specifically, it stipulates that a creditor cannot enforce an international interest in a space asset if doing so would compromise or disrupt the operation of another linked space asset.[37] Nonetheless, there exist two qualifications to the provision outlined in Article XVII(3): (1) The creditor or purchaser of the physically linked space asset must have registered their interest in the space asset before the interest of the enforcing creditor was registered; and (2) Article XVII(3) operates in alignment with any mutually agreed contrary arrangement between the relevant parties, signifying its non-mandatory character.

It is worth noting that this rule does not, strictly speaking, pertain to a priority issue, as it concerns two interests associated with distinct space assets that are physically linked, rather than two interests in the same asset.[38]

3.9 Salvage

Salvage represents a legal or contractual interest tied to a space asset, typically in the case of a full or partial loss of the asset. The asset becomes the property of the insurer upon compensation. In space asset financing, insurance is of great importance, and the Space Protocol aims to ensure minimal disruption in this sector. Article IV(3) explicitly states that neither the Convention nor the Protocol can impact the legal or contractual rights of insurers regarding salvage, as per the relevant law. Consequently, salvage rights, including subrogation rights, remain unaffected by the Convention or Protocol, leaving any priority disputes between salvage and creditor rights to be resolved under the applicable law.[39]

4 Space Protocol and NewSpace

In recent years, the space sector has seen significant growth, particularly in the NewSpace domain, which comprises agile and business-focused companies. This evolution has led to a large influx of satellites being deployed, as well as assets being placed on celestial bodies. This has the potential to create a burgeoning secondary market for space assets, underpinned by advancements in technology. Such a rise would require secured transactions law to play an important role.

The NewSpace sector's emergence has reshaped the space industry landscape. These companies, often lacking extensive on-ground assets or substantial founder support, have revolutionised the sector's dynamics. Simultaneously, technological

[37] Paul B Larsen, 'Berlin Space Protocol: Update' (2015) 64 ZLW 376.
[38] See Goode (n 17) 193.
[39] For more on salvage rights, see Nick Hughes, 'UNIDROIT Draft Space Assets Protocol' (*HFW*, 31 January 2014) <www.hfw.com/UNIDROIT-Draft-Space-Assets-Protocol> accessed 10 October 2023.

innovations have made securing rights over space assets more feasible and economically attractive, resulting in a substantial expansion of the secondary space asset market. Three prominent trends underscore this transformation:

1. **On Orbit Servicing:** The concept of On Orbit Servicing (OOS) has emerged as a transformative paradigm. OOS encompasses a spectrum of activities conducted in the space environment, primarily revolving around the servicing, upgrading, and modification of existing space assets. Notably, it represents a pioneering approach to maximise the utility and longevity of assets in orbit, offering significant advantages for both space industries and broader technological innovation.

   One of the most noteworthy aspects of OOS is its ability to extend the operational lifespan of satellites. Traditionally, satellites had a finite operational duration, and once their fuel was depleted, they became space debris. However, OOS technologies have altered this trajectory. By enabling the refuelling of satellites in orbit, these assets can continue to function beyond their initial mission parameters. This extension of operational life translates into prolonged revenue generation and service provision, significantly enhancing the economic viability of space assets.

   Beyond refuelling, OOS techniques facilitate the retrofitting of space assets for diverse purposes. This adaptability is a game-changer for the space industry as it allows operators to repurpose existing assets, adapting them to new missions, scientific endeavours, or commercial ventures. For example, a satellite initially designed for Earth observation could be retrofitted for telecommunications or space exploration, ensuring that valuable space assets remain relevant and profitable in an ever-evolving market.

   OOS activities can play a pivotal role in invigorating the secondary market for space assets. Traditionally, the space industry had a linear model, with assets being launched, serving their primary mission, and eventually becoming space debris. However, OOS introduces a circular economy to space operations. As assets in orbit can be refuelled, upgraded, and adapted for new tasks, they remain productive and valuable. This not only reduces the economic burden of frequent asset replacements but also fosters a sustainable space ecosystem where used assets continue to contribute to the economy.[40]
2. **In-Space Manufacturing:** The advent of technologies like 3D printing has ushered in a transformative era in space exploration and utilisation. In-space manufacturing represents a paradigm shift, allowing for the creation of space assets directly in orbit around celestial bodies or even on those celestial bodies

[40] Hamza Hameed, 'Secondary Market for Space Assets—The Economic Case for On-Orbit Servicing as a Mechanism to Extend Satellite Life Cycles and Mitigate Space Debris' (72nd International Aeronautical Congress, 2021).

themselves. This approach transcends traditional limitations associated with space missions, enabling the on-demand production of critical assets right within the extra-terrestrial environment.

This emerging frontier in space technology has far-reaching implications. It not only enhances the efficiency of space missions but also expands the scope of economic activities beyond Earth's boundaries. In-space manufacturing has the potential to foster a dynamic marketplace where space assets can be produced and deployed as needed, revolutionising the way we conceive and engage with space ventures. As this new era of in-space manufacturing commences, the role of secured transactions law becomes increasingly critical in shaping the legal framework that governs the rights of various stakeholders in this novel space manufacturing landscape.

In the context of in-space manufacturing, secured transactions law must define and protect the rights of title over assets produced in space. This includes determining whether the manufacturer or the entity financing the manufacturing process holds primary title rights, and how these rights are transferred and recognised under different circumstances. Additionally, the seamless transfer of title to space assets is paramount for the functioning of the burgeoning space economy. Secured transactions law will establish clear rules and procedures for transferring title, ensuring that the transfer process is efficient, transparent, and legally binding.

The financing of in-space manufacturing operations is a complex endeavour, and secured transactions law instruments such as the Space Protocol can provide the necessary legal infrastructure to support it. This includes defining the rights and obligations of financiers, manufacturers, and other parties involved in the financing process. Additionally, it addresses issues related to collateral, default, and dispute resolution, all of which are crucial for maintaining the financial stability of space manufacturing ventures.[41]

3. **Software-Deployed Satellites and Ground Stations as a Service**: Another notable development in space technologies is the convergence of standardised satellite components and the emergence of ground stations as a service. This convergence not only augments communication capabilities for subsequent operators of space assets but also enhances the overall value proposition of these assets within the space ecosystem.

   The proliferation of standardised components in satellite assembly represents a significant milestone in the space industry. This trend has facilitated the streamlined and cost-effective production of satellites. By utilising commercial

[41] See 'In-Space Manufacturing, Servicing, and Transportation Market—A Global Market and Regional Analysis: Focus on End User, Operation and Country Analysis—Analysis and Forecast, 2020–2030' (*BIS Research*, October 2020) <https://bisresearch.com/industry-report/in-space-manufacturing-servicing-transportation-market.html#:~:text=The%20in%2Dspace%20manufacturing%2C%20servicing,with%20a%20share%20of%2045.57%25.> accessed 10 October 2023.

off-the-shelf components and standardised designs, satellite manufacturers can create spacecraft with greater efficiency and reduced production costs. Consequently, this innovation has democratised access to space, allowing more players, including those from emerging space markets, to participate in space missions and ventures.[42]

Ground stations have traditionally been integral components of satellite missions, serving as critical communication hubs that facilitate data transmission to and from orbiting satellites. However, recent advancements have introduced a paradigm shift in the way ground station services are accessed and utilised. With the advent of 'Ground Stations as a Service', satellite operators now have the option to acquire ground station capabilities from service providers, much like a utility service.[43]

This evolution enhances the flexibility and accessibility of ground station services. Satellite operators can leverage these services on a pay-as-you-go basis, optimising their communication networks without the need for significant infrastructure investments. This approach is particularly advantageous for emerging space market players, as it lowers entry barriers and reduces the financial burden associated with establishing and maintaining ground station facilities.

Another transformative aspect of modern space assets is their capacity for dynamic adaptation through software updates. Unlike traditional satellites with fixed functionality, modern space objects equipped with software-defined systems can receive updates and alterations from Earth. This capability empowers operators to modify a space asset's functionality, purpose, or mission parameters remotely.

The ability to deploy software updates from Earth adds a layer of versatility to space assets. Operators can adapt to changing mission requirements, respond to emerging opportunities, or address evolving challenges without the need for physical interventions or replacements. This adaptability significantly contributes to an asset's resilience and long-term relevance, enhancing its attractiveness in the secondary space market.

Collectively, the integration of standardised satellite components, ground stations as a service, and software-defined capabilities amplifies the value of space assets in the secondary market. Space objects that can readily connect to ground stations, receive software updates, and adapt to evolving needs become highly sought-after commodities. For investors and space entrepreneurs, the secondary market offers new avenues for revenue generation and asset utilisation. Rather than considering space assets as static, one-time-use entities,

[42] 'The Space Economy in Figures' (*OECD*, 5 July 2019) <www.oecd.org/innovation/the-space-economy-in-figures-c5996201-en.htm> accessed 10 October 2023.

[43] Martin Krynitz, 'The Ground Station of the Future' (2010) 67 ActaAstro 1252.

the market now envisions them as dynamic, adaptable resources with enduring value. This shift not only fosters a more sustainable space economy but also aligns with the evolving demands of a rapidly advancing space industry.

These transformative developments in the space sector necessitate a robust legal framework governed by secured transactions law. Such laws play a critical role in enabling the efficient functioning of the secondary space asset market by providing clear guidelines for rights, transfers, and financing. They offer the legal certainty required to facilitate transactions in this high-value and evolving industry.[44] Moreover, the growth of these industries has also been motivated by a greater focus on sustainability in space, which necessities consideration of issues such as space debris mitigation.

The potential growth of the secondary market for space assets within the NewSpace sector is closely intertwined with the advancements in technology and the evolving landscape of secured transactions law. As the space industry continues to expand, an effective legal framework becomes indispensable in ensuring the smooth operation of this dynamic marketplace.[45]

5 Conclusion

5.1 The Benefits of Asset-Based Financing for the Space Sector

The Cape Town Convention is rightly described as 'the most successful international secured transactions instrument ever adopted'.[46] The Convention and its Aircraft Protocol have been in force since 2006. The International Registry established under the Aircraft Protocol[47] has recorded over than 1.3 million registrations over an estimated value of collateral of more than US$750 billion.[48] The operators of States that adopt the Convention and the Protocol may receive up to a 10 per cent discount on export credit premiums,[49] and have enjoyed improved

[44] See Frans von der Dunk, 'Transfer of Ownership in Orbit: From Fiction to Problem' in Mahulena Hofmann and Andreas Loukakis (eds), *4th Luxembourg Workshop on Space and Satellite Communication Law* (Nomos Verlagsgesellschaft and Hart Publishing 2017).

[45] Hamza Hameed and Ignacio Tirado, 'Spacecraft Financing: An International Secured Transactions Regime for Space Assets' [2023] ButterworthsJIntlBankFinL 87.

[46] Charles W Mooney, 'Cape Town Convention's Improbable-but-Possible Progeny Part One: An International Secured Transactions Registry of General Application, The Essay' (2014) 55 VaJIntlL 163, 166.

[47] Aircraft Protocol (n 11).

[48] See generally the website of the International Registry for Aircraft Objects at <www.aviareto.aero/> accessed 10 October 2023; also see, for the celebration of reaching 1 million registrations, <www.aviareto.aero/celebrating-one-million-registrations/> accessed 10 October 2023.

[49] 'Arrangement on Officially Supported Export Credits' (*OECD*, 2022) <www.oecd.org/officialdocuments/publicdisplaydocumentpdf/?cote=TAD/PG(2022)1&docLanguage=en> accessed 10 October 2023.

rating in capital market lending operations. For example, it was calculated that the adoption of the Convention and the Aircraft Protocol could enable an airline to save $330,000 on the purchase of a new ATR 72 and $2.5 million on the purchase of an Airbus A380.[50]

Asset-based financing presents several compelling advantages for the space sector, with larger benefits for NewSpace, underpinning its appeal as a financing mechanism for ventures within this dynamic industry. Asset-based financing can serve as an important source of capital for NewSpace endeavours, which frequently necessitate substantial financial resources for research, development, and the launch of space assets. The availability of capital through asset-based financing obviates the need for traditional loans or equity investments, which might involve complex negotiations or higher associated costs, especially for operators based out of emerging space economies.

Moreover, asset-based financing often translates into a lower cost of capital as the risk associated with such an investment for a financier is lower. Lenders, appreciating the inherent value of the collateral offered, are inclined to extend credit at comparatively lower interest rates. This feature can significantly reduce the overall financial burden borne by space ventures. In the context of risk management, asset-based financing emerges as an advantageous choice. Given the intrinsic risks associated with the space sector, lenders find comfort in the collateralisation of assets. In the unfortunate event of a default, lenders possess a means to recover their investment through the seizure and sale of the collateralised assets. Additionally, a financier only needs to monitor its assets to protect its investment, rather than having to monitor the whole company into which it has invested.

Asset-based financing can play a pivotal role in facilitating the continued growth of the space sector. This financing avenue empowers companies, whether startups or established entities, to acquire essential equipment and technologies. Such acquisitions can be instrumental in bolstering a company's capacity to innovate, conduct research, and remain at the forefront of technological advancements. Furthermore, asset-based financing aligns with the spirit of innovation pervasive in the NewSpace sector. By allowing companies to secure funding based on the potential value of their space assets, this financing method supports ventures exploring groundbreaking, unproven technologies. It incentivises the pursuit of novel and ambitious projects by providing a financial framework that can accommodate speculative ventures.

The presence of an international system of secured transactions law, such as that offered by the Space Protocol, has the potential to enable greater access to credit and

[50] David Knibb, 'In Focus: How New Export Credit Rules Will Change Aircraft Finances' (*FlightGlobal*, 22 January 2013) <www.flightglobal.com/in-focus-how-new-export-credit-rules-will-change-aircraft-finance/108522.article> accessed 10 October 2023.

growth for the space sector. Flexibility is another salient characteristic of asset-based financing. Companies operating in the NewSpace sector can benefit from tailored financing structures, including adaptable repayment schedules and bespoke deal structures that can be attuned to the unique requirements of the venture. Additionally, asset-based financing allows companies to preserve equity. Unlike equity financing, which entails relinquishing ownership and control, asset-based financing empowers companies to secure the necessary funding while retaining their autonomy.

An international system of secured transactions law provides a uniform set of rules and regulations for the creation, registration, priority, and enforcement of security rights in space assets. This uniformity ensures legal clarity and consistency across borders, making it easier for space companies to engage in international financing and leasing transactions. As such, the Space Protocol of the Cape Town Convention fosters a harmonised regulatory environment that encourages cross-border investments in the space sector, further enhancing the attractiveness of asset-based financing as a means of supporting this industry's growth and innovation.

Emerging space market companies often engage in cross-border transactions, involving multiple jurisdictions. A standardised legal framework streamlines the complexities associated with such transactions, reducing legal uncertainties and risks. A well-defined and internationally recognised secured transactions law system enhances investor confidence. Lenders and investors are more likely to participate in space industry financing when they have confidence in a transparent and predictable legal framework that protects their interests.

Additionally, a standardised legal framework encourages the growth of the secondary market for space assets. This, in turn, attracts additional investments, fosters liquidity, and allows for the efficient trading of space assets. A uniform secured transactions law enables emerging space market companies to compete globally on an equal footing. They can access international markets with confidence, attracting partnerships, customers, and investors from around the world. This results in growth and economic development, as well as allowing emerging space economies to attract foreign investment and encourage local entrepreneurship.

5.2 Current Status and Next Steps

Despite having received some opposition from large satellite companies upon adoption, the Space Protocol has continued to be looked upon favourably by emerging space economies and NewSpace companies.[51] Several States have

[51] See Peter B de Selding, '"Space Protocol" Widely Opposed by Industry Is on Agenda for ITU Meeting in Korea' *SpaceNews* (14 October 2014) <https://spacenews.com/42185space-protocol-widely-opposed-by-industry-is-on-agenda-for-itu/> accessed 10 October 2023.

introduced domestic law reform in recent years to allow space assets to be the subject of secured transactions.[52]

Access to credit has repeatedly been identified as a key barrier in the space sector. This observation was highlighted in a 2019 report co-published by the European Commission and the European Investment Bank.[53] Furthermore, in a document named 'Orientations on the European contribution in establishing key principles for the global space economy', the European Council emphasised the advantages of lowering financial expenses for satellite operators. They also stressed the importance of offering creditors reliable and easily accessible securities for their loans and raising awareness about international financial tools. This is the main purpose of the Space Protocol.[54]

Presently, the Space Protocol has been signed by four States.[55] Pursuant to Resolution 1 of the Space Protocol Diplomatic Conference, a Preparatory Commission was set up to act as Provisional Supervisory Authority for the establishment of the International Registry for Space Assets.[56] This Preparatory Commission has already developed the first draft of the Registry Regulations and has also done considerable work towards establishing the process required for setting up the International Registry for Space Assets.

In 2017, the Commission's members agreed to constitute a Sub-Group to reassess industry participation in the promotion and development of the Space Protocol. This Sub-Group continues to serve as the body responsible for furthering the development and implementation of the Space Protocol.

The entry into force requirements of the Space Protocol can be found in Article XXXVIII. The Protocol will enter into force when ten instruments of ratification, acceptance, approval, or accession have been deposited with the Depositary. Additionally, for the Protocol to enter into force, a fully functional International Registry must be established. This is confirmed by the deposit by the Supervisory Authority with the Depositary of a certificate confirming that the International Registry is fully operational.[57]

[52] See eg 'Malta National Space Strategy' (*Government of Malta*, January 2020) <https://meae.gov.mt/en/Public_Consultations/MRIC/PublishingImages/Pages/Consultations/MaltaNationalSpaceStrategy/Space%20Strategy%20EN.pdf> accessed 10 October 2023.

[53] See 'The Future of the European Space Sector' (*European Investment Bank*, 2019)<www.eib.org/attachments/thematic/future_of_european_space_sector_en.pdf> accessed 10 October 2023.

[54] See 'Orientations on the European Contribution in Establishing Key Principles for the Global Space Economy: Council Conclusions' (*European Council*, 11 November 2020) <www.consilium.europa.eu/media/46730/st12851-en20.pdf> accessed 10 October 2023.

[55] Germany, Saudi Arabia, Burkina Faso, and Zimbabwe.

[56] For a discussion on the issue of Supervisory Authority of the Space Protocol, see Claudiu Mihai Tăiatu, 'The International Telecommunication Union (ITU) as the Proposed Supervisory Authority of the Future International Registry for Space Assets' (2018) 23 UnifLRev 506.

[57] See Space Protocol (n 13) art XXXVIII.

14
INTERNATIONAL NORMS ON THE SETTLEMENT OF SPACE DISPUTES

Mark J Sundahl[*]

1 Introduction

The rapid expansion of space activity, particularly in the private sector, will inevitably bring in its wake a surge in the number of space-related disputes. One measure of the magnitude of growth in the space industry can be seen in the number of operational satellites in orbit, which may increase, due to the advent of megaconstellations, from the low thousands only a few years ago to as high as 100,000 by the end of this decade.[1] In addition, a number of new industry sectors are taking root, including on-orbit satellite repair and refuelling, private space stations, space tourism, and the extraction of space resources. In light of these new types of activities, the types of disputes that will be seen in the coming years will also be novel and may involve, for example, satellite collisions, injuries to space tourists, and conflicts between rival mining operations.

To handle these disputes (which will involve both governmental and non-governmental litigants in different combinations), the process for dispute resolution must be tailored to meet the particular demands of the case at hand. One of the primary factors in selecting the appropriate method of dispute resolution is the identity of the parties. The different combinations of litigants include (i) state versus state, (ii) private party versus private party, and (iii) private party versus state. However, even within these scenarios, variations among the types of disputes will occur. For example, the types of space-related transactions between private parties may range from satellite construction and launch services agreements to transponder leases and tickets for suborbital tourism flights. In most cases, the private litigants will both be parties to a contract that defines their relationship. In other cases, innocent third parties could seek damages for harm caused by space

* The author would like to thank Abigail Jones and Amy Burchfield for their assistance in the research and formulation of this chapter.

[1] Stephen Young, 'The Meteoric Rise in Satellite Numbers' (*The Equation*, 17 March 2022) <https://blog.ucsusa.org/syoung/the-meteoric-rise-in-satellite-numbers> accessed 3 August 2023.

Mark J Sundahl, *International Norms on the Settlement of Space Disputes* In: *International Space Law in the New Space Era*. Edited by: Sandeepa Bhat B., Dilip Ukey, and Adithya Variath, Oxford University Press. © Mark J Sundahl 2024.
DOI: 10.1093/9780198909415.003.0014

activity. The nationality of the litigants, that is, whether a dispute is international in nature, will also have a dramatic effect on how disputes can be resolved.

Practitioners must be aware of the options for dispute resolution and select the approach most beneficial to the client. The spectrum of available options in some situations may range from the high formality of litigating before the International Court of Justice (ICJ) to engaging in non-binding confidential mediation. In some cases, the best choice for private disputes may be international arbitration at a recognized institution, such as the International Chamber of Commerce (ICC) in Paris, while other disputes may call for litigation in local courts. Sometimes the answer to resolving disputes can be multidimensional and make use of different types of mechanisms. For example, a dispute resolution provision in a contract may provide for a multi-level dispute resolution mechanism, perhaps beginning with informal negotiation between the parties before escalating to binding arbitration or litigation.

This chapter will structure its discussion according to the three litigant combinations described above. In addition to describing these options and the costs and benefits of each, consideration will be given to the way in which space-related disputes are different from other types of disputes and how the peculiar nature of space activity affects dispute resolution.

2 Resolving State v State Disputes

The resolution of disputes between sovereign states is arguably the most critical form of dispute resolution as it may be necessary in order to prevent war. Two types of interstate disputes are addressed here, namely, (i) disputes arising from the violation of an obligation under international law, and (ii) claims for damages for harm caused by another state's space activity.

2.1 Violations of International Law

Under the United Nations (UN) Charter, all states have standing before the ICJ to bring claims against another state for the violation of international law. Some interesting questions of space law that may be brought before the ICJ in coming years include whether the testing of kinetic anti-satellite weapons is a violation of the duty to exercise 'due regard to the corresponding interests of all other States' required by Article IX of the Outer Space Treaty.[2] Another question is whether

[2] Treaty on Principles Governing the Activities of States in the Exploration and Use of Outer Space, Including the Moon and Other Celestial Bodies (adopted 27 January 1967, entered into force 10 October 1967) 610 UNTS 205, art IX (Outer Space Treaty)

the duty to rescue 'personnel' under the Rescue and Return Agreement requires the rescue of paying passengers and not only crew members (as is implied by the ordinary meaning of 'personnel').[3] As to the operations and procedures of the ICJ, they are spelled out in the Statute of the International Court of Justice.[4] The ICJ Statute addresses the composition of the court, its competence, and procedures for matters brought before the court. Perhaps most famously, Article 38 of the ICJ Statute sets forth the acceptable sources of international law, including treaties, customary international law, general practice, and academic commentators. There is not enough space here to cover the procedural details of an ICJ claim, although guidance can easily be found elsewhere.[5]

To clarify what may be obvious, private parties have no standing and cannot bring a case before the ICJ. That said, a state may file a claim to protect the interests of one of its nationals in an act of diplomatic intervention. Moreover, Article VI of the Outer Space Treaty provides that parties to the treaty 'shall bear international responsibility for national activities in outer space . . . and for assuring that national activities are carried out in conformity with the provisions set forth in the present Treaty'.[6] In addition to imposing responsibility on a state for 'national activities', the Outer Space Treaty also imposes liability for harm caused by a 'space object' on all states involved in the launch of the object (including space objects that are privately owned and/or which were launched by a private company):

> Each State Party to the Treaty that launches or procures the launching of an object into outer space, including the Moon and other celestial bodies, and each State Party from whose territory or facility an object is launched, is internationally liable for damage to another State Party to the Treaty or to its natural or juridical persons by such object or its component parts on the Earth, in air space or in outer space, including the Moon and other celestial bodies.

Due to this imputation of responsibility and liability, the misdeeds of private space actors may be remedied by bringing a case against the state of nationality at the ICJ. But what are some potential private actions that might warrant the filing of a case before the ICJ? One question that may come before the court is whether particular activities of a private company (such as lunar landings that create long-lasting dust storms) harmfully interfere with the activities of another state. Under Article IX of the Outer Space Treaty, any potential harmful interference must be

[3] See Mark Sundahl, 'The Duty to Rescue Space Tourists and Return Private Spacecraft' (2009) 35 JSpaceL 163.

[4] Statute of the International Court of Justice (adopted on 24 June 1945, entered into force 24 October 1945) XV UNCIO 355.

[5] See eg Robert Kolb, *The Elgar Companion to the International Court of Justice* (Edward Elgar 2014); Juan José Quintana, *Litigation at the International Court of Justice: Practice and Procedure* (Brill 2015).

[6] Outer Space Treaty (n 2) art VI.

addressed in consultations between the state of nationality and the affected state.[7] Another, ultimately more consequential, question that the ICJ could be asked to answer is whether the duty of 'due regard' provides protection (equivalent to a mining claim) to mining companies that publicize their intention to harvest specific resources on a celestial body. Similar questions are awaiting contemplation by the court as the development of space activity demands clear interpretations of how long-standing treaty obligations should apply to current practices.

2.2 Claims for Damage Caused by Space Objects: The Liability Convention

The Outer Space Treaty, as mentioned above, imposes liability for harm caused by a 'space object' on all states involved in the launch of the object. This obligation was elaborated upon in the 1972 Convention on International Liability for Damage Caused by Space Objects (Liability Convention) which imposes liability on any state that either launches or procures the launch of a space object, as well as any state from whose facility or territory the space object is launched (each of which would qualify as a so-called 'launching State').[8] Among the most important innovations of the Liability Convention are those provisions that (i) clarify the standards of liability, and (ii) create a process for a harmed state to bring a claim for damages.

While the Outer Space Treaty was silent as to the standard for finding a state liable for harm caused by its space activity, the Liability Convention fills this gap by providing not one, but two standards of liability. Which standard applies depends on where the damage takes place (on the ground, in the air, or in space) as well as whether harm is caused to an airplane, a 'space object', another type of property, or a person. If the damage occurs either on the surface of the Earth (even if the damaged asset is a spacecraft) or to an aeroplane in flight, the launching state is strictly liable (or, in the words of the Convention, 'absolutely liable').[9] In contrast, if damage is caused to a space object anywhere other than on the surface of the Earth, a launching state will only be found liable if the damage was 'due to its fault or the fault of persons for whom it is responsible'.[10] What precisely constitutes 'fault' is subject to debate, but may be clarified as case law develops or other clarificatory actions are taken, such as the issuance of a multilateral understanding of the treaty language.

[7] ibid art IX.

[8] Convention on International Liability for Damage Caused by Space Objects (adopted 29 March 1972, entered into force 1 September 1972) 961 UNTS 187 art I (Liability Convention); see, in general, Kirsten Schmalenbach, *Convention on International Liability for Damage Caused by Space Objects* (Springer 2022).

[9] Liability Convention (n 8) art II. If there are multiple launching states, they are jointly and severally liable, ibid art V.

[10] ibid art III.

In addition to elaborating upon the substantive law of liability, the Liability Convention also established a procedure that can be utilized by a state seeking compensation for harm done by another state's space activity. The first step in seeking compensation under the Liability Convention is for the harmed state to present a claim for damages via diplomatic channels directly to the offending state.[11] The claim must be submitted no later than one year after the discovery of the damage and the state at fault (or within a year of when such facts should have reasonably been discovered).[12] There is no requirement to exhaust other remedies—thus making the Liability Convention remedy immediately available to an injured state. However, the claim must be exclusive, that is, no claim may be presented if the injured state is pursuing remedies in another venue.[13]

If a claim is presented and the states fail to reach a resolution in a timely manner, Article XIV calls for the appointment of a Claims Commission to arbitrate the matter. A Claims Commission is composed of three members, with each party selecting the first two members, who then select the third.[14] After the formation of a Claims Commission, the procedural rules are generally left to the agreement of the parties, as is the determination of the location of the proceedings.[15] However, there are some issues that are addressed in the Convention. One of these rules concerns the voting of the Commission: a majority vote is required to take any action. When determining the size of damages to be awarded, Article XIII requires the Commission to award the amount needed 'to provide such reparation in respect of the damage as will restore the person, natural or juridical, State or international organization on whose behalf the claim is presented to the condition which would have existed if the damage had not occurred'. The tribunal is also required to produce a reasoned decision within one year of the Commission's establishment.[16] However, it is crucial to note that a decision handed down by a Claims Commission is only 'final and binding if the parties have so agreed; otherwise, the Commission shall render a final and recommendatory award, which the parties shall consider in good faith'.[17] Obviously, a practitioner would emphasize the need for the parties to agree to the binding nature of any award.

A Claims Commission has never yet been established to resolve a case concerning liability for space activity. However, the Liability Convention did guide the actions of the affected parties in the wake of the 1978 Kosmos 954 incident when a nuclear-powered Soviet spy satellite reentered and disintegrated over Canada. Canada submitted a claim for damages through diplomatic channels as contemplated in Articles VIII of the Liability Convention. Although a Claims Commission was

[11] ibid art IX.
[12] ibid art X.
[13] ibid art XI.
[14] ibid art XV.
[15] ibid art XVI.
[16] ibid art XIX.
[17] ibid art XIX(2).

not established, a Settlement Agreement was reached in 1981 in which the Soviet Union accepted liability and paid C$3 million in compensation.[18] This Settlement Agreement was explicitly based on the parties' international obligations under the Liability Convention (as well as under general international law).[19]

2.3 Arbitration of State v State Disputes

In addition to the mechanism prescribed by the Liability Convention, states may also avail themselves of the modern tools of arbitration which have become popular for the resolution of private disputes, and which offer states the same benefits enjoyed by private parties as described below. In this vein, it should be mentioned that in 1984 the International Law Association adopted the Draft Convention on the Settlement of Disputes related to Space Activities which was inspired by the 1982 United Nations Convention on the Law of the Sea dispute resolution system.[20] However, this Draft Convention has not been ratified by any state nor has it been utilized in any known arbitration.

3 Private Disputes

Since the ICJ and the Liability Convention do not assist with private claims, space companies are left with the remaining terrestrial methods of dispute resolution: domestic courts, mediation, and arbitration.

3.1 Litigation of Private Disputes

In a transactional context, the method of dispute resolution is typically included in the agreement, as can be seen in the following clause taken from a SpaceX Launch Services Agreement:[21]

[18] 'Settlement of Claim between Canada and the Union of Soviet Socialist Republics for Damage Caused by "Cosmos 954"' (2 April 1981) see at <www.jaxa.jp/library/space_law/chapter_3/3-2-2-1_e.html#:~:text=The%20Government%20of%20the%20Union%20of%20Soviet%20Socialist%20Republics%20shall,Cosmos%20954%20in%20January%201978> accessed 3 August 2023.

[19] ibid para 14.

[20] 'Draft Convention on the Settlement of Disputes related to Space Activities' Report of the 61st Conference of the International Law Association (1984). For a concise description of the features of the Convention, see Lotta Viikari, 'International Law Associations Draft Convention on the Settlement of Disputes related to Space Activities' (*Arbitration.ru*, March–April 2021) 14–17 <https://journal.arbitration.ru/upload/iblock/d12/Arbitration_ru_N2_26_March_April2021.pdf> accessed 3 August 2023.

[21] Contract for Launch Services, No IS-10-008 between Iridium Satellite LLC and Space Exploration Technologies Corp <www.sec.gov/Archives/edgar/data/1418819/000141881918000017/irdm10-q093018exx101.htm> accessed 3 August 2023.

18.1 Informal Dispute Resolution: Prior to the initiation of formal dispute resolution procedures, the Parties shall first attempt to resolve their Dispute informally, in a timely and cost-effective manner, as follows:

18.1.1 If, in connection with the performance of this Contract, a Party believes it has a Dispute with the other Party . . . Contractor's Launch Program Manager shall promptly consult with Customer's Launch Program Manager in an effort to reach an agreement to resolve the Dispute.

18.1.2 In the event that the Contractor's and Customer's Launch Program Managers cannot resolve the Dispute within fifteen (15) Business Days . . . the respective positions of the Parties shall be forwarded to the appropriate Executive Vice President (EVP) or equivalent of each Party for resolution of the Dispute.

18.1.3 In the event that the Contractor's EVP and Customer's EVP cannot resolve the Dispute within fifteen (15) Business Days . . . the respective positions of the Parties shall be forwarded to the Chief Executive Officer (CEO) or equivalent of each Party for resolution of the Dispute.

18.1.4 In the event a resolution cannot be reached . . . either Party may proceed in accordance with Section 18.2.

18.2 Litigation: If any Dispute arising between the Parties cannot be settled pursuant to Section 18.1 . . . either Party shall have the right to bring suit.

18.3 Jurisdiction and Venue: Any suit brought shall be brought in the U.S. District Court for the Eastern District of Virginia, and the Parties hereby waive any objection to that venue and that court's exercise of personal jurisdiction over the case.

According to this multi-level dispute resolution clause, the parties agreed to first engage in a series of negotiations leading, ultimately, to an attempt between the CEOs to settle the dispute informally. If no settlement is reached, litigation is the method chosen here to resolve the dispute. It is noteworthy that SpaceX prefers litigation to arbitration which has become a popular alternative method of dispute resolution for a variety of reasons which will be explained in detail in the following section. Despite this trend of selecting arbitration to resolve disputes, a practitioner must keep in mind that the choice of arbitration, while potentially bringing a bevy of benefits, also strips the litigants of certain rights and protections, such as the ability to engage in broad discovery, the ability to seek sanctions against a miscreant opponent, and, perhaps most importantly, the right to appeal. The importance of these rights and protections must be balanced against the costs of litigation, which include time-consuming pre-trial and trial procedures, multiple potential appeals, judges who are unfamiliar with the space industry and space law, and the risk of bad publicity.

One area of litigation that will likely emerge in the coming years will involve tort claims against suborbital tourism companies for harm caused during flight. In anticipation of these claims, the US Congress has taken action to guard against potential diversity of state court opinions regarding the liability of operators by establishing exclusive federal jurisdiction over any claims brought by space flight participants (ie passengers) or third parties on the ground.[22] However, although such claims must be heard in federal court, the appropriate state law will still govern the substance of the claim (as interpreted by the sitting federal judge). In many states, the defendant may face a state law eliminating liability on the grounds of negligence for any harm caused in the course of the suborbital flight provided the passengers sign a waiver. Whether such liability waivers are honoured in court has yet to be tested.

In the case of an international dispute (involving litigants from different countries), the costs of litigation increase dramatically. In addition to the costs previously discussed, international disputes pose challenges arising from the expense and inconvenience of travel, the potential bias of foreign courts, language barriers and costs of document translations, uncertainty as to venue and procedures, uncertainty as to applicable law, and, perhaps most critically, uncertainty as to the enforceability of a judicial award in the country where the defendant's assets are located. Even if you win a case in a Manhattan courtroom and the court issues an award which is ironclad as to substantive law, the value of that award may be questionable given the uncertainty of whether that award can be taken to another country for enforcement against assets located there. For these reasons, the arbitration of international space-related disputes is highly recommended rather than litigating in domestic court.

3.2 Arbitration of Private Disputes

Arbitration has become an increasingly popular form of dispute resolution. The potential benefits of arbitration as compared to litigation are well known and will not be given much space here except to list them:

- Lower expense
- Quicker outcome
- Convenience of location
- Expertise of arbitrators
- Limits on discovery and other controls on procedure

[22] Under this federal statute, 'any claim by a third party or space flight participant for death, bodily injury, or property damage or loss resulting from an activity carried out under the license shall be the exclusive jurisdiction of the Federal courts'. 51 US Code s 50914 (g).

- Finality of award
- Confidentiality

When the dispute is international in character, arbitration also avoids many of the pitfalls of international litigation, such as the bias of the court, uncertainty about the applicable law, and the uncertain enforceability of an award. Parties to an arbitration are able to jointly select arbitrators of good reputation to ensure a lack of bias in the proceedings. Moreover, parties can choose the governing law as well as the seat and venue of the arbitration.

It is crucial to keep in mind that the features listed above are only *potential* benefits of arbitration. None of them manifest automatically but instead require the proper management of the arbitration, beginning with the drafting of an arbitration clause tailored to the transaction at hand. Among the first decisions to be made when drafting an arbitration clause is whether the arbitration will be overseen by an institution (such as the ICC or the many other arbitral institutions available around the world) or will be an ad hoc arbitration where all aspects of the arbitration are privately arranged. If the fees of the arbitral institution are affordable, practitioners may choose to conduct a supervised arbitration to take advantage of the roster of arbitrators, procedural rules, and the ability of the institution to apply pressure on one or more of the parties to keep the process on track.

If an ad hoc arbitration is chosen, the parties must agree on which set of arbitral rules will govern the procedure—and the number of rule sets to choose from is rich. In addition to the rules developed for use at particular institutions, the United Nations Commission on International Trade Law (UNCITRAL) has also published rules that have been broadly utilized in ad hoc arbitrations.[23] Some of the other critical issues that should be addressed in an arbitration clause include (i) the number of arbitrators, (ii) the venue and seat of arbitration, (iii) the choice of governing law, (iv) limits on discovery, (v) timing, and (vi) language. To assist with drafting, model arbitration clauses are provided by the various arbitral institutions, such as the ICC.[24]

Perhaps the most compelling reason to choose arbitration over litigation to resolve an international dispute is the special enforceability of international arbitration agreements and arbitral awards under the 1958 Convention on the Convention on the Recognition and Enforcement of Foreign Arbitral Awards (popularly known as the New York Convention).[25] Pursuant to the New York Convention, the courts of a state party are required (i) to enforce a written agreement to arbitrate, and (ii) to enforce an

[23] UNCITRAL Arbitration Rules [1976] 31 UN GAOR Supp No 17, UN Doc A/31/17.

[24] See eg ICC Rules and Procedures: Arbitration Clause, <https://iccwbo.org/dispute-resolution/dispute-resolution-services/arbitration/rules-procedure/arbitration-clause/#:~:text=Standard%20ICC%20Arbitration%20Clause,accordance%20with%20the%20said%20Rules> accessed 3 August 2023 .

[25] Convention on the Recognition and Enforcement of Foreign Arbitral Awards (adopted 10 June 1958, entered into force 7 June 1959) 330 UNTS 3 (New York Convention); see also Tatiana Sainati and

arbitral award issued in another state. Both requirements are subject to minimal exceptions. With respect to the enforcement of arbitration agreements, a court receiving a complaint subject to a written arbitration clause must dismiss the action and refer the parties to arbitration unless, under applicable contract law, the agreement is 'null and void, inoperative or incapable of being performed'.[26] This exception is rarely triggered in the context of a corporate transaction. With respect to the recognition and enforcement of arbitral awards, Article V of the New York Convention describes the narrow circumstances under which a court can refuse to enforce an award:

- Incapacity of the parties.
- Invalid arbitration agreement.
- A party was unable to present its case (due to lack of notice or otherwise).
- The award concerns an issue outside the scope of the arbitration agreement.
- Procedures followed during arbitration violated the agreement.
- The award is not yet final or has been overturned.[27]
- The award concerns an issue that is not capable of settlement by arbitration (or is 'inarbitrable') under the law of the state where enforcement is sought.[28]
- The enforcement of the award would be contrary to the public policy of the state where enforcement is sought.[29]

Due to the reliable enforcement of arbitral awards, arbitration has become the preferred method of dispute resolution for international transactions.

3.3 The PCA Optional Rules and the Special Nature of Space-Related Disputes

In 2011, the Permanent Court of Arbitration (PCA) adopted a new set of Optional Rules for Arbitration of Disputes Relating to Outer Space Activities.[30] These

Joshua B Simmons, 'The New York Convention in Space: International Arbitration of Satellite Disputes' (*Jus Mundi Blog*, 17August 2022) <https://blog.jusmundi.com/the-new-york-convention-in-space_international-arbitration-of-satellite-disputes/> accessed 3 August 2023.

[26] New York Convention (n 25) art II(1) and (3).

[27] Under US law, a US court can reject an award issued in an arbitration seated in the US if the award was issued in 'manifest disregard' of the law. See *Bergesen v Joseph Muller Corp* [1983] 710 F2d 928, 932 (2d Cir). See also 'The "Manifest Disregard of Law" Doctrine and International Arbitration in New York' (*New York City Bar Committee on International Commercial Disputes*, 2012) <www2.nycbar.org/pdf/report/uploads/20072344-ManifestDisregardofLaw--DoctrineandInternationalArbitrationinNewYork.pdf> accessed 3 August 2023.

[28] Inarbitrable matters in some countries include employment, securities, antitrust, and bankruptcy matters, all of which carry profound implications for the fair treatment of employees and the health and orderliness of the financial system at large.

[29] An example of this exception for an award contrary to public policy would be a court's refusal to impose an award against a member of the press if enforcing such an award would fly in the face of a state's fundamental value of freedom of the press.

[30] Permanent Court of Arbitration Optional Rules for Arbitration of Disputes Relating to Outer Space Activities <https://docs.pca-cpa.org/2016/01/Permanent-Court-of-Arbitration-Optional-Rules-for-Arbi

Optional Rules build on the 2010 UNCITRAL Arbitration Rules, but have been modified to accommodate the 'particular characteristics of disputes having an outer space component'.[31] These 'particular characteristics' of space-related disputes are explained in the introduction to the Optional Rules and include the need for (i) expertise among arbitrators in international space law, (ii) expert testimony regarding space technology and the space industry, and (iii) confidentiality (or other controls) on the disclosure of technical information that is subject to export controls.

In order to accommodate the need for technical and legal expertise, the PCA Secretariat has compiled a roster of arbitrators and technical experts who have been vetted by the Secretariat.[32] When a technical issue arises during an arbitration, the parties can jointly submit to the arbitrator an expert report that explains the issue at hand.[33] To address the need for confidentiality and protection of controlled information, Article 17 of the Optional Rules allows for any party to submit a request that 'special measures' be put in place for the confidential treatment of certain information. The tribunal also has the power to appoint (either unilaterally or by motion) a 'confidentiality adviser' who can maintain the confidentiality of sensitive information by reporting to the tribunal regarding the relevant aspects of the information without disclosing the information in its entirety.[34] Concerns regarding the illegal disclosure of controlled information may also motivate the parties to select arbitrators of a particular nationality (for example, US citizens would be best suited to arbitrate a matter involving technical data subject to US export controls).

Despite being discussed in this section on private dispute resolution, the PCA Optional Rules and the services of the Secretariat are available for use not only in private 'dispute[s] relating to outer space', but also in private party versus state and state versus state space-related disputes. Moreover, although written for space-related disputes, the Optional Rules may be used in any arbitration. Under Article 1, it is sufficient if the parties choose to arbitrate under the Optional Rules regardless of the connection (or lack of connection) to outer space. In fact, the Rules may appeal to parties involved in other disputes involving controlled technology that would benefit from the Rules' provisions regarding the handling of such information. For those parties who choose to arbitrate under the Optional Rules, an annex to the Rules provides a model arbitration clause.

tration-of-Disputes-Relating-to-Outer-Space-Activities.pdf> accessed 3 August 2023 (PCA Optional Rules). For a discussion of the Optional Rules from the perspective of one of the drafters, see Fabio Tronchetti, 'The PCA Rules for Dispute Settlement in Outer Space: A Significant Step Forward' (2013) 29(3) SpacePoly 181–89.

31 PCA Optional Rules (n 30) introduction.

32 The Panels of Arbitrators and Experts can be found at <https://pca-cpa.org/en/about/panels/panels-of-arbitrators-and-experts-for-space-related-disputes/#:~:text=The%20PCA%20Optional%20Rules%20for,have%20expertise%20in%20this%20area> accessed 3 August 2023.

33 PCA Optional Rules (n 30) art 27.

34 ibid art 17(8).

4 Private v State Disputes

One of the trends in the space economy has been the expansion of the number of government contracts for private services, from launching government satellites and transporting astronauts to the International Space Station to providing transponder capacity or remote sensing products for government use.[35] This market in private space services has freed space agencies to use their limited resources to undertake deep space exploration projects (such as exploring the moons of Jupiter) for which the National Aeronautics and Space Administration (NASA) and other space agencies are best suited. However, as these contractual arrangements become more common, they bring with them an increase in the number of contractual disputes between private parties and sovereign states (or their agencies). Any such claims brought by a private party against a state can be pursued in a domestic court (subject to significant barriers) or can be resolved through alternative dispute resolution techniques.

4.1 Litigation of Private v State Disputes

Any time a private company sues a governmental agency, it will face the challenge of sovereign immunity.[36] The concept of sovereign immunity of a foreign state in US courts is embodied in the Foreign Sovereign Immunities Act (FSIA)[37] which prohibits US courts from acquiring personal jurisdiction over a sovereign state—subject to certain enumerated exceptions. The relevant exception for contractual disputes allows the case against the government to proceed when the dispute arises from 'commercial activity' that takes place in the United States or causes a direct effect in the United States.[38] This doctrine is consistent with the 'restrictive' approach adopted by many other countries which attempts to balance the right of private

[35] NASA's expanded use of commercial partners has increased through a series of public-private partnerships including the Commercial Orbital Transportation Service (COTS) programme, the Commercial Resupply Services (CRS) programme, and the Commercial Crew Development (CCDev) programme.

[36] When litigating in the United States, in addition to the challenge of obtaining jurisdiction over a sovereign entity, a plaintiff will also have to overcome the Act of State doctrine—a judge-made rule (based on the international legal principle of comity) that is unique to the United States. As explained by Prof William Dodge, the Act of State doctrine 'is a federal common law doctrine providing that courts in the United States will not question the validity of an official act of a recognized foreign government fully performed within its own territory'. William Dodge, 'A Primer on the Act of State Doctrine' (*Transnational Litigation Blog*, 25 March 2022) <https://tlblog.org/a-primer-on-the-act-of-state-doctrine> accessed 3 August 2023. Professor Dodge goes on to provide an example in the 1964 US Supreme Court case of *Sabbatino* which 'held that the Act of State doctrine required U.S. courts to recognize the validity of Cuba's expropriation of U.S.-owned property even if the taking violates customary international law'. ibid.

[37] Foreign Sovereign Immunities Act, 28 USC s 1602 et seq.

[38] ibid s 1605(a)(2).

citizens to seek redress in domestic courts while respecting the independence of other states.[39]

When determining whether a dispute falls under the 'commercial activity' exception, a US court will inquire whether the action which gave rise to the complaint is a type of activity in which a private party would engage—or whether it is an action that is reserved exclusively for governments.[40] In the past, governments may have been successful in arguing that space activity is 'governmental' by nature in light of the lead role played by government agencies in conducting space missions. However, this claim is becoming increasingly untrue as private activity has taken over many of the roles that were in the past exclusively reserved to governments (such as launching spacecraft). Could a government argue that putting a spy satellite in orbit is not a commercial activity since private parties do not engage in espionage? Perhaps—if the argument is framed in this manner. But the argument could also be made that many private companies engage in remote sensing for a variety of purposes (most notably, to provide the images on Google Earth and other mapping services), and that such remote sensing is therefore not an exclusively governmental activity, thereby allowing the case to proceed under the 'commercial activities' exception. In order to provide greater certainty on this issue, when drafting the documents in a transaction with a government agency, a governmental party should make a representation acknowledging the commercial nature of the contract and all related activity. For ultimate protection against the argument of sovereign immunity, a private party should demand that the governmental partner waive its sovereign immunity outright in the transactional documents.

Once the commercial activity (or another) exception to sovereign immunity is satisfied, an action against the foreign state can move forward. One example of an action between a private operator against a government entity is found in the 2001 case of *Hughes Communications Galaxy Inc v United States* brought before the United States Court of Appeals for the Federal Circuit.[41] In this case, Hughes filed suit claiming that NASA breached a Launch Services Agreement entered into between the parties in 1985 to use its 'best efforts' to place ten of Hughes' HS-393 satellites into orbit through a series of Space Shuttle launches. In 1986, due to the Space Shuttle Challenger tragedy, the Shuttle fleet lost a significant portion of its overall payload capacity. In order to ensure that room remained for government

[39] Amelia L McCarthy, 'The Commercial Activity Exception—Justice Demands Congress Define a Line in the Shifting Sands of Sovereign Immunity' (1994) 77 MarquetteLRev 893, 894 and 899.

[40] In the seminal *Dunhill* case, the US Supreme Court acknowledged that foreign governments acting in their commercial capacities 'do not exercise powers peculiar to sovereigns. Instead, they exercise only those powers that can also be exercised by private citizens.' *Alfred Dunhill of London, Inc v Republic of Cuba* [1976] 425 US 682, 704. In *Nelson v Saudi Arabia*, the Supreme Court provided an example of activity that could only be undertaken by governments. In this case, the abuse of police power by Saudi Arabia causing injury was deemed not to be commercial in nature. Saudi Arabia v Nelson [1993] 507 US 349.

[41] *Hughes Communications Galaxy, Inc v US* [2001] 271 F3d 1060 (*Hughes*).

payloads, President Reagan issued an executive order that restricted NASA's launch of commercial satellites to only those satellites that required the 'unique capabilities' of the Shuttle.[42] The *Hughes* court rejected NASA's claims of sovereign immunity and held that NASA was required to pay damages to compensate Hughes for the additional cost outlay to 'cover' the failure of NASA to provide the promised launch services.[43]

4.2 Arbitration of Private Party v State Disputes

If a private company succeeds in convincing the court to seize jurisdiction over a sovereign defendant, the company will face an opponent with vast resources and limitless time to litigate. Under these circumstances, alternatives to litigation become desirable. For example, a well-managed arbitration with tight timelines can help level the playing field by preventing the governmental party from engaging in dilatory tactics that would, in a courtroom setting, be extremely expensive and time-consuming to the detriment of the private party.

If there is an agreement to arbitrate, parties have many of the same options discussed above in the context of purely private disputes—with the addition of certain venues that are available only for disputes involving sovereign states. As is true for purely private disputes, the parties could choose to arbitrate at the Permanent Court of Arbitration under the Optional Rules or at other institutions such as the ICC (which actively invites the arbitration of state/investor disputes). In addition, the parties can choose to resolve their dispute with the aid of the International Centre for Settlement of Investment Disputes (ICSID), an institution that is available only for state/investor disputes. The option of ad hoc arbitration is not recommended for state/investor disputes because such an informal process is unsuitable when the relative position and resources of the parties are so asymmetric. The ability of an arbitral institution to apply pressure to maintain timelines and to otherwise ensure compliance with the agreed upon procedure is of particular value to a private party facing a sovereign opponent.

4.3 The ICSID

The ICSID was established by the World Bank in 1966 as a venue for resolving disputes that arise between one of its member states and a private investor of another

[42] National Security Decision Directive 254, *United States Space Launch Strategy* (1986).

[43] The court granted relief based on the doctrine of 'cover' contained in s 2-712 of the Uniform Commercial Code on the Sales of Goods, which the court applied to this services contract by analogy. *Hughes* (n 41) 1068.

member state.[44] The ICSID is not a court of arbitration, but instead provides rules and a variety of institutional services to support arbitrations, conciliations, and mediations.[45] The services provided are extensive and include, for example, a panel of recommended arbitrators, expert advice regarding procedure, financial management assistance, and facilities in Washington, DC (and elsewhere) for conducting the arbitration. One benefit of the ICSID rules is that once written consent to arbitration has been given by a party, it cannot be withdrawn.[46] Moreover, awards issued pursuant to the Convention are not subject to appeal or any remedies other than those explicitly provided for in the Convention, which are limited to the following:[47]

Article 49: Supplementation in the event an issue was omitted or a clerical error was made in the original award.

Article 50: Interpretation in the event that a dispute between the parties arises as to the meaning of the award.

Article 51: Revision on the grounds that a fact was discovered of such a nature as 'decisively to affect the award'.

Article 52: Annulment on the grounds that (i) the Tribunal was not properly constituted, (ii) the Tribunal has manifestly exceeded its powers, (iii) corruption of the Tribunal, (iv) there was a serious departure from a fundamental rule of procedure; or (v) that the award has failed to state the reasons on which it is based.

Member states are also required to enforce in their domestic courts awards issued pursuant to the Convention.[48] That said, the official commentary to the

[44] Art 25 of the ICSID Convention states that '[t]he jurisdiction of the Centre shall extend to any legal dispute arising directly out of an investment, between a Contracting State . . . and a national of another Contracting State, which the parties to the dispute consent in writing to submit to the Centre'. Convention on the Settlement of Investment Disputes between States and Nationals of Other States (adopted 18 March 1965, entered into force 14 October 1966) 575 UNTS 159, art 25 (ICSID Convention).

[45] Under the ICSID Convention, there are certain key distinctions between conciliation and mediation. Among the most critical differences is that (i) mediations may involve litigants of the same state, and (ii) there is no requirement for either party to be linked to a member state, thus turning ICSID mediation into a possibility open to all parties. In another example, similar to arbitration under the ICSID Rules, a party may not unilaterally withdraw from a conciliation once consent has been given. In contrast, either party is entitled to withdraw from a mediation at any time. 'Key Differences between Mediation and Conciliation at ICSID' (*World Bank ICSID Website*) <https://icsid.worldbank.org/rules-regulations/mediation/key-differences-between-mediation-and-conciliation> accessed 3 August 2023.

[46] ICSID Convention (n 44) art 25 (stating that '[w]hen the parties have given their consent, no party may withdraw its consent unilaterally').

[47] 'Report of the Executive Directors on the Convention on the Settlement of Investment Disputes between States and Nationals of Other States' (*ICSID Convention, Regulations and Rules*, 2006) para 41 <https://icsid.worldbank.org/sites/default/files/ICSID%20Convention%20English.pdf> accessed 3 August 2023 (Report of the Executive Directors).

[48] ICSID Convention (n 44) art 53.

Convention (the *Report of the Executive Directors*) explains that this does not obligate a state to enforce an ICSID award 'in cases in which final judgments could not be executed', which is then illustrated with the specific example of sovereign immunity as a permitted basis for refusing enforcement.[49] Bilateral investment treaties (BITs) frequently require that investor disputes be resolved through the ICSID processes, although investors may submit a request to arbitrate in the absence of any BIT requirement. For those choosing ICSID arbitration to resolve their disputes, the World Bank provides model arbitration clauses to assist with drafting.[50]

Practitioners should be aware that a request to arbitrate under the ICSID rules can be challenged on the ground that it is not an 'investment' (which is not defined in the ICSID Convention or Rules). The long-standing *Salini* test defines an investment as having four elements: (1) a contribution of money or assets, (2) a certain duration, (3) an element of risk, and (4) a contribution to the economic development of the host state.[51] Meeting this fourth requirement may appear particularly challenging for investments that involve activity only on a celestial body, but the requirement does not demand that the activity take place in the territory of the other party—only that it contribute to the economy, which could be satisfied if, for example, the space operator maintains a subsidiary located in the opposing party's territory and benefits the community through the employment of local talent, purchasing from local vendors, and paying taxes.

4.4 A Note on the Uncertainty of Investor Protections under BITs

On 1 December 2022, a discussion was held on the topic of international investment protection of space assets as part of Washington Arbitration Week.[52] One of the starkest comments of the discussion concerned the uncertainty as to whether space activities are afforded the investor protections provided in BITs in light of the fact that BITs typically require that the protected investment be made in the

[49] Report of the Executive Directors (n 47) para 43.

[50] 'ICSID Model Clauses' (*World Bank Group ISCSID Website*) <http://icsidfiles.worldbank.org/icsid/icsid/staticfiles/model-clauses-en/7.htm#a> accessed 3 August 2023.

[51] *Salini et al v Morocco*, ICSID Case No ARB/00/4, Decision on Jurisdiction (23 July 2001), (2003) 42 ILM 609, para 52. See also, in general, Alex Grabowski, 'The Definition of Investment under the ICSID Convention: A Defense of Salini' (2014) 15 ChicJIntlL 287.

[52] The panel, moderated by Ben Love, included Ernie Chung, Petr Polášek, Laura Yvonne Zielinski, and Philippe Oudinot. See Luc Colin and Gide Loyrette Nouel, 'Washington Arbitration Week 2022: International Investment Protection of Space Assets, Quo Vadis?' (*Kluwer Arbitration Blog*, 28 December 2022) <https://arbitrationblog.kluwerarbitration.com/2022/12/28/washington-arbitration-week-2022-international-investment-protection-of-space-assets-quo-vadis> accessed 3 August 2023. See also Ben Love and Sagar Gupta, 'Investor-State Space Arbitration Rules' (*Space Arbitration Association Blog*, 2023) <https://space-arbitration.com/investor-state-space-arbitration-rules> accessed 3 August 2023.

territory of the parties.[53] The answer to this question of territoriality will depend on the facts of a specific case, but it clearly poses a problem when the activity being funded by the investor takes place on a celestial body. Consider the following hypothetical:

> NASA puts out a call to its international Artemis partners for a private company to build an H_2O processing plant on the Moon for separating the oxygen and hydrogen atoms in mined water-ice for use on the Moon as breathable air and fuel.
>
> A Mexican company responds and contracts with NASA to build the facility. After construction has begun, the U.S. implements new regulations that unduly burden the Mexican company and destroy its expected return-on-investment.

Does this company have the right to seek compensation from the United States under the US-Mexico-Canada Agreement (USMCA) for appropriation (or actions 'tantamount to appropriation')? The question may hinge on the territorial aspect of the USMCA definition of 'covered investment':[54]

> Covered investment means, with respect to a Party, an investment in its territory of an investor of another Party in existence as of the date of entry into force of this Agreement or established, acquired, or expanded thereafter.

If the investment is deemed to be located in the United States, the Mexican company would likely have a strong case for compensation. We can analogize to the ecological decree in the North American Free Trade Agreement (NAFTA) *Metalclad* case which was issued after approval was given to the US investor for the operation of a landfill, and which was seen as 'tantamount to expropriation' in violation of NAFTA.[55] This leaves the critical question of whether the investment in our hypothetical was located in US territory. If space companies want to avail themselves of the special protections of BITs, those companies should take special care to establish a clear territorial presence in the appropriate country, such as setting up a subsidiary on foreign soil to oversee the space operations.

[53] Washington Arbitration Week 2022: International Investment Protection of Space Assets, Quo Vadis? (explaining that '[w]ith respect to investment treaties, Polášek observed that the main issue is that they only apply to investments made in the territories of the signatory States').

[54] US-Mexico-Canada Agreement (agreed to 1 October 2018) art 14(1) <https://ustr.gov/sites/default/files/files/agreements/FTA/USMCA/Text/20_Intellectual_Property_Rights.pdf> accessed 3 August 2023.

[55] The Supreme Court of British Columbia (stating, with respect to a governmental decree establishing a nature preserve that would prevent the operation of the licensed landfill, 'the Tribunal's conclusion that the issuance of the Decree was an act tantamount to expropriation is not patently unreasonable'). *Mexico v Metalclad Corp* [2001] BCSC 644 (BC Sup Ct); see also William S Dodge, 'International Decisions: Metalclad Corp. v. Mexico and Mexico v. Metalclad Corp.' (2001) 95 AJIL 910.

5 Conclusion: Current Initiatives and the Future of Space-Related Disputes

As the space industry grows on multiple fronts, the future promises an increased frequency and diversity of disputes in the sector. At one extreme, one state may sue another state before the ICJ challenging the legality of its mining operations on the Moon, or perhaps an ICJ case will be brought to test the legality of kinetic anti-satellite weapons. At the other extreme, the surviving family members of a star-crossed space tourist may have to file a lawsuit in state court seeking compensation under tort law from a tourism company in the wake of a tragic accident. In between these bookends, there will be a plethora of contractual disputes at various levels of the space economy.

This chapter illustrates the ample existing frameworks for resolving such disputes, ranging from international litigation to various alternatives, including informal negotiation, mediation/conciliation, or binding arbitration. That said, there is a need to strengthen this infrastructure as the demands of the industry grow. Law schools need to teach space law and other relevant courses, such as administrative law and alternative dispute resolution, in order to provide enough lawyers with the needed expertise. Law firms, in turn, need to develop space law practices and gain expertise in the legal issues that will be at the centre of the coming disputes. Academic commentators should give thought to the substance of space law and dispute resolution procedures with an eye to making improvements as the industry evolves. States and corporations should also use their significant resources to support the development of new institutions for dispute resolution, whether in the form of specialized courts or institutional services to assist in the efficient and fair settlement of space-related disputes. In fact, such investments in the resolution of space-related disputes are already taking place around the world in a variety of forms:

- There has been a robust stream of academic publications recently in the form of journal articles, book chapters, and monographs on space-related dispute resolution as a new generation of formally trained space lawyers take a fresh look at these issues.[56]

[56] In addition to the works already cited herein, the author recommends the following publications: Gérardine Goh, *Dispute Settlement in International Space Law: A Multi-Door Courthouse for Outer Space* (Martinus Nijhoff 2007); Tare Brisibe, 'Settlement of Disputes and Resolution of Conflicts' in Ram S Jakhu and Paul Stephen Dempsey (eds), *Routledge Handbook of Space Law* (Routledge 2017); Frans G von der Dunk, 'The Legal Framework for Space Projects in Europe: Aspects of Applicable Law and Dispute Resolution' in Lesley Jane Smith and Ingo Baumann (eds), *Contracting for Space: Contract Practice in the European Space Sector* (Ashgate 2011); Maureen Williams, 'Dispute Resolution Regarding Space Activities' in Frans G von der Dunk (ed), *Handbook of Space Law* (Edward Elgar 2015); Huan Yu and Mingyan Nie, 'Space Dispute Settlement in Asia in the Context of the Belt & Road Initiative: An Appraisal from China's Perspective' (2022) 47(6) Air&SpaceL 553–76 .

- In order to create a hub for practitioners in this specialized field, the Space Arbitration Association (SAA) was created in 2021 by Laura Yvonne Zielinski. The SAA plans, among other things, to create a registry of articles and public documents on space-related arbitration proceedings which promises to be of great value for building the jurisprudence of the field.[57]
- The United Arab Emirates is establishing the *Courts of the Future* with the intent to create specialized courts and dispute resolution services for high-tech industries, including the *Courts of Space*.[58] Near-term plans for the project include the creation of a Space Dispute Guide 'encompassing a set of guidelines to support such space-related disputes' which will be used to assist in the education of the courts' specially trained judges.

Through these initiatives and others like them, the field of space-related dispute resolution is coming of age. As the space industry grows in new directions, dispute resolution professionals must respond with an understanding of space technology, the space industry, and space law. Although existing institutions and rules provide a solid foundation for resolving space-related disputes, a growing industry will demand the continuing evolution of laws and procedures so that participants in the space economy at every level can rely on an efficient and just system of dispute resolution.

[57] 'Our Objective' (*Space Arbitration Association Website*) <https://space-arbitration.com/about-us> accessed 3 August 2023.

[58] 'Courts of the Future Home Page' (*Courts of the Future Website*) <www.courtsofthefuture.org> accessed 3 August 2023.

15
THIRD WORLD APPROACHES TO INTERNATIONAL SPACE LAW

Kiran Mohan Vazhapully[*]

1 Introduction

The genesis of international space law, as we comprehend it today, can be traced back to the Cold War era, a period marked by intense rivalry between the two superpowers, the United States and the Soviet Union. The launch of the first satellites and spacecraft in the late 1950s sparked questions about the existence of laws in outer space and the extent of sovereign rights into this new frontier.[1] This competition culminated in the 1967 Outer Space Treaty (OST),[2] a product of the United Nations Committee on the Peaceful Uses of Outer Space (UNCOPUOS), which was heavily influenced by the interests of its primary architects.[3]

The OST formalised the conditions under which humans use and explore outer space. The dominant powers, leveraging their superior technology, began here, shaping the contours of international space law, crafting a legal system that, while claiming universality, predominantly catered to their interests.[4] The principles of non-appropriation[5] and the promotion of the 'peaceful' use[6] of space therein, although seemingly noble, were underpinned by a bias towards these technologically advanced States. The Third World Approaches to International Law (TWAIL) presents us with the conceptual prism to dissect and critique these realities.

Viewing through a broad lens, TWAIL is a set of perspectives and methodologies that place the experiences and interests of the Global South at the forefront of international law discourses.[7] TWAIL emerged in the late twentieth century as a

* Views expressed are personal. I thank Prof Sandeepa Bhat for his constructive comments and suggestions.

1 Jason Beery, 'Unearthing Global Natures: Outer Space and Scalar Politics' (2016) 55 PolGeo 92.

2 Treaty on Principles Governing the Activities of States in the Exploration and Use of Outer Space, Including the Moon and Other Celestial Bodies (adopted 27 January 1967, entered into force 10 October 1967) 610 UNTS 205 (Outer Space Treaty).

3 Beery (n 1) 95.

4 See Matt Craven, 'Other Spaces: Constructing the Legal Architecture of a Cold War Commons and the Scientific-Technical Imaginary of Outer Space' (2019) 30(2) EJIL 547.

5 Outer Space Treaty (n 2) art II.

6 Outer Space Treaty (n 2) Preamble and art IV.

7 Makau Mutua, 'What is TWAIL?' (2000) 94 ProceASIL 31.

Kiran Mohan Vazhapully, *Third World Approaches to International Space Law* In: *International Space Law in the New Space Era*. Edited by: Sandeepa Bhat B., Dilip Ukey, and Adithya Variath, Oxford University Press. © Kiran Mohan Vazhapully 2024.
DOI: 10.1093/9780198909415.003.0015

counter-narrative—a potent retort to the perceived dominance of Western, particularly Eurocentric, perspectives in international law. The fundamental axiom of TWAIL is the notion that international law is far from neutral; rather, it has been historically shaped by and has catered to the interests of the powerful, typically Western States.[8]

TWAIL scholars seek to understand international law's role in perpetuating global disparities and explore avenues to reimagine and reconstruct legal norms to better represent the interests of the Third World. They engage critically with the history, structure, and consequences of international law, unmasking the power dynamics embedded within and seeking transformative possibilities.[9] By doing so, TWAIL scholars not only challenge the prevalent Eurocentrism in international law, but they also provide a platform for voices from the Global South that have often been silenced or ignored. They endeavour to scrutinise the status quo, questioning the dominant narratives and systems that have shaped the international law landscape as we know it. Their work, thus, has the potential to serve as a catalyst for change, instigating a shift towards a more egalitarian and inclusive legal system that truly respects and incorporates the diversity and plurality of our global community.[10]

Applying a TWAIL perspective to international space law requires us to interrogate the historical and current realities of this legal regime. It requires us to ask: how has the development of space law mirrored and facilitated the geopolitical interests of dominant States? Who does international space law really serve? How does it perpetuate the existing global power structures? How can space law be reconstructed to better represent the aspirations and rights of the Third World? And, most importantly, how can we ensure that the cosmos, a shared heritage of humankind, truly serves all of humanity equitably?

The journey we undertake in this chapter involves exploring the roots of TWAIL (Section 2), before scrutinising the dominance of developed countries in international space law (Section 3). We will then apply a TWAIL analysis to specific principles of international space law, such as the principle of non-appropriation and the principle of peaceful use (Section 4). Subsequently, we will propose reforms to international space law from a TWAIL perspective (Section 5). In conclusion, we will reflect on the challenges and opportunities for TWAIL in the domain of international space law (Section 6). This critique aspires to serve as a compass, guiding us towards a more equitable legal framework that truly encompasses the interests of all States, spacefaring or not.

[8] ibid.

[9] Antony Anghie, 'The Evolution of International Law: Colonial and Post-Colonial Realities' (2006) 27 TWQ 739.

[10] Bhupinder Chimni, 'Third World Approaches to International Law: A Manifesto' (2006) 8 ICLR 3.

2 The Emergence of TWAIL

2.1 Historical Context of TWAIL

Embarking upon the journey of TWAIL demands a voyage through the tumultuous tides of the mid-to-late twentieth century, a period teeming with the liberating waves of decolonisation and the birth of newly independent States, thirsting for equity within the international realm. The nascent States, freshly unshackled from colonial grip, found themselves plunged into an international legal landscape that was far from impartial; it was a realm deeply furrowed with Eurocentric principles favoured by the Western world.[11] This landscape, undeniably influenced by its colonial architects, left a conspicuous impression upon the emerging nations. Consequently, in a collective endeavour to contest this disbalanced status quo, the Third World States formed the Non-Aligned Movement (NAM) and the Group of 77 at the United Nations.[12] However, the depth of the Eurocentric bias within the global legal system was such that a mere contestation proved insufficient. It became increasingly clear that a comprehensive reorientation of international law was needed, one that would not merely question the prevailing biases but would also seek to construct a more equitable legal order.

In this crucible of transformation and challenge, TWAIL sprouted. Although not an organised movement as such, it was a shared consciousness among scholars from the Global South, critically examining international law from their unique post-colonial perspective. This vantage point gave voice to the Third World, unveiling the latent Eurocentric biases and challenging the entrenched power asymmetry in the international legal order.[13] TWAIL's intellectual roots can be traced back to the Bandung Conference of 1955, a significant rendezvous of leaders of newly independent Asian and African nations to discuss peace, cooperation, and the role of the Third World in the Cold War era.[14] These roots grew stronger with resistance movements and liberation struggles, offering a critique of international law from 'below'.[15]

The unfolding of decolonisation brought forth stark economic disparities for the newly sovereign Third World States. These nations were entrenched in a cycle of economic underdevelopment and dependency, remnants of their colonial

[11] Antony Anghie, *Imperialism, Sovereignty and the Making of International Law* (CUP 2007).

[12] Jürgen Dinkel, *The Non-aligned Movement: Genesis, Organization and Politics (1927–1992)* (Brill 2019).

[13] Mutua (n 7) 32.

[14] Luis Eslava and others (eds), *Bandung, Global History, and International Law: Critical Pasts and Pending Futures* (CUP 2017); and see Seng Tan and Amitav Acharya, *Bandung Revisited: the Legacy of the 1955 Asian-African Conference for International Order* (NUS Press 2008).

[15] Balakrishnan Rajagopal, *International Law from Below: Development, Social Movements and Third World Resistance* (CUP 2003).

history.[16] In the 1970s, as a response to these pressing issues, these States initiated a global campaign for a New International Economic Order (NIEO) at the United Nations (UN), aiming to restructure the world economy in a way that would foster their economic development and lessen their dependency on the industrialised countries of the Global North.[17]

The call for the NIEO marked a watershed moment in the evolution of TWAIL. With its focus on economic sovereignty, equitable allocation of global resources, and technology transfer, the NIEO became a beacon for Third World countries to voice their economic goals and challenge the global economic architecture that continually marginalised them. Although the NIEO ultimately fell short of full realisation, owing to a myriad of political and economic factors, its role in shaping the TWAIL discourse cannot be understated. The advocacy for the NIEO by developing States at the UN underscored the intertwined relationship between law and economic structures. It openly and comprehensively discredited the purported objectivity of international law and reaffirmed the critical tenet that inequities, not law alone, were what called for redress and justice.

Starting in the 1950s, TWAIL had grown into a powerful critique of international law by the start of the twenty-first century. Guided by prominent scholars like Antony Anghie and BS Chimni, it laid the groundwork for a significant rethinking of international law from the perspective of the Third World. This challenged the Eurocentric bias and sought to transform the global legal system into a more equitable one.[18]

TWAIL is a continuing endeavour that keeps engaging with important areas of international law, including issues concerning Indigenous peoples, settler colonialism, and labour law. The world is undergoing rapid transformation due to advancements in technology such as artificial intelligence, blockchain, and genomics. These changes are significantly impacting various aspects of life and areas of international law. TWAIL scholars are actively involved in interpreting these changes and their implications for the global legal system.[19] TWAIL thus represents a shared historical context and collective endeavour of the Global South to rectify the historical imbalances in international law and to reconstruct it in a manner that genuinely serves all of humanity. The journey of TWAIL continues, with its scathing critique expanding across multiple facets and branches of modern international law, which is explored in the coming sections.

[16] Rolin Mainuddin, 'The New World Order Transition in the Third World: Implications for the Nation-State' (1995) 12(1) JThrdWldStud 36.

[17] Linus Hoskins, 'The New International Economic Order: A Bibliographic Essay' (1981) 3(3) TWQ 506.

[18] Obiora Okafor, 'Critical Third World Approaches to International Law (TWAIL): Theory, Methodology, or Both?' (2008) 10 ICLR 371.

[19] Antony Anghie, 'Rethinking International Law: A TWAIL Retrospective' (2023) 34 EJIL 106.

Over the years TWAIL has undergone evolution, with a generation of scholars making valuable contributions to its extensive and groundbreaking research. This has given rise to various branches of TWAIL, each offering perspectives and criticisms to the ongoing discourse. The community of TWAIL scholars is vast and diverse and their collective efforts cannot fully capture every development within this movement. Despite its diversity, TWAIL remains united by a shared context and the collective pursuit of deconstructing and reconstructing law to address historical imbalances and truly benefit all of humanity.[20]

2.2 Key Theoretical Underpinnings of TWAIL

TWAIL, as both a movement and an intellectual standpoint, is grounded in a few foundational principles. These principles, shaped by the experiences and viewpoints of the Global South, shed light on and critique the biases deeply embedded within the international legal system while also presenting possibilities for its reconstruction. One essential assumption of TWAIL is the recognition that international law is intrinsically biased. Scholars emphasise that international law in its form is far from neutral or benign; it has been influenced by and continues to serve the interests of powerful Western States. TWAIL underscores that the principles, structures, and norms of international law, though presented as universal, are essentially Eurocentric, born out of the historical experiences of Europe and the West.

Building on this understanding, the second core assumption of TWAIL is the notion that international law has been a significant tool in the perpetuation of colonial exploitation and continues to be implicated in the neo-colonial realities that characterise the relationships between the Global North and South. The structures and institutions of international law are not innocent bystanders in the global scenario; rather, they have actively facilitated the global disparities we see today.[21] As Chimni rightly points out, 'it is not just that international law has failed to address these disparities, but rather it has played a role in creating and sustaining them'.[22] Therefore, the critique and transformation of international law from a TWAIL perspective involve an unflinching interrogation of these historical and ongoing injustices.

Another key theoretical underpinning of TWAIL is its approach to history.[23] It views history not as a linear progression of events but as a complex interplay

[20] Okafor (n 18).

[21] Balakrishnan Rajagopal, 'International Law and Third World Resistance: A Theoretical Inquiry' in Antony Anghie and others (eds), *The Third World and International Order: Law, Politics and Globalization* (Martinus Nijhoff 2003) 145.

[22] Chimni (n 10).

[23] James Gathii, 'Promise of International Law: A Third World View' (2020) 114 ProceASIL 165.

of power dynamics that continues to influence the present. It probes into the historical origins of international law, tracing it back to the colonial encounters between the European powers and the rest of the world, thereby highlighting the colonial remnants in contemporary legal structures. Its critique is also rooted in the intersectionality of various forms of oppression. It recognises that the Global South's experiences are shaped by myriad intersecting structures of domination and imperialism—including race, class, gender, and geography.[24] This recognition deepens TWAIL's critique of the global power structure, allowing for a more nuanced understanding of the disparities in international law.

Indeed, the TWAIL perspective is conscious of the heterogeneity of the experiences of the Global South. Each region, country, and community within the Third World is uniquely positioned at the crossroads of a multitude of differences. These differences are deeply embedded in their societies, resulting in diverse forms of struggle and resistance, and contributing to their unique experiences and interactions with the international legal discourse. Therefore, TWAIL's commitment to intersectionality compels it to consider these multiple layers of experience, understanding that these identities and struggles cannot be analysed separately. It also acknowledges that these differences within the Global South must be respected and included in the process of envisioning and implementing change in international law. In this context, TWAIL's critique of international law gains significant depth and breadth, covering not just the critique of North–South disparities, but also the internal disparities within the Global South, including gender, racial, ethnic, and economic injustices. This approach amplifies the multitude of voices from the Global South and strengthens TWAIL's advocacy for a more inclusive and equitable international legal order.

TWAIL is not limited to the 'Third World'—it offers a universal vision of international law and justice, evolving over decades with its origins, development, and basic intellectual orientations forming a part of its history. In fact, there are multiple TWAILs, reflecting the diversity within the movement and the breadth of scholarship it encompasses. This diversity is evidence of its richness and its ability to resonate with and accommodate a wide range of experiences and aspirations manifested in the Global South. Furthermore, the heart of TWAIL's theoretical framework lies in its commitment to social justice. This commitment drives the movement's critical engagement with international law and its search for transformative possibilities. It is not just about exposing the disparities and biases in international law, but also about reimagining and reconstructing it in a way that is more equitable and representative of the diverse experiences and aspirations of the global community.

[24] James Gathii, 'Writing Race and Identity in a Global Context: What CRT and TWAIL Can Learn from Each Other' (2021) 67 UCLALRev 1610; Mosope Fagbongbe, 'The Future of Women's Rights from a TWAIL Perspective' (2008) 10 ICLR 401.

Finally, the core of the TWAIL discourse is its emphasis on the role of the empire. This attention is crucial in understanding the effects of international law on the Third World. TWAIL incorporates a broad spectrum of traditions and thinkers, from Mahatma Gandhi and Karl Marx to post-colonial and subaltern studies scholars such as Edward Said and Gayatri Spivak. This allows TWAIL to offer a comprehensive and nuanced critique of the imperialistic leanings in international law.[25] As an ongoing initiative, TWAIL is continually growing and widening its scope. It seeks to engage more deeply with crucial areas of international law, from Indigenous peoples and settler colonialism to labour law and the global commons. This ongoing evolution reflects TWAIL's commitment to illuminating overlooked dimensions of international law and contributing to its transformation. Indeed, one does not have to be from the Third World to be a 'TWAIL scholar'. TWAIL offers a different vision not just of the Third World, but also of Europe and the West itself, challenging conventional perspectives and pushing for a more inclusive understanding of international law.

2.3 Application of TWAIL in Various Fields of International Law

Since its inception, TWAIL scholarship has been instrumental in critiquing, illuminating, and challenging the biases embedded in various fields of international law. From the traditional fields such as international humanitarian law and international trade law, to emerging areas like international environmental law and space law, TWAIL's lens has proven to be indispensable.

International humanitarian law (IHL) exemplifies the structural biases critiqued by TWAIL. IHL, largely created by Western powers, has often been accused of legitimising violence against peoples of the Global South under the guise of 'civilising missions' or 'humanitarian interventions'. TWAIL critiques of IHL challenge this narrative and argue for the need to recognise the colonial origins and implications of these laws.[26] TWAIL perspectives have been applied to international trade law, which is often touted as a tool for global development. However, TWAIL scholars argue that international trade law often exacerbates disparities between the Global North and South. The World Trade Organization (WTO), for instance, despite its avowed goal of promoting fair trade, often perpetuates inequities through its dispute resolution mechanism and policy structures.[27]

[25] Anghie (n 19).

[26] James Gathii and Henry Richardson, 'Introduction to Symposium on TWAIL Perspectives on ICL, IHL, and Intervention' (2015) 109 AJILUnbound 252; Anne Orford, *Reading Humanitarian Intervention* (CUP 2003).

[27] BS Chimni, 'The WTO, Democracy and Development: A View from the South' in C Deere Birkbeck (ed), *Making Global Trade Governance Work for Development: Perspectives and Priorities from Developing Countries* (CUP 2011) 261.

Likewise, in the context of emerging fields like international environmental law and space law, TWAIL has emerged as a valuable perspective. As the world grapples with climate change, the role of international environmental law and its impact on the Global South is being scrutinised through a TWAIL lens. The impacts of climate change disproportionately burden the Global South, despite these nations contributing the least to global emissions, underscoring the need to incorporate TWAIL perspectives in shaping international environmental policies.[28]

As humanity looks to the stars, international space law is coming into sharper focus. TWAIL critiques question whether space law will replicate the inequalities of terrestrial international law, or whether it presents an opportunity for more equitable governance. Given the increasing commercialisation of space and the consolidation of control by a few countries and multinational corporations, the need for TWAIL's critique is more urgent than ever. The potency of the TWAIL methodology, its capacity to question, illuminate, and challenge the ingrained biases in international law, is what this chapter seeks to harness in the analysis of space law. The exploration and use of outer space, a realm beyond national jurisdictions, presents both opportunities and challenges. The underlying question is whether the legal framework that governs space exploration will reflect the disparities and inequalities that mar terrestrial international law, or whether it will rise above these limitations and offer an equitable and inclusive legal order.

This chapter, drawing on TWAIL's rich theoretical heritage, seeks to answer these questions. By interrogating space law through a TWAIL lens, we aim to uncover the latent biases, question the prevailing power structures, and contribute to the ongoing discourse on creating a more equitable framework for humanity's cosmic endeavours. The transformative potential of TWAIL, in this context, is more significant than ever. Indeed, the space law frontier invites not just a physical exploration but also an intellectual one, and TWAIL is our chosen instrument of inquiry.

3 The Dominance of Developed Countries in International Space Law

3.1 Tracing the Historical Development of International Space Law

The origins of international space law are anchored in the geopolitical rivalries of the twentieth century. The post-World War II era bore witness to the escalating tensions of the Cold War and the compelling narrative of space exploration. The launch of Sputnik 1 by the Soviet Union in 1957 served as the symbolic opening

[28] Malavika Rao, 'A TWAIL Perspective on Loss and Damage from Climate Change: Reflections from Indira Gandhi's Speech at Stockholm' (2022) 12 AsianJIntlL 63.

of this new frontier, and the consequent 'Space Race' between the United States and the Soviet Union largely dictated the nascent trajectory of international space law.[29] Just as TWAIL emerged in the aftermath of decolonisation, the development of space law was precipitated by momentous political transformations; however, the drivers of these transformations were diametrically opposed. While TWAIL was born out of the struggles of the Global South, space law was primarily the product of the technologically advanced superpowers of the North, who projected their interests and ideologies onto this new legal landscape.

The OST laid the cornerstone for the contemporary architecture of international space law. The Treaty, formulated amidst the strategic chessboard of the Cold War, was the result of negotiations primarily between the United States and the Soviet Union, with minor inputs from a select group of technologically advanced countries. The Treaty enshrined foundational principles such as the non-appropriation of outer space by any State, the peaceful uses of outer space, and the freedom of exploration and use. However, these principles were arguably shaped by the interests and perspectives of the spacefaring nations, which were then merely a handful of Western powers and the Soviet Union.

The OST, while embedding principles that seemed universal and equitable, subtly reflected the power structures that dominated its formulation. Although it prohibited national appropriation of outer space, it did not address the issue of the commercial or private exploitation of space resources. This absence, which may seem innocuous at first glance, bears the hallmarks of a system dominated by technologically advanced States that envisioned potential commercial interests in space. This foresight has allowed a privileged few to enjoy the fruits of this legal lacuna today, as we witness the advent of private space exploration and space mining.[30] Furthermore, the principles of non-appropriation and freedom of exploration and use were more or less abstract and idealistic, perhaps intentionally so. Their interpretation and application were left largely to the discretion of the spacefaring States who held the keys to the realm of space. The resulting imprecision of these principles and their consequent manipulation exposed the developing countries to the whims of the dominant powers.[31]

The complexity of space law cannot be understood without considering the international legal framework that governs it. As Perpetua Akoth Adar posits, 'International Space Law needs to develop to respond to current and future activity'.[32] This development should take into account the evolving dynamics of space activities, with emerging and aspirant space actors from Latin America, the Middle

[29] Albert Lai, *The Cold War, the Space Race, and the Law of Outer Space* (Routledge 2021); Linda Billings, 'Ideology, Advocacy, and Space Flight—Evolution of a Cultural Narrative' in Steven Dick and Roger Launius (eds), *Societal Impacts of Space Flight* (NASA 2007) 483.

[30] Perpetua Adar, 'Space and the Future of Humanity: A TWAIL Critique of International Space Law and Space Discourse' (2022) 3 TWAILRev 92.

[31] ibid 95.

[32] ibid.

East, Africa, and Southeast Asia joining the fold. The existing space law framework, shaped by the interests and perspectives of technologically advanced States, does not adequately address the needs and concerns of these emerging actors. The growing influence and involvement of non-state actors, particularly private entities, in space activities further complicate this scenario.

These subtle manifestations of power asymmetry in the formulation of the OST unveil the complex tapestry of international space law. The legal order, while ostensibly founded on universal principles, served to bolster the strategic interests of a select group of States, thereby revealing the potent influence of power structures in shaping international law. This historical perspective elucidates the skewed power dynamics underlying international space law, making a compelling case for a TWAIL analysis of the same.

The 1972 Liability Convention,[33] the 1975 Registration Convention,[34] and the 1979 Moon Agreement[35] followed, further solidifying the legal framework governing space activities. Yet, much like the OST, these agreements were formulated within the geopolitical dynamics of the spacefaring powers, reflecting their techno-strategic priorities and anxieties. Still, the discourse surrounding the shared human heritage of outer space and the celestial bodies was largely dominated by the States possessing the technology to access them. A significant majority of the world, particularly the developing countries, had limited participation in these crucial negotiations. Therefore, while space law, on its face, appears to be a universal legal order espousing the common interests of humanity, it is fundamentally a product of the strategic interests of the few technologically capable States. This historical trajectory provides a crucial backdrop against which we can further explore the role and influence of developed States in international space law in the subsequent sections.

3.2 The Role and Influence of Developed Countries in Shaping International Space Law

In the absence of any robust global governance structure for outer space, the technologically advanced countries have largely driven the formulation of norms and rules in the space domain. For instance, the American and European space agencies have developed detailed regulations for their space activities, shaping the

[33] Convention on International Liability for Damage Caused by Space Objects (adopted 29 March 1972, entered into force 1 September 1972) 961 UNTS 187 (Liability Convention).

[34] Convention on Registration of Objects Launched into Outer Space (adopted 14 January 1975, entered into force 15 September 1976) 1023 UNTS 15 (Registration Convention).

[35] Agreement Governing the Activities of States on the Moon and Other Celestial Bodies (adopted 18 December 1979, entered into force 11 July 1984) 1363 UNTS 3 (Moon Treaty).

practical application and interpretation of the broad principles outlined in international treaties.[36] These nations have also dominated international forums like the United Nations Committee on the Peaceful Uses of Outer Space (UNCOPUOS), asserting their influence on the development of international norms. This preponderance of developed nations in the establishment and application of space law manifests itself in numerous ways.

One of the most apparent is in the arena of satellite communication. Most of the orbital slots in the geostationary orbit, a valuable outer space real estate, have been occupied by the developed nations, leaving limited options for the countries of the Global South.[37] This control over critical space resources further extends their influence and power over the governance of outer space. Furthermore, the significant influence of these technologically advanced countries also reverberates in their ability to shape the discourse around emergent issues in space law. Be it the debates on space mining, militarisation of space, or the issues of space debris, the positions and interests of the developed nations often take centre stage, subtly shaping the contours of these discussions. The rules surrounding these frontier issues are in flux, and the influential voices of the technologically advanced countries undeniably hold sway over the law's future trajectory.

This dynamic becomes even more nuanced when we consider the increasing role of private entities in space activities. Many of the leading commercial players in the space industry, such as SpaceX and Blue Origin, are based in developed countries. The ability of these countries to regulate and support these entities inevitably influences the development of norms and practices related to commercial space. Given the increasing role of private entities in space activities, it is important to consider the socio-economic and political conditions of all nations while formulating space laws. This becomes particularly crucial in an era of space commercialisation, where private companies, predominantly based in developed nations, lead in space exploration and exploitation. Rather than a reactive process to technological advancements or the interests of a select few, the development of space law should be a proactive endeavour. The goal should be to ensure equitable sharing of space benefits and the preservation of outer space for future generations.

These realities underscore the critical need for an inclusive and democratic process based on multilateralism in the formation of international space law, a process that acknowledges and accommodates the interests of all States irrespective of their space capabilities. By questioning these power structures and advocating for

[36] Ram Jakhu, 'Legal Issues Relating to the Global Public Interest in Outer Space' (2004) 32 JSpaceL 31.

[37] Tanja Masson-Zwaan and Stephan Hobe (eds), *The Law of Outer Space: An Experience in Contemporary Law-Making, by Manfred Lachs* (2nd edn, Brill 2010) 130; Nandasiri Jasentuliyana, *International Space Law and the United Nations* (Kluwer Law International 1999) 130.

a more equitable space order, TWAIL opens the door to this transformative possibility. It highlights the urgency of re-envisioning a space law that moves beyond the interests of the powerful few and genuinely embodies the common interest of humankind. The domination of developed countries in shaping international space law raises critical questions about its equity and inclusiveness. If the norms and rules of outer space are predominantly determined by the perspectives and interests of the developed nations, can they truly serve the collective interests of humanity? A TWAIL analysis of international space law invites us to question these power dynamics and critically explore the possibilities of a more equitable and inclusive legal order for outer space.

4 A TWAIL Analysis of Specific Principles of International Space Law

The journey into the heart of international space law, viewed through the discerning lens of TWAIL, necessitates a critical inspection of its bedrock principles, principles that have been ensconced in the very fabric of space law. This investigation naturally leads us to the doctrines of non-appropriation, peaceful uses, and the freedom of exploration and use. Each of these principles, enshrined in the OST, serves as a testament to the pervasive influence of the dominant powers of the era.

Imbued with the spirit of TWAIL, we engage in a deconstruction of these principles, to expose the veiled biases and deeply entrenched power structures that they sustain. Such an exercise is indispensable if we are to unmask the Eurocentric biases that subtly pervade these principles. It also facilitates a broader discourse, one that seeks to integrate the often-neglected perspective of the Global South, thereby fostering a more inclusive dialogue. At the same time, this dissection paves the way for a constructive engagement with international space law, where we seek not merely to critique but to contribute to its evolution. This endeavour stems from a critical recognition of the potential of space law to transcend its past and present, to move beyond the constrictions of power politics and assert its truly universal character.

The challenge here is substantial, for it requires that we scrutinise these principles not in isolation, but in the broader context of their interpretation and application, by state and non-state actors alike. The doctrines of non-appropriation and peaceful uses of space, for instance, must be critically evaluated in the light of the commercialisation and militarisation of space. Similarly, the principle of freedom of exploration and use must be juxtaposed with the environmental and ethical considerations, particularly in the context of the Global South, whose space capabilities and aspirations often remain marginalised in the face of the Western-centric space agenda. Such a comprehensive critique, embedded in the TWAIL philosophy, illuminates the systemic inequities that underpin international space law and prompts a radical reimagining of it.

4.1 The Principle of Non-Appropriation

Delving into the intricacies of the principle of non-appropriation, we encounter its facade of neutrality. The principle, enshrined in Article II of the OST,[38] forbids States from asserting sovereignty over celestial bodies. This principle is presented as a benign provision, positing an equal playing field for all nations—regardless of their technological prowess. Therein, however, lies a deception that requires a critical gaze. The genesis of this principle, like most tenets of international space law, is steeped in the power dynamics of the Cold War era. It emerged as a strategic move by the superpowers to prevent a celestial scramble that could disrupt the strategic balance. Thus, from its very inception, the principle was a product of power politics, serving the interests of the spacefaring nations which, while agreeing not to claim territorial sovereignty over celestial bodies, had the requisite capabilities to make use of space resources without doing so.

Further probing into the application of this principle unravels yet another layer of partiality. Several States and scholars argue that the non-appropriation principle, while preventing territorial claims, does not expressly preclude the use or extraction of space resources.[39] Consequently, it grants an undue advantage to technologically advanced countries, equipped to exploit space resources without claiming territorial sovereignty. In effect, the ostensibly neutral principle of non-appropriation bolsters the de facto monopolisation of space resources by a select few. A TWAIL critique of the non-appropriation principle therefore necessitates a radical reappraisal of its presumed neutrality. It urges us to question the narrative of universality and benignity, laying bare the Eurocentric biases and power imbalances that underlie the principle. It calls for an acknowledgment of the political realities and the power dynamics that birthed and continue to shape the application of this principle, thus challenging the myth of neutrality that it enjoys. International space law, as it exists today, is ill-equipped to navigate the increasing complexities of space activities, a situation further complicated by the emergence of new space actors from the Global South. This inadequacy is particularly glaring in the face of escalating private sector involvement in space activities. The existing

[38] Outer Space Treaty (n 2) art II: 'Outer space, including the moon and other celestial bodies, is not subject to national appropriation by claim of sovereignty, by means of use or occupation, or by any other means.'

[39] Abigail Pershing, 'Interpreting the Outer Space Treaty's Non-Appropriation Principle: Customary International Law from 1967 to Today' (2019) 44(1) YaleJIntlL 149; Paliouras Zachosa, 'The Non-Appropriation Principle: The Grundnorm of International Space Law' (2014) 27 LJIL 37.

international legal framework falls short of effectively regulating these entities, which further deepens the inherent disparities in the non-appropriation principle.

The technological advancements and our increasing dependency on space technology amplify these imbalances. The countries with the means to utilise these technologies gain an unfair edge, while those from the Global South are left to confront the fallout of their technological and economic limitations. The historical and political biases entrenched in international space law have contributed to this state of inequality. These biases have predominantly neglected the interests of countries from the Global South, perpetuating a narrative that favours the interests of technologically advanced nations. In light of these stark imbalances, there is an immediate demand for reform in international space law.[40] Such reform must centre on equity and inclusivity, ensuring that space, as the shared heritage of mankind, does not become the exclusive territory of a select few. This demand for reform aligns with the TWAIL critique of the non-appropriation principle and urges us to conceive a more equitable interpretation of the principle, one that resonates with the realities and ambitions of the Global South.

The task is not merely to deconstruct but also to envisage a more equitable interpretation of the principle that aligns with the realities and aspirations of the Global South. To illuminate the inequities implicit in the non-appropriation principle, one can look to a host of recent developments that underscore its biased application. These case studies from the world of space exploration and exploitation provide us with a vantage point to critically examine the principle and its effect on the Global South.

Take, for example, the much-celebrated Artemis Accords, spearheaded by NASA, which seek to set 'norms of behaviour' for spacefaring nations.[41] While the Accords affirm the non-appropriation principle, they subtly manoeuvre around it by endorsing the extraction and use of space resources, including those on the moon, by private entities. This allows technologically advanced nations, which are home to these private entities, to monopolise space resources, in clear violation of the spirit of the non-appropriation principle.

Moreover, the formulation of national space laws, such as the US Commercial Space Launch Competitiveness Act of 2015 and the Luxembourg Law on the Exploration and Use of Space Resources of 2017, further underscores the inequity inherent in the non-appropriation principle.[42] These laws grant private entities the right to own, sell, and profit from the space resources they extract. By doing so, they facilitate the de facto appropriation of space resources by private actors from

[40] Saadia Pekkanen, 'Governing the New Space Race' (2019) 113 AJILUnbound 92; Dave Baiocchi and William IV Welser, 'The Democratization of Space' (2015) 94 ForeignAff 98.

[41] Artemis Accords: Principles for Cooperation in the Civil Exploration and Use of The Moon, Mars, Comets, and Asteroids for Peaceful Purposes (adopted 13 October 2020) (Artemis Accords).

[42] US Congress, 'Commercial Space Launch Competitiveness Act' (2015); Luxembourg Parliament, 'Law on the Exploration and Use of Space Resources' (2017).

technologically advanced nations, effectively sidelining countries from the Global South, most of whom lack the technological capability to do so.

The case of the geostationary orbit (GEO), a finite resource crucial for satellite communication, provides yet another testament to the skewed implementation of the non-appropriation principle. The positions on the GEO are mostly occupied by satellites from developed nations, restricting access for developing countries. The International Telecommunication Union, tasked with allocating slots on the GEO, has been unable to ensure equitable distribution, thereby compounding the disadvantage for the Global South.[43]

These instances vividly illustrate the exploitative application of the non-appropriation principle, validating the TWAIL critique. They expose the systemic structures that allow the developed nations to navigate around the principle, cementing their hegemony in space while the Global South grapples with the stark realities of their technological backwardness and economic constraints. In light of these stark realities underscored by the TWAIL critique, it becomes imperative to search for alternative frameworks that can address the inherent inequities of the non-appropriation principle. Transitioning from an exposition of the problem, we turn our gaze to potential solutions gleaned from international law. Indeed, the essence of TWAIL lies not only in critiquing the existing structures of international law but also in harnessing its potential to foster solutions. By delving into the vast corpus of international legal principles and practices, we can unearth transformative ideas that can reshape our understanding of space law in a manner that aligns with the ethos of justice, equity, and inclusivity.

When we probe further into the possibilities of reconceptualising the principle of non-appropriation, we find guiding lights in the realm of international law. The Law of the Sea Treaty, for example, offers an instructive parallel. This treaty operationalises the principle of the common heritage of mankind to the deep seabed, establishing an international regime to oversee the extraction of resources and ensuring their equitable distribution among all nations.[44] Drawing from this, we could envision a similar regime in the context of space law, which would address the exploitation of space resources while also ensuring a fair distribution of space-derived benefits.

Likewise, the concept of intergenerational equity, rooted in environmental law, holds significant implications for space law. This principle mandates that our generation's actions should not compromise the ability of future generations to meet their own needs, particularly in relation to environmental resources.[45] When

[43] Cris Van Eijk, 'Unstealing the Sky: Third World Equity in the Orbital Commons' (2022) 47 Air&SpaceL 25.

[44] Convention on the Law of the Sea (adopted 10 December 1982, entered into force 16 November 1994) 1833 UNTS 397, vol 1833, 3; vol 1834, 3; vol 1835, 3 Part IX s 2 (Convention on Law of the Seas).

[45] Edith Weiss, *In Fairness to Future Generations: International Law, Common Patrimony, and Intergenerational Equity* (Transnational Publishers 1989).

applied to space, this principle could discourage the rush for space resources by a select few countries, thereby preserving these resources for future generations globally.

Furthermore, the doctrine of 'Permanent Sovereignty over Natural Resources', delineated in General Assembly Resolution 1803 (XVII) of 1962, has the potential to reshape our comprehension of space resource appropriation. This resolution strengthens the entitlement of peoples and States to wield sovereignty over their indigenous resources. It simultaneously accentuates their duty to ascertain the equitable redistribution of advantages, thereby infusing the discourse on space resource management with a nuanced complexity that resonates with the broader aspirations for justice and inclusivity within the international legal framework. This principle, steeped in historical context and reflective of evolving global norms, serves as a beacon guiding the ongoing dialogue on the ethical and legal dimensions of space exploration and utilisation. In the context of space law, this principle could guide a new approach to space resource appropriation that reflects the collective aspirations of humanity, rather than the interests of a technologically determined elite.[46] Together, these principles from other domains of international law—common heritage, intergenerational equity, and permanent sovereignty over natural resources—offer valuable insights into how the principle of non-appropriation in space law could be reimagined. This would foster an interpretation that aligns more closely with the ideals of global justice and equity that TWAIL embodies.

Turning to existing frameworks within space law, the Moon Agreement provides a significant, albeit underutilised, point of departure. Despite its noble intentions of articulating a benefit-sharing regime for lunar resource exploitation, the Agreement has been hampered by a lack of detailed implementing provisions and the reluctance of major spacefaring nations to ratify it.[47] One possible reformative path could involve revisiting and revitalising the Moon Agreement. By defining terms like 'equitable sharing' and operationalising benefit-sharing mechanisms, as well as creating a robust international body to ensure compliance, the Agreement could be significantly strengthened. This would not only breathe new life into the Agreement but also give practical form to the aspirations of TWAIL.

In line with the principle of Permanent Sovereignty over Natural Resources, we could envisage a regime that recognises the rights of all nations, including those of the Global South, over space resources. Regulatory mechanisms could include the establishment of a global trust, where a portion of profits from space resource exploitation would be deposited. These funds could then be used to foster the development of space capabilities in the Global South, thereby addressing the current inequities.[48]

[46] Nico Schrijver, *Sovereignty over Natural Resources: Balancing Rights and Duties* (CUP 1997)

[47] Fabio Tronchetti, *The Exploitation of Natural Resources of the Moon and Other Celestial Bodies: A Proposal for a Legal Regime* (Brill 2009).

[48] Jakhu (n 36).

Indeed, the challenges surrounding space resource exploitation necessitate a careful balancing of diverse interests. While it is important to incentivise space exploration, these ambitions must be balanced against the ideals of equity and justice. An inclusive approach, inspired by TWAIL, could serve as a powerful tool in reshaping space law to better reflect the shared aspirations of humanity.

4.2 The Principle of Peaceful Use

The 'peaceful use' principle as enshrined in the ÓST is one of the key pillars of international space law, but ambiguity pervades its interpretation and application. The Treaty asserts, in Article IV,[49] that space and celestial bodies shall be used exclusively for 'peaceful purposes', yet the exact meaning of this phrase remains elusive and contentious. This vagueness has left the term susceptible to a range of interpretations that, in turn, shape its implementation. The interpretations largely differ among countries, with spacefaring States often steering the direction.[50]

These varying interpretations have substantial impacts on the implementation of the principle, often creating a disparity between the stated intent of the Treaty and the actual outcomes. For instance, the United States and Union of Soviet Socialist Republics, during the Cold War, maintained that 'peaceful' merely indicated 'non-aggressive', thus allowing military use of space as long as it did not involve direct aggression. This interpretation enabled the proliferation of military satellites and reconnaissance systems in space, arguably deviating from the intent of peaceful exploration.[51] Importantly, these interpretations and implementations seem to disproportionately favour technologically advanced nations. Such nations, due to their technical prowess and political influence, have been able to mould the interpretation of the principle in ways that allow them to exploit the strategic and economic potential of space while simultaneously curbing the opportunities for less developed countries. For example, surveillance and reconnaissance satellites, while arguably breaching the spirit of the 'peaceful use' principle, have been tacitly accepted due to the influential positions of the countries deploying them.

[49] Outer Space Treaty (n 2) art IV:

> States Parties to the Treaty undertake not to place in orbit around the earth any objects carrying nuclear weapons or any other kinds of weapons of mass destruction, install such weapons on celestial bodies, or station such weapons in outer space in any other manner. The moon and other celestial bodies shall be used by all States Parties to the Treaty exclusively for peaceful purposes. The establishment of military bases, installations and fortifications, the testing of any type of weapons and the conduct of military manoeuvres on celestial bodies shall be forbidden. The use of military personnel for scientific research or for any other peaceful purposes shall not be prohibited. The use of any equipment or facility necessary for peaceful exploration of the moon and other celestial bodies shall also not be prohibited.

[50] Bin Cheng, *Studies in International Space Law* (OUP 1997) 514; Edward Finch, 'Outer Space for 'Peaceful Purposes' (1968) 54(4) ABAJ 365.

[51] ibid, Bin Cheng.

This biased application, apart from skewing the benefits of space activities in favour of the powerful nations, also creates an environment where non-spacefaring nations are rendered passive observers rather than active participants. This undercuts the egalitarian vision that the OST aspired to uphold, once again demonstrating the structural biases ingrained in international space law. Further, when we delve deeper into the intricacies of the 'peaceful use' principle, we encounter the complex challenge posed by dual-use technologies. Dual-use technologies, in the context of space exploration, refer to those technologies that can be used for both civilian and military purposes. This includes, but is not limited to, communication satellites, Global Positioning Systems (GPS), and remote sensing technologies. The inception of these technologies has introduced a nuanced layer of complexity to the principle of 'peaceful use' enshrined in international space law.

The ambiguity of 'dual-use' lies in its potential for both benign and malicious applications. A communication satellite can enable international connectivity and information sharing, yet it can also be used for military communication and cyber espionage. This dual nature blurs the lines between peaceful and military use, casting a long shadow of uncertainty over the interpretation and implementation of the 'peaceful use' principle.[52]

For the Global South, these ambiguities have profound implications. While dual-use technologies have the potential to benefit these countries in terms of development and connectivity, they also expose them to new vulnerabilities. Technologically advanced nations have the means to deploy and exploit these technologies for strategic and economic advantages, thereby escalating the asymmetries and security dilemmas in the international arena. The existing legal framework of international space law appears to be inadequately equipped to tackle these challenges. The principle of 'peaceful use', due to its inherent ambiguity, fails to effectively regulate the use of dual-use technologies. This regulatory gap exposes the structural biases of international space law, which tend to privilege the interests of technologically advanced nations while leaving the concerns of the Global South inadequately addressed. A TWAIL-inspired critique of this situation once again illuminates the disparities and inequities inherent in the current architecture of international space law.

As an illustrative example of the issues raised thus far, consider the deployment of satellite technology by technologically advanced nations. Satellite technology is quintessentially dual-use, with potential applications ranging from climate monitoring and GPS to military surveillance and potential offensive capabilities. One of the most controversial cases concerning this issue was the 2007 Chinese anti-satellite (ASAT) test, in which China used a missile to destroy one of its defunct weather satellites.[53]

[52] Sandeepa Bhat B. and Kiran Mohan, 'Anti-Satellite Missile Testing: A Challenge to Article IV of the Outer Space Treaty' (2009) 2 NUJSLRev 205.

[53] '2007 Chinese Anti-Satellite Test Fact Sheet' (*Secure World Foundation*, 23 November 2010) <https://swfound.org/media/9550/chinese_asat_fact_sheet_updated_2012.pdf> accessed 2 October 2023.

While China maintained that the test was conducted for peaceful purposes, many Western nations interpreted it as a military demonstration of space capabilities, thereby escalating the tensions surrounding the militarisation and weaponisation of outer space.[54] From a Global South perspective, such instances raise serious security concerns. Despite their limited participation in the space domain, these nations are by no means immune to the repercussions of space militarisation. The weaponisation of space by technologically advanced nations poses an existential threat to international peace and security, to which countries of the Global South are particularly vulnerable due to their limited defensive capabilities in space.

Recognising the impact of these technologies, especially on less developed nations, calls for an enhanced regulatory approach. This entails creating a balance that safeguards against misuse without stifling innovation. To achieve this, restrictions on the deployment of such technologies, rigorous oversight mechanisms created under the aegis of UN-mandated bodies like UNCOPUOS or the Conference on Disarmament (CD), and transparent dispute resolution mechanisms are needed. Negotiating new treaty provisions to explicitly address dual-use technologies and safeguard less developed nations' security interests is another imperative step. It can lead to a more inclusive and equitable legal framework for outer space. These provisions could explicitly address the challenges posed by dual-use technologies and ensure the safeguarding of the security interests of the Global South.

4.3 The Principle of Freedom of Exploration and Use

The principle of freedom of exploration and use, as stated in Article I of the OST,[55] delineates that the exploration and use of outer space should be carried out for the benefit and in the interests of all countries. It emphasises that space is the province of all humankind, irrespective of the state of economic or scientific development. This tenet envisions a scenario where space, being the common heritage of humanity, is accessible and usable by all nations equitably. However, the

[54] Anthony Mastalir, *The U.S. Response to Chinese ASAT Test* (Drew Paper No 8, Air University Press 2009).

[55] Outer Space Treaty (n 2) art I:

> The exploration and use of outer space, including the moon and other celestial bodies, shall be carried out for the benefit and in the interests of all countries, irrespective of their degree of economic or scientific development, and shall be the province of all mankind.
>
> Outer space, including the moon and other celestial bodies, shall be free for exploration and use by all States without discrimination of any kind, on a basis of equality and in accordance with international law, and there shall be free access to all areas of celestial bodies.
>
> There shall be freedom of scientific investigation in outer space, including the moon and other celestial bodies, and States shall facilitate and encourage international co-operation in such investigation.

interpretation of this principle varies significantly among the developed and developing nations. Predominantly, developed nations, with their advanced space capabilities, interpret this principle as carte blanche to conduct their space activities, often driven by commercial or strategic interests. On the other hand, developing nations perceive this principle in the light of equal opportunities to access and use space resources and technologies, viewing it as a tool to bridge the technological divide. The divergent interpretations reflect the varying capacities and aspirations of nations, subsequently shaping the dynamics of international space law.[56]

In the race to explore and harness space, it is crucial to juxtapose the freedom of exploration and use with responsibility, specifically considering environmental and ethical aspects. Unchecked exploration and use of space pose substantial environmental challenges. Satellite constellations, while instrumental in advancing communication technologies, have raised concerns over light pollution affecting astronomical observations. Additionally, the potential exploitation of celestial bodies raises the risk of irreversible environmental damage in space.

Ethically, the principle must grapple with issues surrounding resource exploitation, space debris, and implications for extra-terrestrial life. The idea of space mining, though scientifically exciting, is fraught with ethical issues, particularly relating to the ownership and distribution of resources. It brings forth the question of whether a select few should be allowed to exploit space resources for commercial gain.[57] The growing issue of space debris, fuelled by rapid advancements in space technology, is yet another ethical dilemma. With every new satellite launched, the risk of collision and consequent debris generation increases, posing threats to both manned and unmanned space missions.[58] Lastly, as we tread further into space, the ethical implications for potential extra-terrestrial life must also be part of the discourse. Any form of life, if found, would need to be treated with respect and caution, adhering to planetary protection measures.[59] This delicate balance between freedom of exploration and use, and environmental and ethical considerations, must guide the future discourse of international space law.

The freedom of exploration and use of space, as enshrined in the OST, has not translated into equal opportunities for all nations. There exists a marked disparity in space capabilities between the technologically advanced Global North and the Global South. A majority of the countries in the Global South are yet to fully realise

[56] David Myers, '"Common Interest" and "Non-Appropriation" in Outer Space: Political Interpretation of Legal Principles' (1979) 6(3) IntlRel 529.

[57] Reed Loder, 'Asteroid Mining: Ecological Jurisprudence beyond Earth' (2018) 36(3) VaEnvtLJ 275.

[58] Lauryn Hallet, 'Rise of Mega Constellations: A Case to Adapt Space Law Through the Law of the Sea' in Annette Froehlich (ed), *Legal Aspects Around Satellite Constellations* (Springer 2021).

[59] John Rummelet and others, 'Ethical Considerations for Planetary Protection in Space Exploration: A Workshop' (2012) 12(11) Astrobio 1017.

their space capabilities due to a myriad of reasons ranging from financial constraints to lack of technological know-how. This effectively excludes them from the benefits that could accrue from the exploration and use of outer space.

The 1996 UN Declaration on 'Space Benefits',[60] however, was a significant attempt to address this imbalance. Stemming from extensive debates in UNCOPUOS, driven by both developing and developed nations, the declaration was a culmination of diplomatic efforts by key players such as Germany, France, and Brazil. It sought to reconcile the principles of free exploration of space with the necessity for equitable benefits from such exploration. Marietta Benko and Kai-Uwe Schrogl illuminate the political tightrope UNCOPUOS had to tread in this exercise.[61] In the initial stages, the developing countries, striving for a more equitable stake in space benefits, faced resistance from the Western powers. This push-and-pull dynamic led to various proposals, including a rather radical approach that sought automatic transfer of resources from North to South.

However, it was clear that such an aggressive stance would not gain the consensus needed for a lasting solution. The eventual merging of texts from both the developing countries[62] and the German-French proposal[63] resulted in a declaration which, while preserving the freedom of space exploration, also emphasised cooperation that is beneficial for all. The declaration's emphasis on efficiency, intellectual property rights, and the role of both States and commercial entities in space exploration marked a paradigm shift from confrontation to collaboration. Although it did not compel the spacefaring nations to actively share technology or resources, it created a conducive environment for the 'demand-pull' from developing nations, encouraging them to actively participate and integrate in the global space community. This new approach encouraged the pooling of resources, potentially on a regional basis, thereby opening up new avenues for budding space powers of the Global South.

Despite this modest success, the Western-centric space agenda continues to marginalise the space aspirations of the Global South. The agendas, policies, and frameworks governing space exploration have been predominantly formulated by and favour the interests of the spacefaring nations, the majority of which are from the Global North. These frameworks often overlook or undervalue the unique interests, needs, and aspirations of the Global South. To redress this imbalance, there is an urgent need for inclusivity in the interpretation and implementation of the principle of freedom of exploration and use. Greater participation of the Global

[60] Declaration on International Cooperation in the Exploration and Use of Outer Space for the Benefit and in the Interest of All States, Taking into Particular Account the Needs of Developing Countries, UNGA Res 51/122 (13 December 1996) UN Doc A/RES/51/122.

[61] Marietta Benko and Kai-Uwe Schrogl, 'History and Impact of the 1996 UN Declaration of Space Benefits' (1997) SpacePoly 139.

[62] UN Doc A/AC.105/C.2/L.182/Rev.1 (31 March 1993).

[63] UN Doc A/AC.105/C.2/L.197 (27 March 1995).

South in the space discourse, and the framing of policies and regulations that encourage and facilitate their engagement in space activities, are crucial. The normative vision of international space law must shift from a Western-centric perspective to a more equitable and inclusive one, embodying the voices and aspirations of all nations, irrespective of their level of technological advancement. The critique offered by TWAIL of the current space governance framework emphasises the need for a shift in perspective. This shift should recognise the right of the Global South to participate in the benefits of space exploration and its use and ensure their equal access to space resources. Correspondingly, active participation of the Global South in the space law-making processes could bring a more equitable dimension to these structures, thereby preserving the principle of space as a common heritage of humankind.

Considering the rapid advances in space technologies and the increasing enthusiasm from developing nations in space exploration, it is time for the global community to jointly re-evaluate and modify its stance on space governance. The evolution of space law must guarantee that every country, no matter its developmental stage, has the same chances to explore, utilise, and profit from space and its assets. A prime example of this is the creation of large satellite constellations by private corporations. Projects like SpaceX's Starlink, while offering the prospect of worldwide broadband access, also raise substantial concerns about the environment and fairness. The deployment of thousands of satellites increases the risk of space debris and the associated dangers of collisions. Furthermore, it raises concerns about the 'tragedy of the commons' in outer space and whether all nations, irrespective of their technological capabilities, will have equitable access to these services.

Having underscored the centrality of the principle of freedom of exploration and use in international space law, it is equally crucial to emphasise that such freedom should not become a licence for unbridled exploitation or cause the degradation of the space environment. In fact, balancing freedom and responsibility becomes an imperative. Lessons from environmental law are instructive here. Principles such as the precautionary principle and the principle of sustainable development provide potent paradigms for developing an ecocentric approach in space law. The precautionary principle, for instance, mandates that if an action or policy has the potential to cause harm to the public or the environment, in the absence of a scientific consensus the burden of proof falls on those advocating for the action or policy.[64] This principle could effectively be transposed to the realm of space law, ensuring that potentially harmful activities in space are subject to thorough examination and validation before being permitted.

[64] Bittencourt Neto 'Preserving the Outer Space Environment: The "Precautionary Principle" Approach to Space Debris' (2013) 14 ProceIntlAstroCongress 11213.

Similarly, the principle of sustainable development, which seeks to reconcile economic development with the preservation of the environment, could serve as a valuable reference for space law.[65] Given that space activities have both economic dimensions and potential environmental consequences, sustainable development's core tenets could help guide space exploration and use towards an equilibrium. This could include promoting practices that limit space debris, encouraging the use of space technology for Earth's environmental challenges, and ensuring that space resource exploitation does not harm the space environment or compromise future generations' ability to benefit from space.

Moreover, the principle of 'inter-generational equity', another cornerstone of environmental law, could also provide guidance for a reformed interpretation of the freedom of exploration and use principle.[66] This principle insists that each generation should pass on the Earth and its natural resources in no worse condition than when they received it, ensuring that future generations have the same opportunities for development as the current one. Transposing this principle to space law would mean that our generation's exploration and use of space should not preclude future generations from similar opportunities. Thus, by intertwining environmental law principles with the freedom of exploration and use, international space law could strike a balance between enabling space activities and preserving the space environment. It is through such a reformulation that space law could align better with TWAIL's objectives of equity, justice, and sustainability.

Further, as space technology rapidly advances, a new frontier of ethical considerations unfolds before us. The TWAIL approach, grounded in principles of justice and equity, provides a critical perspective for exploring this emerging landscape. A potent example is the ethical conundrum surrounding astrobiology. The possibility of encountering extra-terrestrial life forms, be it microbial or more complex organisms, demands a robust ethical framework within space law. The precautionary principle, borrowed from environmental law, may be an effective tool here, urging us to err on the side of caution when the potential for harm exists, even if it is not fully understood. Furthermore, drawing inspiration from bioethics and animal rights law, we may need to formulate principles for the potential rights of extra-terrestrial life. We must consider not just our right to explore space, but the rights of potential life forms that we might encounter or disrupt. A respectful, considerate approach to exploration could lay a robust ethical foundation for future space activities, aligning with the TWAIL ethos of justice and respect for all forms of life. Another crucial component of fostering inclusivity is technology transfer. However, the technological gap between the Global North and South is widening,

[65] Virginie Barral, 'Sustainable Development in International Law: Nature and Operation of an Evolutive Legal Norm' (2012) 23(2) EJIL 377.

[66] Louis Sohn and Edith Weiss, 'Intergenerational Equity in International Law' (1987) 81 ProceASIL 126.

with space technology being a prime example. An equitable regime that facilitates the transfer of space technology could enable the Global South to participate more fully in space exploration.

In the final analysis, the potential democratisation of space exploration may be realised through the strategic implementation of cooperative frameworks. Regional collaborative entities, exemplified by organisations such as the African Space Agency, stand poised to synergise both intellectual capital and tangible resources; this amalgamation, in turn, could engender a collective ascension in space-faring capabilities.[67] The United Nations, an institution imbued with the solemn responsibility of safeguarding international peace and security, could orchestrate a pivotal role in catalysing such cooperation. The intrinsic value of inclusivity and equity within the multifaceted sphere of space activities is not merely a theoretical construct. Rather, it manifests tangibly in the tapestry of successful cooperative endeavours, ranging from the monumental achievement of the International Space Station to the nuanced collaborations within regional space agencies. Conversely, the examination of instances wherein the absence of inclusivity precipitated conflict or the forfeiture of opportunities serves as a sombre testament. Such instances underscore the profound significance of these principles, illuminating the intricate interplay between cooperation, equity, and the broader implications for the future of space exploration.

5 Concluding Thoughts: Challenges and Opportunities in International Space Law

In the complex terrain of international space law, scholars and practitioners alike are confronted with the Herculean task of assimilating and operationalising a TWAIL perspective. This endeavour, far from being a mere intellectual exercise, is beset with several obstacles. These barriers are not simply substantial in their magnitude; they are inextricably woven into the very sinews of the global political and economic structure. The complexity is further exacerbated by the historical, sociological, and philosophical dimensions that underpin the existing legal frameworks, rendering the pursuit of a TWAIL vision both a profound challenge and an imperative in the contemporary discourse on space governance. A concerted effort is required to circumvent these barriers, to instigate a meaningful dialogue that acknowledges and addresses the stark realities of power dynamics that have long governed the realm of space law.

Incorporating a TWAIL perspective demands a critical examination and reconfiguration of the existing legal and institutional structures. It requires an

[67] 'AUC Inaugurates the African Space Agency' (*Space in Africa*, 26 January 2023) <https://africanews.space/auc-inaugurates-the-african-space-agency> accessed 2 October 2023.

interrogation of the entrenched norms shaped by the world's leading powers and a broad-based consensus among nations with often disparate interests and worldviews. The application of TWAIL to international space law carries the potential to advance global justice and enhance the sustainability of space activities. It posits a new vision for space law—one that is not an extension of terrestrial power dynamics but a manifestation of a truly global community with shared interests and responsibilities.[68]

The role of emerging space powers, particularly those from the Global South, is crucial in shaping this future. These countries, by advocating for their interests and perspectives, have the power to challenge the status quo and contribute to a more balanced and diverse international space regime.

In the pursuit of a governance structure for outer space that embodies fairness, inclusiveness, and sustainability, the prospects for legal and institutional alterations in harmony with Third World Approaches to International Law are substantial. Such transformations might include an augmented presence of the Global South in authoritative entities, alterations to the tenets of non-acquisition and pacific utilisation, and stratagems to guarantee a just apportionment of celestial resources and advantages. Notwithstanding the intrinsic difficulties, the amalgamation of an augmented acknowledgement of the significance of inclusiveness and sustainability, together with the burgeoning proficiencies of Global South nations in cosmic endeavours, furnishes a singular occasion. This intersection invites us to reformulate international space law into a structure that epitomises fairness and justice.

The exigency for metamorphosis within the sphere of international space law is discernible. The prevailing structure, in spite of its seminal assurances, has proven deficient in suitably addressing the apprehensions and requisites of the Global South. Positioned at the brink of a novel era in cosmic exploration, typified by unparalleled technological innovations and the increased involvement of non-sovereign participants, the call for a more encompassing, fair, and enduring cosmic law structure is more urgent than ever. Adopting the philosophy of TWAIL empowers us to unveil the disparities of the current legal regime and conceptualise alternatives. It beseeches us to re-evaluate the core principles of space law, to re-contemplate our communal aspiration for our communal celestial domain, and to magnify voices that have been overshadowed in these dialogues. Moreover, the prospective trajectory of space law, when perceived through the lens of TWAIL, heralds a path towards fairness and inclusiveness. It foresees a future wherein the ambitions of all States, regardless of their fiscal or political stature, are not merely recognised but vigorously cultivated. This undertaking, although undeniably

[68] Anghie (n 11).

arduous, is not merely preferable but a vital stride towards a more just future in outer space.

As our attention is increasingly drawn to the pursuit of equity and true democracy in outer space, the sagacious utterances of Antony Anghie reverberate profoundly: 'the "universal" was actually the "particular" of Europe.'[69] This revelation compels us to face the historical wrongs that have sculpted the extant international legal regime, reminding us that the entrenched 'universal' standards were predominantly a manifestation of the inclinations and viewpoints of a limited number of pre-eminent States. Thus, the future of space law necessitates a fundamental reassessment, a shift from merely reinterpreting treaties and principles to confronting the underlying political, economic, and social dynamics that have been at play since the dawn of space exploration. As Upendra Baxi posits, '[t]he cruellest lie of the existing structures of international law is the lie of formal equality of all nation-states.'[70] This critical perspective, prevalent in TWAIL and other critical approaches to international law, underscores the imperative to address systemic imbalances.

If this journey is navigated successfully, we may arrive at a legal framework that aligns more closely with the ethos of TWAIL, fostering cooperation over competition and inclusivity over exclusion. It could truly embody the principle of space as the province of all humankind, echoing Makau Mutua's call to decentre the hegemonic discourses and structures of the traditional human rights corpus.[71] Thus, the task before us is not merely to reform, but to transform—to envision and create an equitable and inclusive cosmos that is reflective of our shared humanity, a cosmos that transcends terrestrial power dynamics and truly belongs to us all. This is the promise of a future guided by TWAIL principles and the legacy we strive to leave for the generations to come.

[69] ibid.

[70] Upendra Baxi, *The Future of Human Rights* (OUP 2006).

[71] Makau Mutua, 'Savages, Victims, and Saviors: The Metaphor of Human Rights' (2001) 42 HarvIntlLJ 201.

Index

For the benefit of digital users, indexed terms that span two pages (e.g., 52–53) may, on occasion, appear on only one of those pages.